Corrosion and Mechanical Behavior of Metal Materials (2nd Edition)

Corrosion and Mechanical Behavior of Metal Materials (2nd Edition)

Guest Editor

Ming Liu

Basel • Beijing • Wuhan • Barcelona • Belgrade • Novi Sad • Cluj • Manchester

Guest Editor
Ming Liu
Materials Science and
Engineering
Xi'an University of Technology
Xi'an
China

Editorial Office
MDPI AG
Grosspeteranlage 5
4052 Basel, Switzerland

This is a reprint of the Special Issue, published open access by the journal *Materials* (ISSN 1996-1944), freely accessible at: www.mdpi.com/journal/materials/special_issues/AJ1CG6FMP0.

For citation purposes, cite each article independently as indicated on the article page online and using the guide below:

Lastname, A.A.; Lastname, B.B. Article Title. *Journal Name* **Year**, *Volume Number*, Page Range.

ISBN 978-3-7258-3238-5 (Hbk)
ISBN 978-3-7258-3237-8 (PDF)
https://doi.org/10.3390/books978-3-7258-3237-8

Contents

Article

Application of Galvanostatic Non-Linear Impedance Spectroscopy to the Analysis of Metallic Material Degradation

Pawel Slepski [1],*, Husnu Gerengi [2], Dominika Parasinska [1] and Lukasz Gawel [1]

[1] Department of Electrochemistry, Corrosion and Materials Engineering, Faculty of Chemistry, Gdansk University of Technology, 11/12 G. Narutowicza Str., 80-233 Gdansk, Poland; dominika.parasinska@pg.edu.pl (D.P.); lukasz.gawel@pg.edu.pl (L.G.)

[2] Corrosion Research Laboratory, Department of Mechanical Engineering, Faculty of Engineering, Düzce University, 81620 Düzce, Turkey; husnugerengi@duzce.edu.tr

* Correspondence: pawsleps@pg.edu.pl

Abstract: This study presents a novel application of Non-Linear Electrochemical Impedance Spectroscopy (NLEIS) in galvanostatic mode for the rapid, non-destructive assessment of metal degradation. By using galvanostatic mode instead of traditional potentiostatic methods, polarization-related challenges are mitigated, enabling more accurate and reliable analysis. The technique allows for the determination of corrosion rates (corrosion current) and material susceptibility to oxidation (Tafel coefficient) through a single measurement with a modulated AC perturbation signal. Theoretical assumptions of the method were validated through tests on both a non-linear model system and an experimental system. The proposed research methodology is highly effective for monitoring the condition of metallic materials in various environments, covering both anodic and cathodic processes.

Keywords: degradation of metallic materials; non-linear electrochemical impedance spectroscopy; non-destructive corrosion monitoring

Citation: Slepski, P.; Gerengi, H.; Parasinska, D.; Gawel, L. Application of Galvanostatic Non-Linear Impedance Spectroscopy to the Analysis of Metallic Material Degradation. *Materials* **2024**, *17*, 4985. https://doi.org/10.3390/ma17204985

Academic Editor: Geoffrey D. Will

Received: 10 September 2024
Revised: 8 October 2024
Accepted: 10 October 2024
Published: 12 October 2024

1. Introduction

The degradation of metallic materials is a primary cause of various failures in structures and industrial equipment. The global cost associated with this phenomenon exceeds 3% of the global GDP [1]. Solutions involving the appropriate selection of materials and protective techniques are not always feasible due to technological or economic constraints.

In such cases, the most effective approach is the preventive monitoring of material degradation. The deterioration of metallic materials is predominantly related to electrochemical corrosion [2]. Consequently, electrochemical techniques based on current–voltage relationships are commonly used to analyze and monitor the progression of this process [3]:

$$\Delta I = i_{corr}[exp(b_a \Delta E) - exp(-b_c \Delta E)] \qquad (1)$$

where: ΔI—the response of the generated current [A], i_{corr}—corrosion current [A], $\Delta E = E - E_{corr}$—polarization (difference of the actual (E) potential and the corrosion potential (E_{corr}) [V], $b_a = \frac{\ln(10)}{\beta_a}$, $b_c = \frac{\ln(10)}{\beta_c}$, β_a—Tafel slope coefficient of the anodic partial process, β_c—Tafel slope coefficient of the cathodic partial process.

The simplest technique for determining the corrosion rate (i_{corr}) and analyzing the anodic (β_a) and cathodic (β_c) processes is the Tafel analysis [4].

However, the deep polarization applied in this method (approximately ±250 mV) leads to irreversible changes in the material, preventing ongoing degradation monitoring. Using low polarization for linear current–voltage characteristic analysis (polarization resistance technique) avoids material damage but provides only indirect information [5,6].

These limitations are overcome by methods that use variable excitation signals, with sinusoidal perturbation being the most widely used.

Among the aforementioned methods we can distinguish:

- A method utilizing single sinusoidal excitation in the form of harmonic analysis (HA) proposed by Meszarosz et al. [7,8]. This technique has been applied in various corrosion studies [9–11].
- A method employing dual-frequency excitation, known as Electrochemical Frequency Modulation (EFM) [12–14].
- A method utilizing multiple frequency signals, generated either sequentially [15,16] or simultaneously [17–19].

The variable current method that provides the most comprehensive information about metal degradation is Non-Linear Electrochemical Impedance Spectroscopy (NLEIS) [20]. Its limitations related to long measurement time have recently been addressed through appropriately modified excitation signals [21,22].

Virtually all corrosion monitoring methods used to date operate in potentiostatic mode, assuming constant potential during electrochemical degradation. However, the decay of metallic materials is a non-stationary process where the corrosion potential undergoes significant fluctuations. A better alternative is the use of a method operating in galvanostatic mode.

In this paper, the authors introduce a novel approach to investigating corrosion processes by employing Non-Linear Electrochemical Impedance Spectroscopy in the current mode. This method is proposed as a promising approach for corrosion monitoring and a viable alternative to conventional techniques, offering a unique and valuable perspective on corrosion phenomena. By utilizing current perturbation signals, this approach facilitates a more nuanced and dynamic analysis of corrosion processes, which is crucial for advancing the field of materials science. The potential impact of this method on industrial and research applications is considerable, offering enhanced precision in corrosion monitoring and contributing to the development of more robust materials.

2. Method Description

The main problem, which occurs in corrosion analysis for measurements obtained in galvanostatic mode is the relation, which easily connects the voltage response signal with the current perturbation signal. Equation (1), which presents relation $\Delta I = f(\Delta E)$ could not be directly transformed into the form $\Delta E = f(\Delta I)$. The solution requires the introduction of one of three assumptions:

- $b_c = b_a = b$:

$$\Delta E = \frac{1}{b} ln \left(\frac{\Delta I + \sqrt{\Delta I^2 + 4 i_{corr}^2}}{2 i_{corr}} \right) \tag{2}$$

- $b_c \gg b_a$:

$$\Delta E = \frac{1}{b_c} ln(i_{corr}) - \frac{1}{b_c} ln(i_{corr} - \Delta I) \tag{3}$$

- $b_a \gg b_c$:

$$\Delta E = \frac{1}{b_a} ln(\Delta I + i_{corr}) - \frac{1}{b_a} ln(i_{corr}) \tag{4}$$

The first solution pertains to situations in which the Tafel coefficients, both anodic and cathodic, are exactly the same. In practice, such cases are rather rare. The next solution concerns corrosion processes controlled by the anodic reaction. Such a situation, for example, occurs with materials that undergo passivation. The third case describes the corrosion process in which the rate of the phenomenon is determined by the cathodic reaction. This is the most commonly occurring form of the corrosion process and, for this reason, it has been chosen for further description.

2.1. Galvanostatic NLEIS for Cathodic Control Processes

If the current perturbation signal of Equation (4) will be changed to an AC signal in the form:

$$\Delta I = \Delta I_0 cos(\omega t) + I_{DC} \tag{5}$$

where: $I_{DC} = 0$—conditions of general corrosion.

The polyharmonic voltage response signal will take the form, described in Equation (6):

$$\Delta E = \Delta E_0 + \Delta E_1 cos(\omega t + \varphi) + \Delta E_2 cos(2\omega t + 2\varphi) + \Delta E_3 cos(3\omega t + 3\varphi) + \Delta E_4 cos(4\omega t + 4\varphi) + \cdots \tag{6}$$

where the amplitudes ΔE_n of successive harmonics will be a function of the parameters b_a, i_{corr}, and ΔI_0 (Equation (7a–c)):

$$\Delta E_1 = \frac{1}{b_a i_{corr}}\Delta I_0 + \frac{1}{4 b_a i_{corr}{}^3}\Delta I_0{}^3 + \frac{1}{8 b_a i_{corr}{}^5}\Delta I_0{}^5 + \cdots \tag{7a}$$

$$\Delta E_2 = -\frac{1}{4 b_a i_{corr}{}^2}\Delta I_0{}^2 - \frac{1}{8 b_a i_{corr}{}^4}\Delta I_0{}^4 - \frac{5}{64 b_a i_{corr}{}^6}\Delta I_0{}^6 - \cdots \tag{7b}$$

$$\Delta E_3 = \frac{1}{12 b_a i_{corr}{}^3}\Delta I_0{}^3 + \frac{1}{16 b_a i_{corr}{}^5}\Delta I_0{}^5 + \cdots \tag{7c}$$

The EIS technique is based only on the analysis of the fundamental harmonics of the response signal described by Equation (7a). Higher harmonics were not analyzed in the presented work. Based on this equation, the dependence of the Faradaic resistance R_{ct} as a function of the amplitude of the perturbation signal ΔI_0 can be described as Equation (8a,b):

$$R_{ct} = \frac{\Delta E_1}{\Delta I_0} \tag{8a}$$

$$R_{ct} = \frac{1}{b_a i_{corr}} + \frac{1}{4 b_a i_{corr}{}^3}\Delta I_0{}^2 + \frac{1}{8 b_a i_{corr}{}^5}\Delta I_0{}^4 + \cdots \tag{8b}$$

The presented Equation (8b) indicates that the Faradaic resistance of the corrosion system, controlled by the cathodic process, depends on the amplitude of the current perturbation signal used in galvanostatic mode. The higher the value of the perturbation amplitude used, the higher the value of the Faradaic resistance R_{ct} that will be obtained. This dependency can be used to determine the values of the corrosion parameters, such as the corrosion current and the anodic process Tafel slope coefficient. For this purpose, an analysis of the dependency between the Faradaic resistance and the amplitude of the perturbation signal $R_{ct} = f(\Delta I)$, based on at least the first two terms of the polynomial Equation (8b) should be carried out.

2.2. Galvanostatic NLEIS for Model System

The verification of the correctness of research methods and measurement systems often relies on measurements using model systems. For impedance measurements of linear systems, dummy cells are commonly used, consisting of a resistor (in series) connected to a capacitor and resistor connected in parallel, as shown in Figure 1a [23].

In the case of simulating non-linear systems, the resistor is replaced with a diode, which is an element with a non-linear characteristic, as shown in Figure 1b [24]. The current–voltage relationship for a diode can be represented by a modification of the Shockley diode equation, which closely resembles the current–voltage behavior of an electrochemical system (Equations (9) and (10)):

$$\Delta I = i_s exp(b_d \Delta E) \tag{9}$$

$$\Delta E = \frac{1}{b_d} ln\left(\frac{\Delta I}{i_s}\right) \tag{10}$$

where: i_s—saturation current, b_d—Tafel slope coefficient for diode.

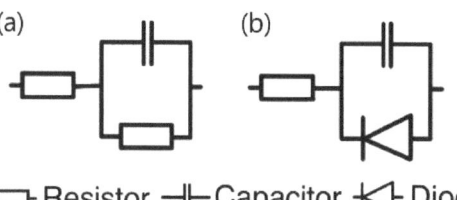

Figure 1. (a) Dummy cell with a resistor connected to a capacitor and resistor connected in parallel. (b) Dummy cell with a diode.

In this context, i_s is the equivalent of the corrosion current, while b_d corresponds to the inverse of the Tafel slope.

In the case of current excitation with a variable nature (Equation (5)), the amplitude of the voltage response for the fundamental harmonic can be described by a polynomial (Equation (11)):

$$\Delta E_1 = \frac{1}{b_d I_{DC}}\Delta I_0 + \frac{1}{4b_d I_{DC}^3}\Delta I_0^3 + \frac{1}{8b_d I_{DC}^5}\Delta I_0^5 + \cdots \tag{11}$$

The above expression can be used to determine the diode resistance as a function of: excitation amplitude, current bias and voltage coefficient (Equation (12)):

$$R_d = \frac{1}{b_d I_{DC}} + \frac{1}{4b_d I_{DC}^3}\Delta I_0^2 + \frac{1}{8b_d I_{DC}^5}\Delta I_0^4 + \cdots \tag{12}$$

It should be noted that the above equation is remarkably similar to Equation (8b) describing a corroding system under cathodic control conditions. This means that a substitute circuit based on the non-linear element, such as a diode, serves as an ideal means of validating the presented method.

3. Experimental

In the research, a dummy cell was used, consisting of a signal switching diode (part number: 1N414) connected in parallel with a 3.3 µF nominal value foil capacitor, with both components connected in series with a 10 Ω resistor. The current–voltage characteristic of the diode was determined through a linear voltage scan in the range of 0 to 0.7 V.

The measurements of the corrosion system under cathodic control were carried out in a 1M KCl solution (Sigma Aldrich, St. Louis, MO, USA) at a temperature of 32 °C stabilized by a LT ecocool 150 thermostat (Shepreth, Cambridgeshire, UK). Electrochemical measurements were made after conditioning the testing system for 24 h at the working temperature. The working electrode was composed of carbon steel (Fe 99.29% Mn 0.49% Ni 0.07% Cu 0.06% Cr 0.02%) in the form of a disk with a surface area of 1 cm². The elemental analysis was performed by a Bruker (Brillerica, MA, USA) handheld XRF spectrometer (S1 Titan 600). The sample was immobilized in epoxy resin. Before measurements were taken, the surface was wet ground using a set of silicon carbide abrasive papers with a gradation from 400 up to 2500 grit, washed with deionized water, dried out and placed in an electrochemical cell. The reference electrode was a saturated calomel electrode (SCE) and the counter electrode was a platinum mesh.

The same card was used to record the total voltage and current signals from the investigated system. Dedicated software was developed using the LabView 2020 SP1 graphical programming environment by Emerson to control the measurement process, including signal generation, acquisition, signal recording, and subsequent analysis of the obtained impedance data.

The current–voltage characteristic, which served for the comparative determination of the corrosion current and Tafel coefficient β_a, was obtained through linear polarization of the working electrode in the range of ± 150 mV vs. E_{oc} at a scan rate of 1 mV/s (Figure 2).

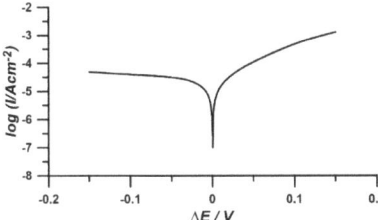

Figure 2. Current–voltage characteristics for determining corrosion current and Tafel coefficient via linear polarization at ± 150 mV vs. E_{oc} with a 1 mV/s scan rate.

Impedance measurements were conducted using a signal composed of elementary sinusoids with linearly varying amplitudes (Figure 3).

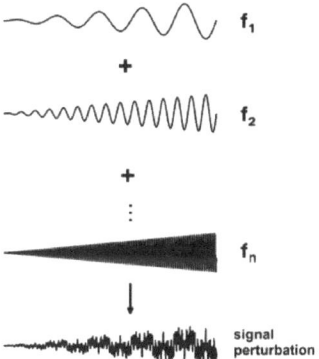

Figure 3. Scheme of generation of an AC signal perturbation.

Impedance measurements for the non-linear model system were conducted at ten frequencies ranging from 1.5 kHz to 3 Hz. A constant current excitation amplitude of 32 μA was used, along with a linear continuously changing perturbation amplitude ranging from 0 μA to 15 μA at the rate of 1 μA/s. Variable current measurements for the corrosion system were performed at nine frequencies in the range of 100 Hz to 30 mHz, with a linear continuously changing perturbation amplitude ranging from 0 μA to 50 μA at the rate of 50 nA/s. The DC component, I_{DC}, was set to 0 A (measured at the corrosion potential). Both the memory effects and hysteresis do not occur in this case.

4. Results and Discussion

4.1. Model System-Verification of the Method

The impedance of the equivalent circuit obtained at a constant current $I_{DC} = 32$ μA and with variable components' amplitudes, ΔI_0 ranging from 0 to 15 μA is presented in Figure 4.

The individual spectra exhibit semicircular shapes, which gradually enlarge as the amplitude of the variable excitation signal increases. The most significant changes in impedance are observed in the range of the lowest analyzed frequencies. By analyzing the individual impedance spectra using the Randles circuit R (CR), the variation in the diode resistance as a function of amplitude was obtained (Figure 5).

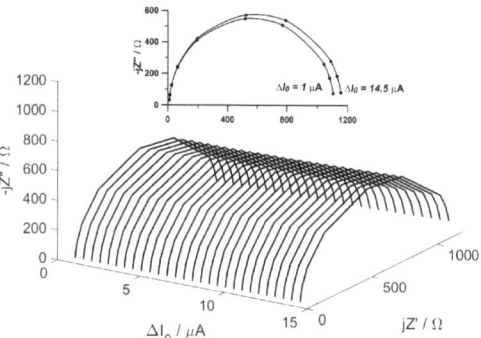

Figure 4. Changes in the impedance of the model system as a function of the amplitude of variable current excitation.

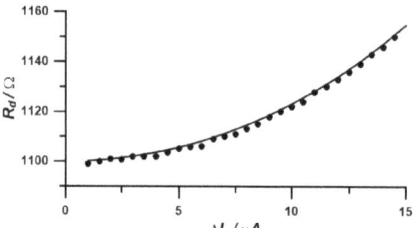

Figure 5. Changes in diode resistance as a function of the amplitude of variable current excitation.

An increase in the variable current excitation from 0 μA to 15 μA results in an exponential increase in the obtained diode resistance, from approximately 1100 Ω to around 1160 Ω. This means that, according to Equation (12), by increasing the current amplitude of the variable excitation, we increase resistance values obtained based on impedance changes.

Based on the obtained current-resistance relationships, a non-linear regression was performed using the first two and the first three terms of Equation (12). The fitting results are presented in the table below.

The experimental values (Table 1), both in the case of the polarization current and the diode voltage coefficient, are close to the real values. The magnitude of the error, which is approximately 5% in the case of fitting to the first two polynomial terms and around 10% in the case of fitting to three polynomial terms, largely depends on the range of amplitudes of the variable signal used for impedance measurement. In the analyzed case, the range of changes in the variable current single frequency amplitude was a maximum of 30 μA peak to peak (Figure 6).

The maximum voltage response generated as a result of the lowest single frequency is less than 37 mV peak to peak, a value which, in many impedance measurements, is considered as the linear range.

Table 1. Comparison of real values to experimentally obtained values.

Method	i_{DC} μA	b_d 1/V
Real value	32.0	28.06
NLEIS-2 terms of the polynomial	33.7	26.98
NLEIS-3 terms of the polynomial	35.1	25.9

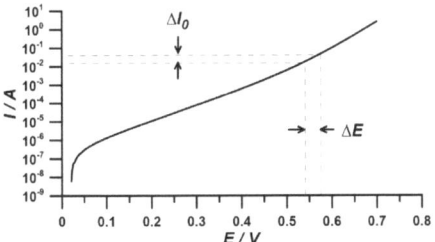

Figure 6. Current–Voltage characteristic of the diode used in the non-linear system model.

4.2. Cathodic Control-Real System

The impedance changes for current perturbation amplitudes ΔI_0 in the corrosion system under cathode control are presented in Figure 7. The current amplitude excitation was varied linearly from 0 μA to 50 μA. The individual spectra have the typical shape of a fragmentary semicircle. There is a marked increase in impedance as the amplitude of the variable excitation increases.

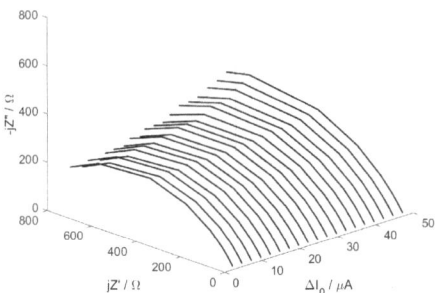

Figure 7. Impedance changes as a function of current perturbation amplitude ΔI_0 for corrosion system under cathodic control.

In order to use Equation (8a) to determine the corrosion current and anodic Tafel coefficient, it must be kept in mind that the R_{ct} value is the ratio of the effective amplitude of the first harmonic appearing on the Faradaic element ΔE_{ef} to the current perturbation signal flowing through the element ΔI_f. In the tested system, an electrical double layer Z_c and an electrolyte resistance R_e can also be distinguished. An electrical equivalent circuit showing this dependency is shown in Figure 8.

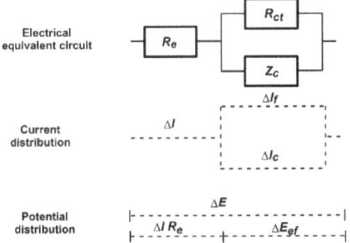

Figure 8. Electrical equivalent circuit, dependency of current distribution and dependency of the potential distribution of the tested system.

In the first element, the resistance of the electrolyte (R_e) of the electrical equivalent circuit influences the value of the voltage response signal on the Faradaic element R_{ct}. In the second element, capacitance C_{dl}, according to Kirchhoff's law, influences the real value of the current perturbation signal on the Faradaic element R_{ct}. Such a situation strongly implies harmonic analysis measurements when a single sinusoidal perturbation is used. In the case of using spectral measurements, the analysis of the obtained impedance spectrum gives the result in the form of the R_{ct} value, not under the influence of the Faraday current. Only in the case of significant influence of the electrolyte R_e, when its value is close to the value of the R_{ct}, should an appropriate correction of the values be included [21,24–28]

The spectra shown in Figure 9 display slightly flattened semicircles instead of ideal ones. Therefore, to determine the Faradaic resistance (R_{ct}), a simple three-element equivalent circuit R (QR) was employed. A constant phase element was used instead of capacitance due to both the energetic heterogeneity of the surface, as the material under investigation is an iron alloy containing additional elements, and the geometric irregularities resulting from the surface preparation, which does not produce a perfectly flat surface. Detailed information regarding surface distribution related to this phenomenon has been explained by Hirschorn et al. [29].

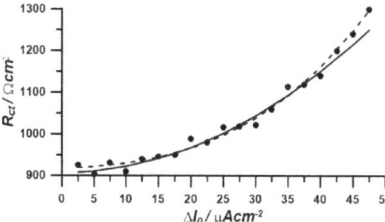

Figure 9. Characteristic of faradaic element R_{ct} values changes in a function of amplitude changes of the current perturbation signal: (●)—experimental data (—)—fitting to the 2-term polynomial of Equation (11b) (- -)—fitting to the 3-term polynomial of Equation (11b).

The obtained characteristic of R_{ct} changes in a function of perturbation amplitude changes $R_{ct} = f (\Delta I_0)$ is shown in Figure 7. In line with the earlier expectations, the course of the changes has the character of an increasing function.

To determine the corrosion parameters values from obtained data, such as corrosion current (i_{corr}) and anodic Tafel coefficient (β_a), it is required to perform a non-linear regression based on at least two terms of Equation (8b). Corrosion parameters were determined using the first two and the first three terms of the polynomial presented by Equation (8b). The results, along with the values obtained from classical polarization measurement are presented in Table 2.

Table 2. Extrapolated corrosion rates and anodic Tafel slope for NLEIS and classic Tafel extrapolation for corrosion system under cathodic control.

Method	i_{corr} μA/cm^2	β_a mV
Tafel extrapolation	47.5	95
NLEIS-2 terms of the polynomial	38.6	81
NLEIS-3 terms of the polynomial	45.8	97

The values obtained by the NLEIS technique in the galvanostatic mode are similar to the values obtained in the classical Tafel extrapolation technique (see Table 2). The slight differences are probably the result of very low AC current perturbation Figure 10.

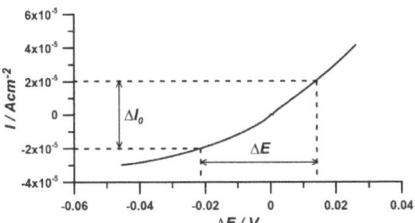

Figure 10. Amplitude-response signal relationship.

The current perturbation amplitude signals generate a small variable voltage response with an amplitude of less than 40 mV peak to peak. Such conditions mean that the results obtained by the polarization method relate to the measured sample significantly deviate from the equilibrium state.

5. Conclusions

Numerous electrochemical techniques exist for monitoring corrosion rates, but achieving reliable measurements while minimizing impact on the system remains challenging. Direct measurements of corrosion current often cause irreversible alterations or are skewed by electrical double layers and electrolyte resistance. Non-destructive methods usually provide only indirect, hard to correlate, results.

To address these limitations, the novel application of Non-Linear Electrochemical Impedance Spectroscopy (NLEIS) in galvanostatic mode offers a promising solution. This approach enables real-time, detailed insights into corrosion processes, making it a powerful tool for advancing the understanding and management of corrosion-related challenges. The use of NLEIS with current perturbation signals provides a unique and valuable perspective on corrosion phenomena. Moreover, the application of galvanostatic NLEIS with amplitude modulation shows great potential in scenarios where rapid changes in corrosion rates or non-stationarity of corrosion potential are critical factors. The adaptability and precision of this method make it an attractive option for researchers and practitioners seeking robust corrosion monitoring solutions. The proposed method represents a significant advancement in the field of corrosion science and technology, offering a novel approach that could lead to more accurate and reliable corrosion monitoring systems. By contributing to a deeper understanding of the relationships between material structure, properties, and functions, this technique aligns with the broader goals of materials research, facilitating the development of more durable and resistant materials across various applications.

Author Contributions: Conceptualization, P.S. and L.G.; methodology, P.S.; software, P.S.; validation, L.G., H.G. and D.P.; formal analysis, L.G.; investigation, P.S., D.P. and H.G.; resources, P.S. and L.G.; data curation, D.P.; writing—original draft preparation, P.S., L.G. and D.P.; writing—review and editing, H.G.; visualization, P.S., L.G. and D.P.; supervision, P.S. and L.G. All authors have read and agreed to the published version of the manuscript.

Funding: This research received no external funding.

Data Availability Statement: The raw data supporting the conclusions of this article will be made available by the authors on request.

Conflicts of Interest: The authors declare no conflicts of interest.

References

1. Kania, H. Corrosion and Anticorrosion of Alloys/Metals: The Important Global Issue. *Coatings* **2023**, *13*, 216. [CrossRef]
2. Revie, R.W. (Ed.) *Uhlig's Corrosion Handbook*; John Wiley & Sons: Hoboken, NJ, USA, 2011.
3. Marcus, P.; Oudar, J. (Eds.) *Corrosion Mechanisms in Theory and Practice*; Marcel Dekker: New York, NY, USA, 2002; pp. 243–286.
4. *ASTM Standard G3-14(2019)*; Standard Practice for Conventions Applicable to Electrochemical Measurements in Corrosion Testing. Annual Book of ASTM Standards. ASTM International: West Conshohocken, PA, USA, 2019; Volume 3.02, p. 9.

5. *ASTM Standard G102-89(2015)e1*; Standard Practice for Calculation of Corrosion Rates and Related Information from Electrochemical Measurements. Annual Book of ASTM Standards. ASTM International: West Conshohocken, PA, USA, 2015; Volume 3.02, p. 7.

6. *ASTM Standard G59-97(2020)*; Standard Test Method for Conducting Potentiodynamic Polarization Resistance Measurements. ASTM International: West Conshohocken, PA, USA, 2020; Volume 3.02, p. 4.

7. Mészáros, L.; Lengyel, B.; Jánászik, F. Study of the rate of underpaint corrosion by a faradaic distortion method. *Mater. Chem.* **1982**, *7*, 165–182. [CrossRef]

8. Mészáros, L.; Mészáros, G.; Lengyel, B. Application of harmonic analysis in the measuring technique of corrosion. *J. Electrochem. Soc.* **1994**, *141*, 2068. [CrossRef]

9. Dong, Z.; Torbati-Sarraf, H.; Poursaee, A. Determining the optimal frequency and perturbation amplitude for AC electrical resistance measurements of cement-based materials using harmonic analysis. *Adv. Civ. Eng. Mater.* **2022**, *11*, 339–353. [CrossRef]

10. Zhang, X.; Moosbauer, D. Application of Harmonic and Total Harmonic Distortion Instrumentation in Corrosion. In *Advances in Electrochemical Techniques for Corrosion Monitoring and Laboratory Corrosion Measurements*; ASTM International: West Conshohocken, PA, USA, 2019; pp. 160–179. [CrossRef]

11. Burczyk, L.; Darowicki, K. Determination of local corrosion current from individual harmonic components. *J. Electrochem. Soc.* **2017**, *164*, C796. [CrossRef]

12. Bosch, R.W.; Hubrecht, J.; Bogaerts, W.F.; Syrett, B.C. Electrochemical frequency modulation: A new electrochemical technique for online corrosion monitoring. *Corrosion* **2001**, *57*, 60–70. [CrossRef]

13. Belkheiri, A.; Dahmani, K.; Aribou, Z.; Kharbouch, O.; Nordine, E.; Allah, A.E.M.A.; Galai, M.; Touhami, M.E.; Al-Sadoon, M.K.; Al-Maswari, B.M. In-depth study of a newly synthesized imidazole derivative as an eco-friendly corrosion inhibitor for mild steel in 1 M HCl: Theoretical, electrochemical, and surface analysis perspectives. *Int. J. Electrochem. Sci.* **2024**, *19*, 100768. [CrossRef]

14. Al Jahdaly, B.A. Electrochemical and DFT insights into 2-amino-4-(4-hydroxy-3-methoxyphenyl)-7-methyl-4 H-chromene-3-carbonitrile: An innovative strategy for antibacterial activity and corrosion protection of carbon steel. *RSC Adv.* **2024**, *14*, 24039–24054. [CrossRef] [PubMed]

15. Lasia, A. Electrochemical Impedance Spectroscopy and Its Applications. In *Modern Aspects of Electrochemistry*; Springer: Boston, MA, USA, 2002; pp. 143–248. [CrossRef]

16. Wang, S.; Zhang, J.; Gharbi, O.; Vivier, V.; Gao, M.; Orazem, M.E. Electrochemical impedance spectroscopy. *Nat. Rev. Methods Primers* **2021**, *1*, 41. [CrossRef]

17. Koster, D.; Du, G.; Battistel, A.; La Mantia, F. Dynamic impedance spectroscopy using dynamic multi-frequency analysis: A theoretical and experimental investigation. *Electrochim. Acta* **2017**, *246*, 553–563. [CrossRef]

18. Darowicki, K.; Krakowiak, S.; Ślepski, P. Evaluation of pitting corrosion by means of dynamic electrochemical impedance spectroscopy. *Electrochim. Acta* **2004**, *49*, 2909–2918. [CrossRef]

19. Bolat, G.; Mareci, D.; Iacoban, S.; Cimpoesu, N.; Munteanu, C. The estimation of corrosion behavior of NiTi and NiTiNb alloys using dynamic electrochemical impedance spectroscopy. *J. Spectrosc.* **2013**, *2013*, 714920. [CrossRef]

20. Fasmin, F.; Srinivasan, R. Review—Nonlinear Electrochemical Impedance Spectroscopy. *J. Electrochem. Soc.* **2017**, *164*, H443–H455. [CrossRef]

21. Slepski, P.; Szocinski, M.; Lentka, G.; Darowicki, K. Novel Fast Non-Linear Electrochemical Impedance Method for Corrosion Investigations. *Measurement* **2021**, *173*, 108667. [CrossRef]

22. Katırcı, G.; Zabara, M.A.; Ülgüt, B. Methods—Unexpected Effects in Galvanostatic EIS of Randles' Cells: Initial Transients and Harmonics Generated. *J. Electrochem. Soc.* **2022**, *169*, 030527. [CrossRef]

23. Slepski, P.; Szocinski, M.; Majcherczak, J.; Gerengi, H.; Darowicki, K. New Method of Non-Linear Electrochemical Impedance Spectroscopy with an Amplitude-Modulated Perturbation Signal. *J. Electrochem. Soc.* **2019**, *166*, C559–C563. [CrossRef]

24. Battistel, A.; La Mantia, F. Nonlinear Analysis: The Intermodulated Differential Immittance Spectroscopy. *Anal. Chem.* **2013**, *85*, 6799–6805. [CrossRef] [PubMed]

25. Darowicki, K.; Majewska, J. Harmonic Analysis Of Electrochemical and Corrosion Systems—A Review. *Corros. Rev.* **1999**, *17*, 383–400. [CrossRef]

26. Diard, J.-P.; Le Gorrec, B.; Montella, C. Deviation from the Polarization Resistance Due to Non-Linearity I—Theoretical Formulation. *J. Electroanal. Chem.* **1997**, *432*, 27–39. [CrossRef]

27. Diard, J.-P.; Le Gorrec, B.; Montella, C. Deviation of the Polarization Resistance Due to Non-Linearity II. Application to Electrochemical Reactions. *J. Electroanal. Chem.* **1997**, *432*, 41–52. [CrossRef]

28. Diard, J.-P.; Le Gorrec, B.; Montella, C. Deviation of the Polarization Resistance Due to Non-Linearity. III—Polarization Resistance Determination from Non-Linear Impedance Measurements. *J. Electroanal. Chem.* **1997**, *432*, 53–62. [CrossRef]

29. Hirschorn, B.; Orazem, M.E.; Tribollet, B.; Vivier, V.; Frateur, I.; Musiani, M. Determination of effective capacitance and film thickness from constant-phase-element parameters. *Electrochim. Acta* **2010**, *55*, 6218–6227. [CrossRef]

Article

Electrochemical Noise Analysis: An Approach to the Effectivity of Each Method in Different Materials

Jesús Manuel Jáquez-Muñoz [1], Citlalli Gaona-Tiburcio [2,*], Ce Tochtli Méndez-Ramírez [3,*],
Cynthia Martínez-Ramos [2], Miguel Angel Baltazar-Zamora [3], Griselda Santiago-Hurtado [4],
Francisco Estupinan-Lopez [2], Laura Landa-Ruiz [3], Demetrio Nieves-Mendoza [3]
and Facundo Almeraya-Calderon [2]

[1] Universidad Autónoma de Ciudad Juárez, Ciudad Juárez 32315, Mexico; jesus.jaquez@uacj.mx
[2] Centro de Investigación e Innovación en Ingeniería Aeronáutica (CIIIA), Universidad Autónoma de Nuevo León FIME, San Nicolás de los Garza 66455, Mexico; francisco.estupinanlp@uanl.edu.mx (F.E.-L.); facundo.almerayacld@uanl.edu.mx (F.A.-C.)
[3] Facultad de Ingeniería Civil, Universidad Veracruzana, Xalapa 91000, Mexico; mbaltazar@uv.mx (M.A.B.-Z.); lalanda@uv.mx (L.L.-R.); dnieves@uv.mx (D.N.-M.)
[4] Facultad de Ingeniería Civil, Universidad Autónoma de Coahuila, Torreón 27276, Mexico; santiagog@uadec.edu.mx
* Correspondence: citlalli.gaonatbr@uanl.edu.mx (C.G.-T.); cmendez@uv.mx (C.T.M.-R.)

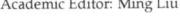
Citation: Jáquez-Muñoz, J.M.;
Gaona-Tiburcio, C.; Méndez-Ramírez,
C.T.; Martínez-Ramos, C.;
Baltazar-Zamora, M.A.;
Santiago-Hurtado, G.;
Estupinan-Lopez, F.; Landa-Ruiz, L.;
Nieves-Mendoza, D.;
Almeraya-Calderon, F.
Electrochemical Noise Analysis: An
Approach to the Effectivity of Each
Method in Different Materials.
Materials **2024**, *17*, 4013. https://
doi.org/10.3390/ma17164013

Academic Editor: Ming Liu

Received: 14 July 2024
Revised: 1 August 2024
Accepted: 2 August 2024
Published: 12 August 2024

Abstract: Corrosion deterioration of materials is a major problem affecting economic, safety, and logistical issues, especially in the aeronautical sector. Detecting the correct corrosion type in metal alloys is very important to know how to mitigate the corrosion problem. Electrochemical noise (EN) is a corrosion technique used to characterize the behavior of different alloys and determine the type of corrosion in a system. The objective of this research is to characterize by EN technique different aeronautical alloys (Al, Ti, steels, and superalloys) using different analysis methods such as time domain (visual analysis, statistical), frequency domain (power spectral density (PSD)), and frequency–time domain (wavelet decomposition, Hilbert Huang analysis, and recurrence plots (RP)) related to the corrosion process. Optical microscopy (OM) is used to observe the surface of the tested samples. The alloys were exposed to 3.5 wt.% NaCl and H_2SO_4 solutions at room temperature. The results indicate that HHT and recurrence plots are the best options for determining the corrosion type compared with the other methods due to their ability to analyze dynamic and chaotic systems, such as corrosion. Corrosion processes such as passivation and localized corrosion can be differentiated when analyzed using HHT and RP methods when a passive system presents values of determinism between 0.5 and 0.8. Also, to differentiate the passive system from the localized system, it is necessary to see the recurrence plot due to the similarity of the determinism value. Noise impedance (Z_n) is one of the best options for determining the corrosion kinetics of one system, showing that Ti CP2 and Ti-6Al-4V presented 742,824 and 939,575 $\Omega \cdot cm^2$, while R_n presented 271,851 and 325,751 $\Omega \cdot cm^2$, being the highest when exposed to H_2SO_4.

Keywords: electrochemical noise; statistical analysis; power spectral density (PSD); wavelets; Hilbert–Huang transform (HHT); recurrence plots (RPs)

1. Introduction

Different conventional electrochemical techniques have been used to determine the corrosion kinetics and reaction mechanisms, such as potentiodynamic polarization (PP), electrochemical impedance spectroscopy (EIS), and linear polarization resistance (LPR). However, these techniques can alter the electrochemical system with external signals in electrochemical measurements [1–6]. Using the electrochemical noise (EN) technique for investigation and corrosion monitoring has allowed many advances in recent years in corrosion science. A particular advantage of EN measurements is that they detect and analyze the early stages of localized corrosion.

Electrochemical noise describes the spontaneous low-level potential and current fluctuations during an electrochemical process. The EN can be used to monitor different corrosion processes, but it is specialized in localized processes; the type of analysis is linked to the type of signal present in the system. One advantage of EN is its efficiency in localized processes, which is a non-perturbative technique [7–11].

The EN analysis can be classified into time-domain, frequency-domain, frequency–time, and chaotic systems. In the first years of the technique, the signal was analyzed by the visual method; the next analysis realized the EN signal by realizing the statistical method, where authors such as Mansfeld, Cottis, Turgosse, Eden, and Bertocci [11–18] related the type of corrosion with statistical parameters such as localization index (LI) and pitting index (based on the standard deviation of ECN and EPN). Furthermore, the noise resistance (R_n) parameter was obtained and is employed as a homolog of R_p, relating to a kinetic variable. Furthermore, some authors [15–18] used kurtosis and skewness to improve the LI to determine the corrosion type.

The components that make up the EN signal include DC, random, and stationary. DC must be separated from stationary and random components to analyze EN data. DC introduces false frequencies and interferes with statistical, visual, and PSD assessments. This way, corrosion data presented at low frequencies are maintained when DC is removed. EN can be expressed via Equation (1) [15,19,20]:

$$x(t) = m_t + s_t + Y_t \tag{1}$$

The noisy signal (x_n) and polynomial of "n" grade (p_o) at n-th term (a_i) in "n" time are defined by the polynomial approach, as stated in Equation (2), to produce a signal devoid of trend (y_n) [1,11,21,22]:

$$y_n = x_n - \sum_{i=0}^{p_o} a_i n^i \tag{2}$$

To determine noise resistance (R_n), obtaining a standard deviation from time series values (EPN and ECN) is necessary. See Equation (3); these statistical values give information about corrosion kinetics and mechanisms. Turgoose and Cottis [11] found a relationship between the increase in variance and standard deviation and an increase in corrosion rate.

$$R_n = \frac{\sigma_v}{\sigma_I} * A \tag{3}$$

Kurtosis and skewness were used in this study to identify the type of corrosion. It was not considered because Mansfeld and Sun [22] decided in 1995 that the localization index (L.I.) exhibits limits and should be used with prudence. According to a patent developed in 2001 by Reid and Eden [23], corrosion type can be identified using statistical moments with skewness and kurtosis (Equations (4) and (5)), which are the third and fourth statistical moments, respectively [24–28]:

$$skewness = \frac{1}{N}\sum_{i=1}^{N} \frac{(x_i - \bar{x})^3}{\sigma^3} \tag{4}$$

$$skewness = \frac{1}{N}\sum_{i=1}^{N} \frac{(x_i - \bar{x})^3}{\sigma^3} \tag{5}$$

For PSD analysis, it is necessary to transform the time-domain EN to the frequency-domain by applying a fast Fourier transform (FFT). Since there is a correlation with the EN signal (with a polynomial filter applied), the spectral density is calculated with Equations (6) and (7) [29,30].

$$R_{xx}(m) = \frac{1}{N}\sum_{n=0}^{N-m-1} x(n) \cdot x(n+m) \text{ when values are from } 0 < m < N \tag{6}$$

$$\Psi_x(k) = \frac{\gamma \cdot t_m}{N} \cdot \sum_{n=1}^{N} (x_n - \overline{x}_n) \cdot e^{\frac{-2\pi k n^2}{N}} \qquad (7)$$

The PSD is interpreted using the limit frequency to cut frequency as a basis. The cut frequency indicates the start and end of the slope, which is useful in determining the corrosion mechanism. Information regarding sample representation following pitting is provided by cut frequency [31]. The slope is represented by Equation (11) and is defined by β_x:

$$\log \Psi_x = -\beta_x \log f \qquad (8)$$

Because the power PSD is correlated with the overall amount of energy in the system, the frequency zero limit (ψ^0) provides information on material disintegration [32]. Material dissolution is limited to the current PSD [33,34]. In 1998, Mansfeld et al. [35] suggested determining the corrosion phenomena on the material surface using the intervals of β (adjusted to decibels) [36,37]. It is important to emphasize that some values are the same for two types of corrosion; this could create another way of studying the slope along frequencies [38,39].

In trying to develop the analysis, different authors used the wavelet transform to make a signal decomposition generating an energy plot, where the signal is divided on crystal, usually from 1 to 8. The authors related the first crystal energy, D1 to D3, to metastable pitting. The crystals D4 to D6 are associated with localized corrosion, while D7 and D8 are linked to diffusion or controlled processes (uniform corrosion) [40]. The S8 crystal is associated with the DC signal from the EN signal [15,16,18,41]. With wavelets, a signal is broken down using a high-low filter; high frequencies are referred to as detail, and low frequencies are approximations [42]. Equation (9) [43] provides the total energy of an N number of data points.

$$E = \sum_{n-1}^{N} x_n^2 \qquad (9)$$

Additionally, Equation (10) provides the energy fractions of details and approximations:

$$ED_j^d = \frac{1}{E}\sum_{n=1}^{N} d_{j,n}^2 \; ED_j^s = \frac{1}{E}\sum_{n=1}^{N} s_{j,n}^2 \qquad (10)$$

Equation (11) states that the total energy evaluated is equal to the energy of each wavelet transform component:

$$E = ED_j^s \sum_{j=1}^{j} ED_j^d \qquad (11)$$

Another type of analysis is the time-frequency domain, which is presented by Hilbert–Huang transform (HHT) [31]. This method is based on an empirical decomposition (EMD) that permits analysis of the non-stationary signal obtained from the intrinsic function of the signal. The graphic obtained is the Hilbert specter, which helps to determine the corrosion process that occurs in the system and the corrosion mechanism; authors such as Mol, Zhao, and Homborg [30,31,39,40] suggest that the energy accumulated at low frequencies is related to uniform and diffusion processes, but if the energy is accumulated at middle and high frequencies, it is related to localized processes.

The HHT is an additional sophisticated technique for identifying the kind and process of corrosion; it assists in eliminating DC from the original signal [43]. Furthermore, using the HHT suggested by Huang et al. [44] in 1998 to investigate non-stationary signals, this method can also localize the frequency and time at which energy interchange occurs. This energy is known as instantaneous energy and is calculated by an empirical method of decomposition (EMD) to obtain intrinsic functions (IMF). It is possible to localize the collected energy by generating a spectrum with the time-frequency-energy distribution [45,46]. Huang's proposed EMD is represented by Equation (12):

$$x(t) = \sum_{i=1}^{N} h^{(i)}(t) + d(t) \qquad (12)$$

where $h^{(i)}(t)$ is the *i*-th term of the IMF that is generated; these numbers must satisfy that the extreme and cross numbers are equal or differ in one at maximum and that each point using the local maximum and minimum must be zero [47]. $d(t)$ is the average trend at a low frequency in the time series $x(t)$ and cannot be decomposed. Equation (13) of HHT is controlled by the following:

$$y_j(t) = \frac{1}{\pi} p \int_{-\infty}^{\infty} \frac{h_j(\tau)}{t - \tau} d\tau \tag{13}$$

The Hilbert transform is denoted by $y_j(t)$, and the IMF is represented by h_j; p is associated with the Cauchy principle and the IMF average [48].

One of the methods to analyze a non-linear (chaotic) system can be interpreted by recurrence plots (RPs). RPs help distinguish if the processes in the system are deterministic recurrence processes. If the system is deterministic (D), it is related to localized processes, and if the recurrence is the system's domain, the process is associated with uniformity [49,50]. A two-dimensional binary diagram known as a recurrence plot (RP) encodes the temporal pattern of a single recorded time series of an observable, such as the current I in this study. It depicts recurrences of a trajectory $x_i \in$ Rm at distinct periods *i*, *j*, that occur in m-dimensional phase space and within a given threshold limit ε. Specifically, an RP is a picture of a two-dimensional square matrix with black and white dots on two axes of time, t_i and t_j, representing ones and zeros, respectively. Each black dot at a position (t_i, t_j) denotes the state $x(t_i)$ reoccurring at time j. The mathematical expression (Equation (14)) populates the matrix [51].

$$R_{ij} = \Theta\left(\varepsilon - \left\|\vec{x_i} - \vec{x_j}\right\|\right), \; i, j = 1, \ldots, N \tag{14}$$

where N is the number of measured points in this case. x_i, epsilon represents the threshold distance, θ_x denotes the heavy function, and $\|.\|$ is normal and can be either Euclidian or maximum. The threshold epsilon constraint is responsible for the binary black-and-white appearance of recurrence graphs and for enabling the quantification of certain values [52].

Authors Ren et al. [53] conducted research to identify corrosion types in electrochemical noise; they used an Adaboost system based on statistical analysis and shot noise theory, and they found that it is possible to identify the corrosion type based on shot noise theory and statistical results, with the shot noise being the complement for statistical results. Montoya-Rangel et al. [54] employed the EN to determine the corrosion mechanism in DP steels. When exposed to chlorides, the PSD analysis detected a galvanic couple in the steel phases. Conventional EN methods cannot obtain that result. Ye et al. [55] did similar research; however, they employed the Wavelets method to determine the galvanic corrosion that occurs on DSS 2205, obtaining good results for this analysis, which was more precise than other methods. The wavets method is used to analyze more sophisticated systems, as Shahri et al. [56] analyzed localized corrosion of PEO coatings. The EN technique analyzed using the wavelet method helped to determine the localized corrosion system and its weaknesses. Homborg et al. [57] realized the importance of employing EN in different systems, such as heat treatment, corrosion inhibition, and the behavior of alloys with different phases. Also, it mentioned the importance of DC drift from the original signal. They concluded that the transient analysis can be treated as an image classification problem and that a wavelet transform helps determine the corrosion system.

Also, the authors work on using HHT to determine processes in dynamic systems, and stochastic behavior is one of the most important limitations of other EN methods. The ability to detect spontaneous reactions is an important advantage over other methods, as is the regeneration of some passive layers [58–60]. Ortíz-Corona et al. [61] employed recurrence plots to determine the chemical reactions that occur in Ag–Cu alloys, and they concluded that in the time series, the transitions of corrosion types could not be distinguished by statistical methods, and with the RPs analysis, the different transitions that occur in the system can be observed, obtaining a better performance than statistical analysis. The use of RPs is highly important in non-linear systems, according to different authors, due to the precision and diversity of results that can be analyzed because it does not present limitations for some signals [62–64].

This research aimed to study, employing electrochemical noise, the effectiveness of each analysis method in different aeronautical alloys (AISI 1018 CS, 304 SS, 316 SS, Inconel 718, Al 2024, AA 2055, AA 6061, Custom 450, Ti-6Al-4V, Ti CP2, Ultimet, and Waspaloy) related to the corrosion process. Optical microscopy (OM) is used to observe the surface of the tested samples. The alloys were exposed to 3.5 wt.% NaCl and H_2SO_4 solutions at room temperature. Electrochemical characterization of these alloys could find potential in aeronautical applications such as fuselage, turbine blades, aircraft landing gear, and structural components. The alloys of aircraft are susceptible to localized or general corrosion when they are exposed to different atmospheres: industrial [acid rain (H_2SO_4)] and marine (NaCl).

2. Materials and Methods

2.1. Materials

The materials used in this work were aeronautical alloys AISI 1018 CS, 304 SS (Austenitic), 316 SS (Austenitic), Inconel 718 (Nickel-based), AA 2024 (Aluminum-Copper), AA 2055 (Aluminum-Lithium), AA 6061, AM 350 (semi-austenitic), Custom 450 (Martensitic), Ti-6Al-4V, Ti CP2 (Pure Titanium), Ultimet (Cobalt-based), and Waspaloy (Nickel-based), used in the received condition.

2.2. Microstructural Characterization

The sample preparation was realized using metallographic techniques according to ASTM E3 [65]. The polishing was performed using different SiC grit papers until 600 grades, followed by ultrasonic cleaning in ethanol (C_2H_5OH) and deionized water for 10 min each.

The microstructural analysis was carried out by optical microscopy (OM, Olympus, Hamburg, Germany) to identify samples' morphology at a magnification of $100\times$.

2.3. Electrochemical Testing

The electrochemical measurements were made at room temperature using a conventional three-electrode cell (Figure 1). The working (aeronautical alloys) and auxiliary electrodes were similar electrodes, and a saturated calomel electrode was used as reference [66–69]. EN measurements were carried out according to the ASTM G199-09 standard [66]. In each experiment, 2048 data points were measured with a scanning speed of 1 data/s. The current and potential time series were visually analyzed to interpret the signal transients and define the behavior of the frequency and amplitude of fluctuations as a function of time. The electrolytes used were 3.5 wt.% NaCl and H_2SO_4 solutions. Tests were performed in duplicate. The electrochemical noise measurements were recorded simultaneously using a Gill-AC potentiostat/galvanostat/ZRA (Zero Resistance Ammeter) from ACM Instruments (Manchester, UK). The tests were realized in triplicate.

Data analysis was processed using a program made in MATALB 2018a software (Math Works, Natick, MA, USA). In the time-domain analysis, the DC trend signal was removed from the original EN signal by the polynomial method, and from the signal without DC, statistical data (R_n, kurtosis, and skewness) were obtained. For frequency-domain analysis of PSD (power spectral density) data, a Hann window was applied before being transformed to the frequency domain by an FFT (fast Fourier transform). Frequency–time-domain analysis energy dispersion graphs were made (EDP), where the orthogonal wavelet transform was applied to the original signal (with DC) because this method separates DC from EN signal. EN analysis with Hilbert–Huang transform (HHT) was necessary to obtain the intrinsic functions (IMF) of EN signal by an empirical decomposition method (EMD), and finally, the instantaneous frequencies were plotted with a Hilbert spectrum.

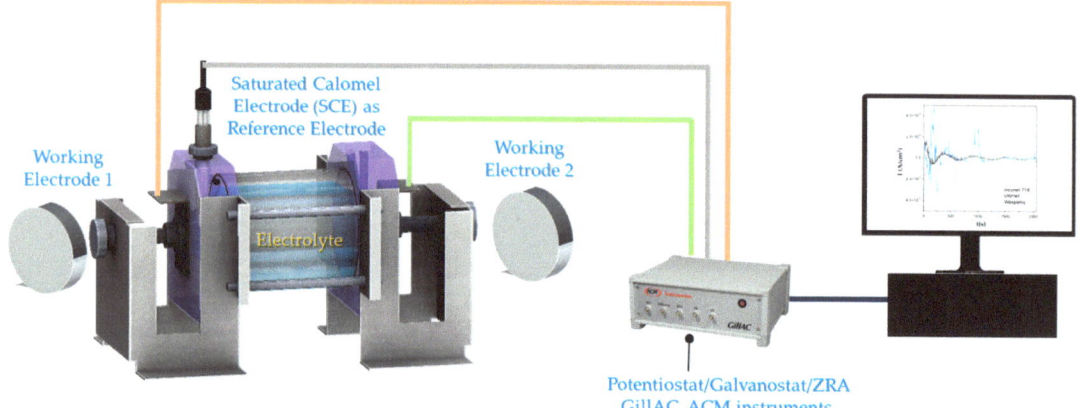

Figure 1. Experimental setup for electrochemical noise (EN) measurements.

3. Results and Discussion

3.1. Electrochemical Noise (EN)

3.1.1. Time-Domain Analysis

Figures 2–5 shows the time series recorded for all the alloys' potential and current. Figure 2a shows the electrochemical potential noise (EPN) when alloys were exposed to NaCl. Figure 2b shows the low amplitude and frequency transients for the Ultimete alloy; these transients are characteristic of generalized corrosion. The Ti-6Al-4V, Figure 2c, presented transients indicating the breaking and regeneration of the passive layer. On the other hand, AA 2024, Figure 2d, presented a higher potential amplitude of 4×10^{-4} V vs. SCE; those fluctuations are related to a mixed corrosion process. The behavior that presented steels, Figure 2e, is associated with a localized process that presented many transients with amplitudes.

Figure 3a shows the electrochemical current noise (ECN); it presents a high amplitude signal for AISI 1018 CS, indicating a possible higher corrosion kinetic with values of 6×10^{-6} A/cm^2, and a cathodic transient of 8×10^6 A/cm^2, indicating a possible localization process. In Figure 3b, superalloys presented anodic transients (8×10^{-7} A/cm^2), and Inconel 718 presented more of this behavior, associating with a localized process for this superalloy. On the other hand, Ultimet and Waspaloy presented a fluctuation with a transient system of low amplitude, relating that process to a possible passivation system. The Ti-6Al-4V presented in Figure 3c shows the behavior of a passive system with a very low signal. In the case of steels, all presented anodic transients of 1×10^{-7} A/cm^2, indicating that a localized process occurs on the surface. The aluminum alloys' behavior differs for each (see Figure 3d); the AA 2024 presented high fluctuations, 1×10^{-6} A/cm^2, indicating a higher corrosion kinetic mean, while the AA 6061 presented lower fluctuations, indicating low corrosion kinetic. Those results are shown in Table 1, where the AA 6061 presented 170,057 Ω·cm^2 and the AA 2024 43,859 Ω·cm^2.

Figure 4a shows the EPN of alloys exposed to H$_2$SO$_4$, and Figure 5a shows the ECN of alloys exposed to H$_2$SO$_4$. In almost all EPN alloys, Figure 5b–e presents a behavior related to the generation of a passive layer. For this reason, it is important to analyze Figure 5 of the ECN to determine if passivation occurs or if another process exists. Figure 4a shows how the AISI 1018 CS presented anodic transients of 2×10^{-5} A/cm^2, indicating a possible localized process. Also, it presented the highest amplitude and AA 2024 (1×10^{-4} A/cm^2), indicating that corrosion kinetics are higher (see Figure 5b). The results correspond to Table 2, where AISI 1018 CS and AA 2024 presented R$_n$ values of 130 and 249 Ω·cm^2. The signal of Ti-6Al-4V and Ti CP2 is related to a passive system (Figure 5c). On the other hand, Inconel 718 (Figure 5d) and Custom 450 (Figure 5e) observed anodic transients of 4×10^{-7} A/cm^2, indicating a localized corrosion process.

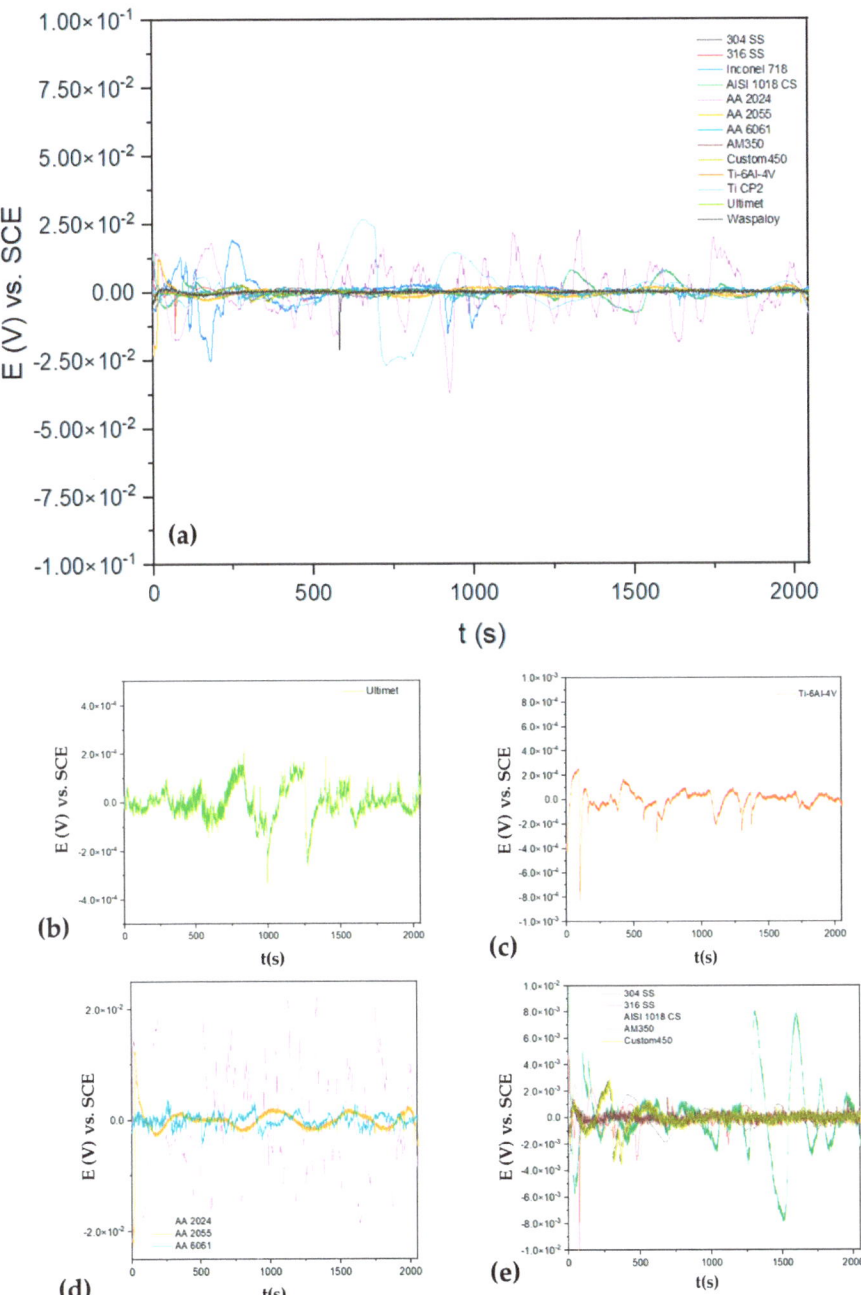

Figure 2. Electrochemical potential noise-time series for alloys in NaCl (**a**) and windowing for (**b**) Ultimet, (**c**) Ti6Al-4V, (**d**) AA2024, AA 2055 and AA6061 (**e**) 304 SS, 316 SS, AISI 1018 CS. AM 350 and Custom 450.

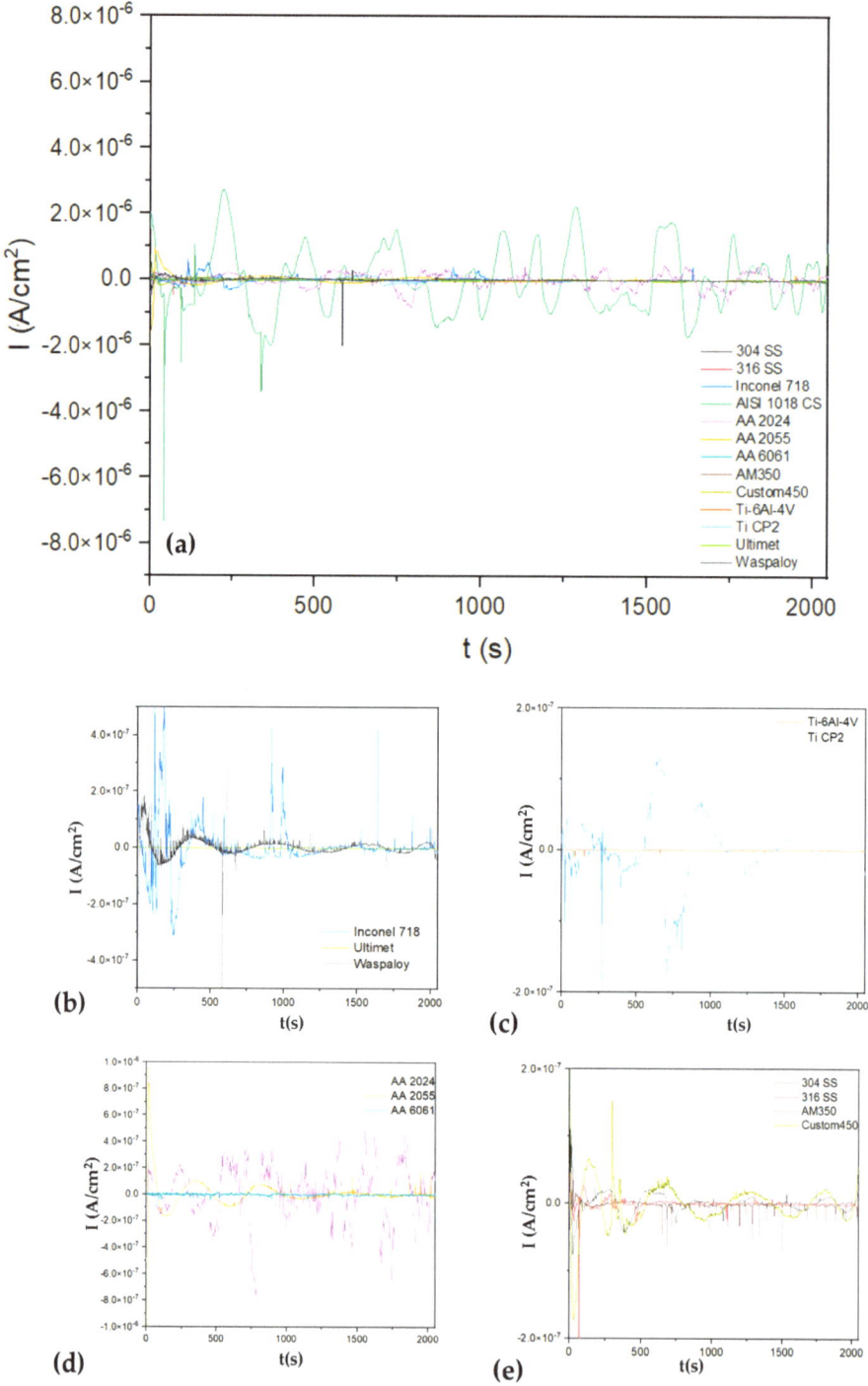

Figure 3. Electrochemical current noise-time series for alloys in NaCl (**a**) and windowing for (**b**) Inconel 718, Ultimet and Waspaloy, (**c**) Ti-6Al-4V and Ti CP2, (**d**) AA 2024, AA 2055 and AA 6061 (**e**) 304 SS, 316 SS, AM 350 and Custom 450.

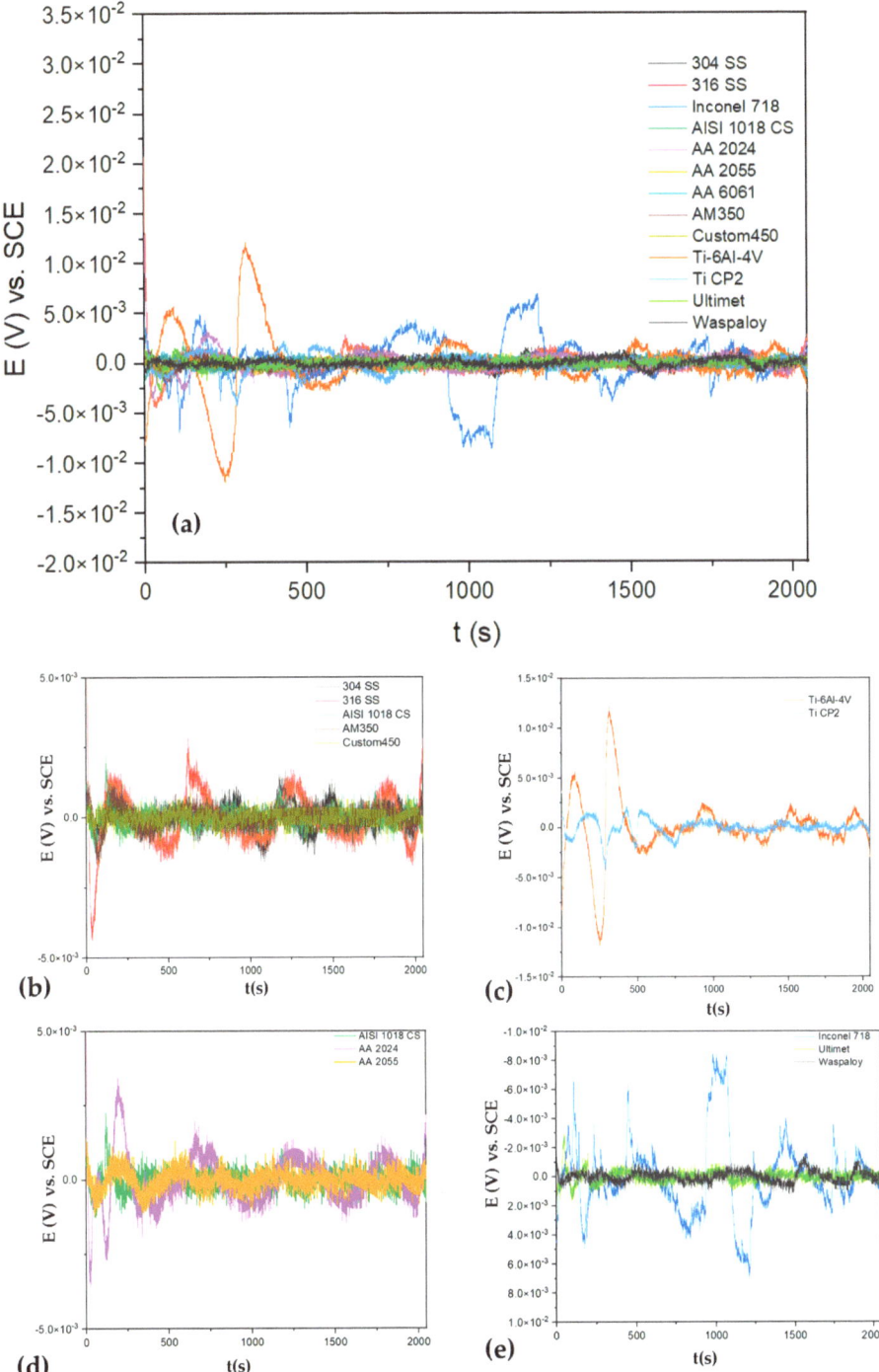

Figure 4. Electrochemical potential noise-time series for alloys in H_2SO_4 (**a**) and windowing for (**b**) 304 SS, 316 SS, AISI 1018 CS, AM 350 and Custom 450, (**c**) Ti-6Al-4V and Ti CP2, (**d**) AA2024, AA 2055 and AISI 1018 CS, (**e**) Inconel 718, Ultimet and Waspaloy.

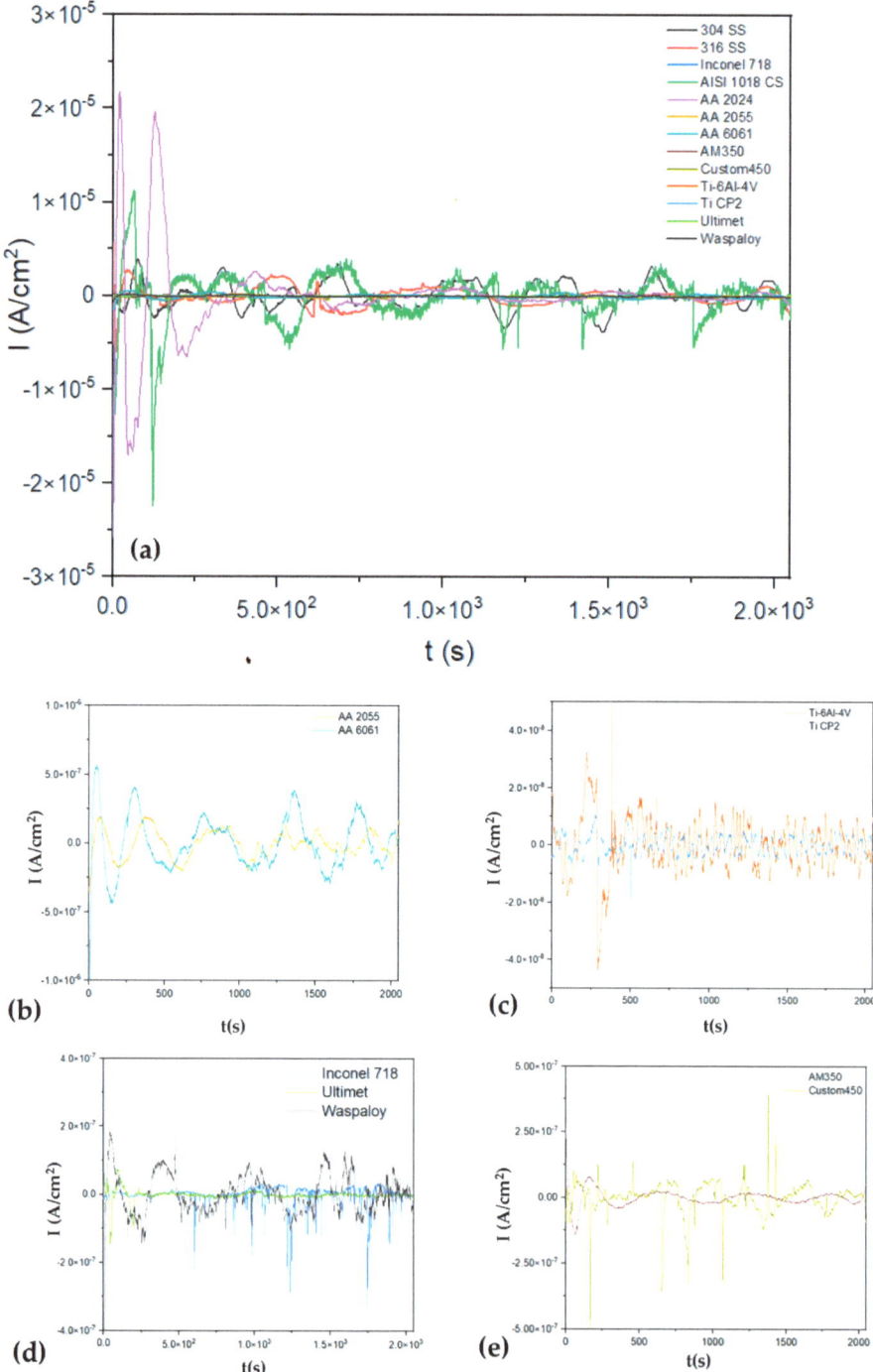

Figure 5. Electrochemical current noise-time series for alloys in H_2SO_4 (**a**) and windowing for (**b**) AA2024 and AA 2055, (**c**) Ti-6Al-4V and Ti CP2, (**d**) Inconel 718, Ultimet and Waspaloy (**e**) AM350 y custom450.

Table 1. Statistical parameters from time series of EN exposed to NaCl solution.

Alloys	R_n ($\Omega \cdot cm^2$)	LI	Corrosion Type	Kurtosis	Corrosion Type	Skewness	Corrosion Type
304 SS	54,772 ± 11	0.07 ± 0.02	Mix	4.5 ± 0.8	Loc	−1 ± 0.2	Uni
316 SS	66,621 ± 10	0.2 ± 0.03	Loc	26 ± 2	Loc	−29 ± 0.8	Loc
Inconel 718	57,745 ± 8	0.12 ± 0.08	Loc	12 ± 1.2	Loc	1 ± 0.5	Loc
1018 CS	2817 ± 14	0.03 ± 0.01	Mix	4 ± 1.0	Loc	0 ± 0.7	Uni
AA 2024	43,856 ± 9	0.28 ± 0.1	Loc	4 ± 0.6	Loc	−1 ± 0.3	Uni
AA 2055	15,212 ± 20	0.18 ± 0.07	Loc	72 ± 3	Loc	−4 ± 0.8	Loc
AA 6061	170,057 ± 24	0.06 ± 0.02	Mix	22 ± 2.6	Loc	−2 ± 0.9	Loc
AM350	76,329 ± 13	0.02 ± 0.02	Mix	69 ± 4	Loc	−4 ± 1.2	Loc
Custom450	17,038 ± 24	0.16 ± 0.12	Loc	87 ± 6	Loc	5 ± 0.9	Loc
Ti-6Al-4V	175,051 ± 29	0.01 ± 0.01	Mixt	188 ± 5	Loc	−0 ± 0.3	Uni
Ti CP2	204,605 ± 24	0.36 ± 0.08	Loc	6 ± 1.5	Loc	−0 ± 0.6	Uni
Ultimet	75,342 ± 26	0.13 ± 0.03	Loc	538 ± 7	Loc	13 ± 1.1	Loc
Waspaloy	14,451 ± 14	0.05 ± 0.02	Mix	10 ± 1.3	Loc	−17 ± 1.4	Loc

Table 2. Statistical parameters from time series of EN exposed to H_2SO_4 solution.

Alloys	R_n ($\Omega \cdot cm^2$)	LI	Corrosion Type	Kurtosis	Corrosion Type	Skewness	Corrosion Type
304 SS	326 ± 5	0.01 ± 0.02	Mix	3 ± 0.8	Uni	0 ± 0.2	Uni
316 SS	1208 ± 15	0.06 ± 0.02	Mix	6 ± 0.2	Loc	−0 ± 0.7	Uni
Inconel 718	90,112 ± 20	0.4 ± 0.1	Loc	32 ± 1.5	Loc	−4 ± 0.3	Loc
1018 CS	130 ± 8	0.06 ± 0.02	Mix	14 ± 1.1	Loc	−2 ± 0.1	Loc
AA 2024	249 ± 12	0.01 ± 0.01	Mix	15 ± 1.7	Loc	1 ± 0.1	Uni
AA 2055	3564 ± 29	0.009 ± 0.001	Uni	3 ± 1.3	Uni	−0 ± 0.3	Uni
AA 6061	2384 ± 20	0.4 ± 0.1	Loc	6 ± 0.8	Loc	−0 ± 0.3	Loc
AM350	12,563 ± 24	0.02 ± 0.004	Mix	14 ± 0.7	Loc	0 ± 0.6	Uni
Custom450	4484 ± 33	0.03 ± 0.02	Mix	15 ± 0.5	Loc	−2 ± 0.2	Loc
Ti-6Al-4V	325,751 ± 16	0.4 ± 0.02	Loc	6 ± 0.3	Loc	−0 ± 0.5	Uni
Ti CP2	271,851 ± 123	0.04 ± 0.005	Mix	4 ± 1.2	Loc	0 ± 0.2	Uni
Ultimet	25,560 ± 37	0.5 ± 0.04	Loc	28 ± 2.5	Loc	−3 ± 0.7	Loc
Waspaloy	6356 ± 57	0.01 ±	Mix	16 ± 1.1	Loc	−1 ± 0.6	Loc

Table 1 shows the statistical results, which will be correlated with the time-series analysis. The R_n shows a higher resistance for Ti-alloys (17.5×10^4 and 20.4×10^4 $\Omega \cdot cm^2$); meanwhile, the AISI 1018 CS presented a lower corrosion resistance of 2.8×10^4 $\Omega \cdot cm^2$. Comparing the localization index in Table 3, any one of the samples presented a uniform process; all presented localized or mixed processes. When kurtosis analyzed samples, all samples presented localized corrosion in NaCl solutions. However, when analyzed by skewness, 304 SS, AISI 1018 CS, AA 2024, Ti-6Al-4V, and Ti CP2 presented uniform corrosion. This marks a difference in the analysis and generates uncertainty for the analysis.

For the samples analyzed on H_2SO_4, the results (see Table 2) were very similar; Ti-6Al-4V and Ti CP2 presented higher values of R_n (32.5×10^4 and 27.1×10^4 $\Omega \cdot cm^2$), and the AISI 1018 and AA 2024 presented lower corrosion resistance (130, 249, and 326 $\Omega \cdot cm^2$). When corrosion type is analyzed by localization index, only AA 2055 (0.09) presented uniform corrosion, and the rest of the samples presented mixed and localized corrosion. When those samples were analyzed by kurtosis, 304 SS and AA 2055 presented uniform corrosion. On the other hand, skewness showed that 304 SS, AA 2024, AA 2055, AM350, Ti-6Al-4V, and Ti CP2 presented values related to uniform corrosion. Therefore, it is necessary to validate the results with other analysis methods.

Table 3. PSD parameters when alloys are exposed to NaCl and H_2SO_4 solutions.

Alloys	Slope (dBi)	Limit Frequency	Z_n ($\Omega \cdot cm^2$)	Alloys	Slope (dBi)	Limit Frequency	Z_n ($\Omega \cdot cm^2$)
	NaCl Solution				**H_2SO_4 Solution**		
304 SS	-14 ± 1	-130 ± 12	$63{,}123 \pm 70$	304 SS	-15 ± 1.2	-109 ± 5	612 ± 12
316 SS	-12 ± 0.8	-140 ± 15	$183{,}756 \pm 122$	316 SS	-17 ± 1	-100 ± 3	1129 ± 67
Inconel 718	-13 ± 1.1	-130 ± 9	$85{,}956 \pm 130$	Inconel 718	-10 ± 1.1	-133 ± 8	$268{,}894 \pm 50$
1018 CS	-17 ± 1.2	-106 ± 14	1342 ± 27	1018 CS	-9 ± 0.5	-106 ± 4	136 ± 8
AA 2024	-15 ± 0.7	-122 ± 11	$109{,}304 \pm 159$	AA 2024	-11 ± 0.4	-107 ± 8	2493 ± 36
AA 2055	-2 ± 0.5	-119 ± 8	$36{,}459 \pm 98$	AA 2055	-15 ± 0.3	-121 ± 2	2567 ± 38
AA 6061	-6 ± 0.9	-145 ± 10	$162{,}779 \pm 211$	AA 6061	-10 ± 0.7	-115 ± 9	5983 ± 69
AM350	-5 ± 0.3	-138 ± 14	$46{,}864 \pm 136$	AM350	-7 ± 0.4	-123 ± 10	$11{,}072 \pm 96$
Custom450	-2 ± 0.2	-123 ± 13	3242 ± 67	Custom450	-19 ± 1.3	-122 ± 9	1793 ± 82
Ti-6Al-4V	-12 ± 0.4	-155 ± 13	$223{,}794 \pm 177$	Ti-6Al-4V	-12 ± 0.9	-140 ± 11	$939{,}575 \pm 241$
Ti CP2	-13 ± 0.8	-128 ± 10	$222{,}411 \pm 389$	Ti CP2	-9 ± 0.4	-154 ± 13	$742{,}824 \pm 265$
Ultimet	-4 ± 0.5	-151 ± 17	$59{,}992 \pm 450$	Ultimet	-4 ± 0.2	-135 ± 7	$34{,}595 \pm 76$
Waspaloy	-5 ± 0.8	-123 ± 15	9656 ± 87	Waspaloy	-11 ± 0.3	-119 ± 9	2328 ± 53

3.1.2. Frequency-Domain Analysis

Power Spectral Density and Noise Impedance (Z_n)

Figures 6 and 7 show the PSD and noise impedances of alloys. Table 3 presents the results of the parameters obtained in the frequency domain analysis. The AISI 1018 CS presented the lower Z_n (1.3×10^4 $\Omega \cdot cm^2$) in NaCl solution, followed by Custom 350, 304 SS, and 316 SS when exposed to NaCl and H_2SO_4 solutions. That behavior matches the results obtained by statistical analysis with the noise resistance (R_n). The PSD slope results for alloys exposed to NaCl presented values of localized corrosion for SS 304, SS 316, Inconel 718, AISI 1018 CS, AA 2024, Ti-6Al-4V, and Ti CP2; the rest of the alloys presented values of uniform corrosion.

When the alloys were exposed to H_2SO_4, only the Ultimet and AM350 alloys presented slope values of uniform corrosion; the rest of the samples presented localized corrosion according to the slope parameters. The AISI 1018 CS presented 136 $\Omega \cdot cm^2$ of resistance, followed by the 304 SS with 612 $\Omega \cdot cm^2$.

Limit frequency to cero showed results that were divergent with Z_n; this indicated a variation between potential and current results. Some authors consider this to be correct. However, when the system is microbial, using the PSD in current as the direct parameter to determine corrosion kinetics is better; on the other hand, the slope shows values related to localization and uniform corrosion; however, those values do not match those obtained by statistical analysis.

3.1.3. Time-Frequency Domain Analysis

Wavelets Analysis

Eight details and one estimate comprise the crystal numbers that need to be analyzed for this study. A metastable pitting process is thought to be responsible for energy accumulation on the first crystals (D1–D3). Localized corrosion is linked to major energy presented in crystals D4–D6, while diffusion, generalized, or controlled processes are thought to be responsible for energy in crystals D7 and D8 [24,30,35]. The DC from the EN signal is associated with the approximate crystal S8. Equation (15) [70] is employed to ascertain each crystal time range.

$$\left(c_1^j, c_2^j \right) = \left(2^{-j} \Delta t, 2^{j-1} \Delta t \right) \tag{15}$$

where Δt is the time display and c is the crystal. Every crystal scale ranges in both Hz and seconds. The initial crystals are high-frequency, whereas the latter display low-frequency phenomena.

Figure 8 shows the energy dispersion plot calculated by the wavelet method. Figure 7a,b shows the results when exposed to NaCl solution. Figure 7 shows how almost all processes are dominated by uniform corrosion or diffusion. Only the Ti-6Al-4V and Ti- CP2 presented on passivation due to their low energy in the crystal. The 316 SS presented energy accumulation at the middle crystals, meaning a localized process; also, the energy at the last crystals presented more energy due to a pitting diffusion.

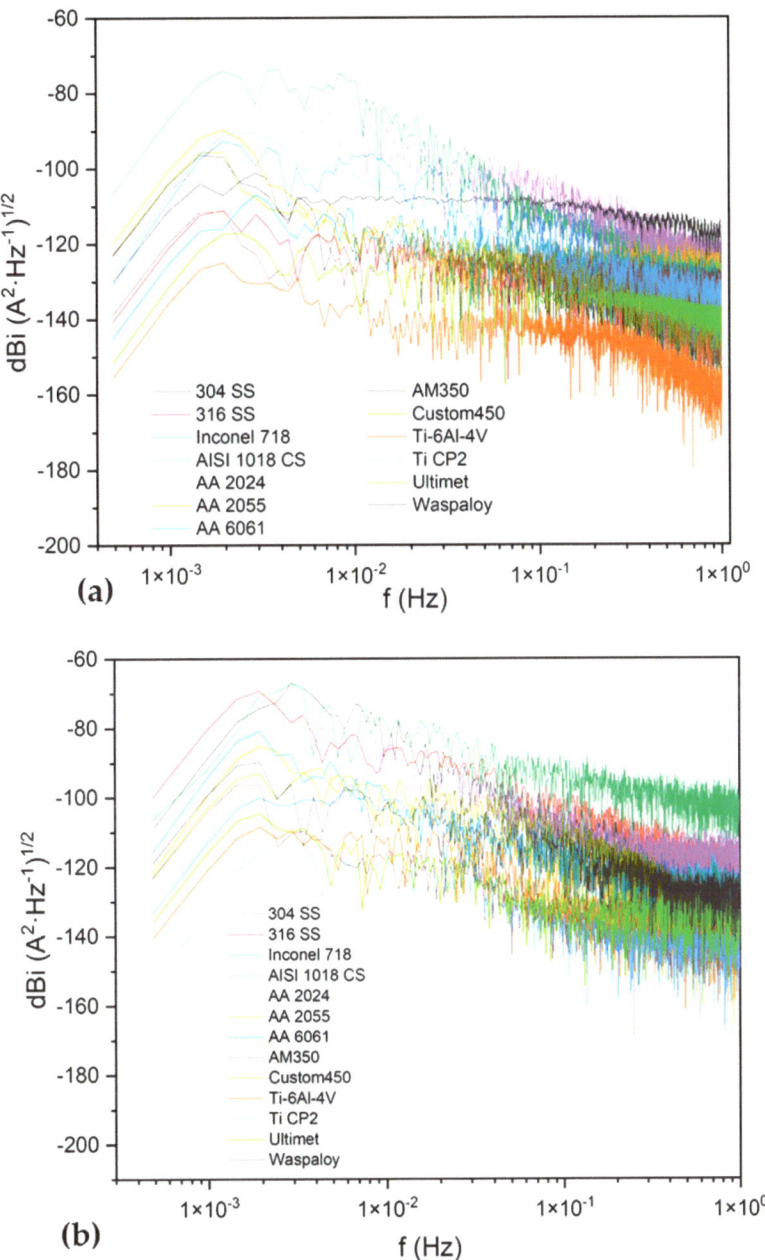

Figure 6. Power spectral density (PSD) in current for samples in NaCl (**a**) and H_2SO_4 (**b**).

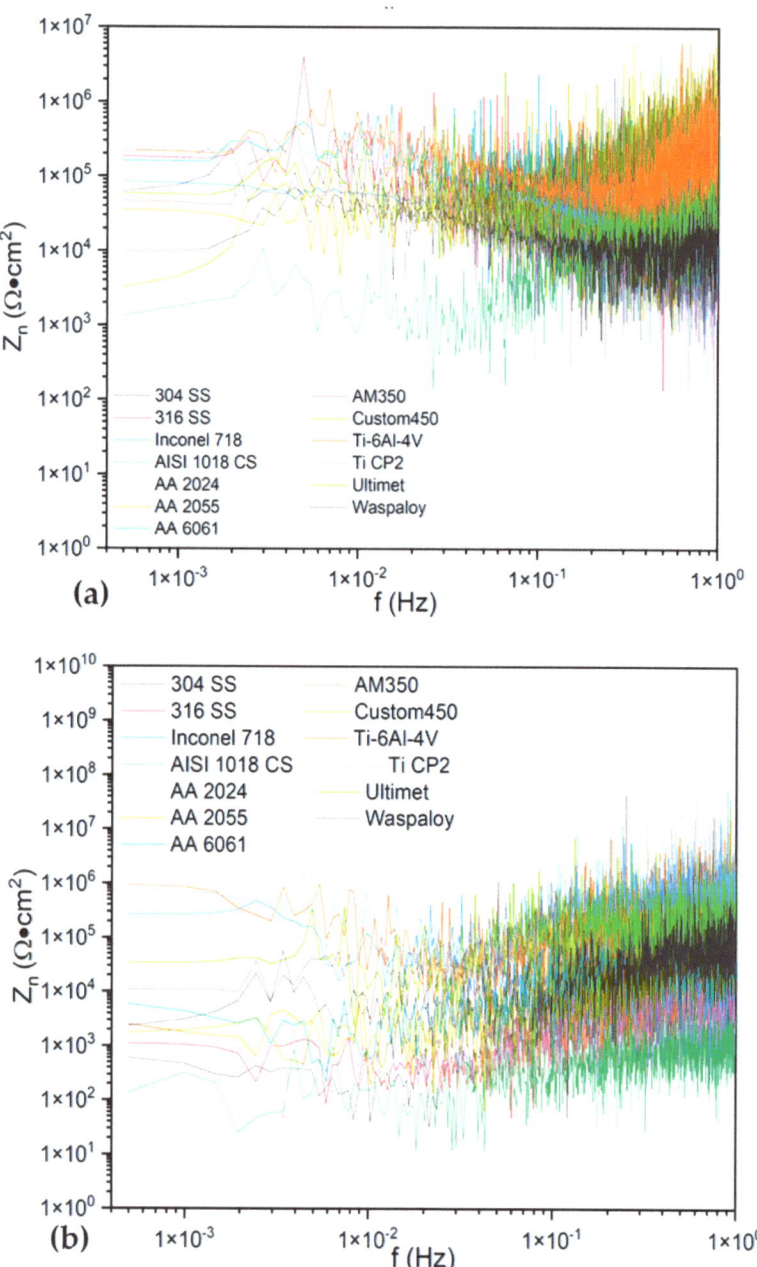

Figure 7. Noise impedance (Z_n) for samples in NaCl (**a**) and H_2SO_4 (**b**).

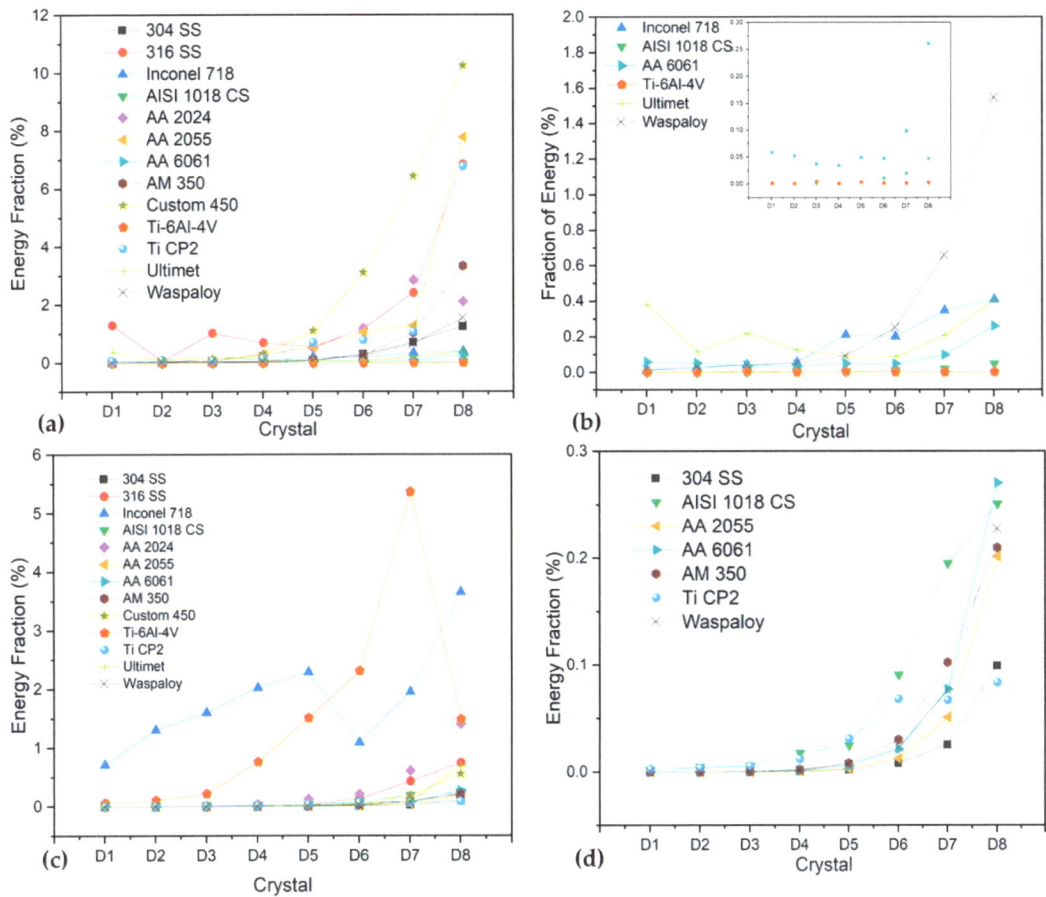

Figure 8. Energy dispersion plot in NaCl (**a**,**b**) and H_2SO_4 (**c**,**d**).

In Figure 8c,d, alloys were exposed to H_2SO_4. The AISI 1018 CS presented a similar behavior in H_2SO_4, where the pitting process is observed. The 316 SS is susceptible to pitting only in NaCl solutions.

Hilbert–Huang Transform Analysis and Recurrence Plots

Only the more significant samples were considered for this analysis due to the graphic number. Figure 9a,c shows a uniform corrosion process; the results obtained by the Hilbert specter and recurrence plot showed congruence, and the microscopy figure helps to show the congruence of the method. The HHT methods show energy accumulation at low frequencies, indicating that a long-term process is occurring. The RP shows a system with a difference in its repeatability, resulting in more recurrence. Also, the determinism value (DET) is higher than when samples present pitting (0.9856 and 0.9295). The value of AA 2025 is lower than the value of 304 SS due to the system being governed by pitting (uniform corrosion), and 304 SS is a uniform oxide layer created.

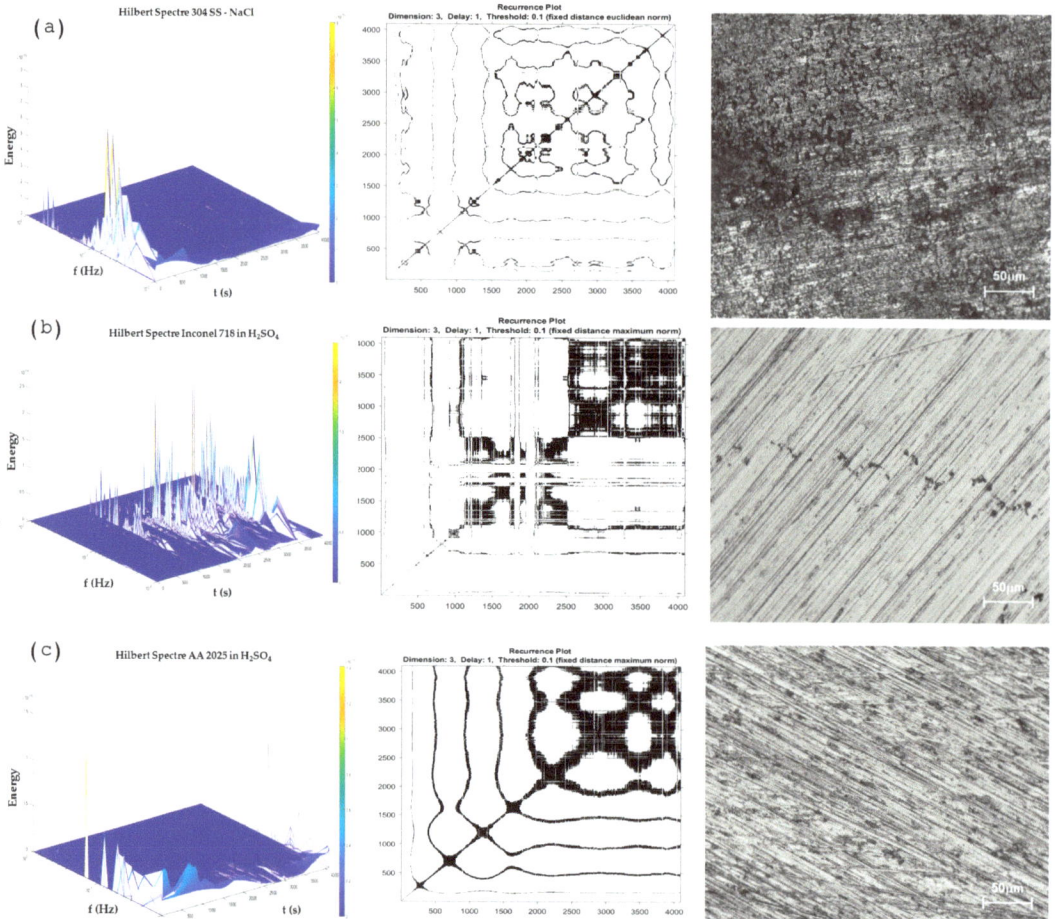

Figure 9. The Hilbert spectra with recurrence plot and morphology by optical microscopy. (**a**) 304 SS in NaCl, (**b**) Inconel 718 in H_2SO_4, (**c**) AA2025 in H_2SO_4.

On the other hand, Figures 9b and 10a, shows a system domain by pitting, where the energy accumulation in the Hilbert specter showed energy at middle frequencies, and the RP presented a high presence of horizontal and vertical lines, indicating that a process is being repetitive; this occurs when localization is present on the surface. When it occurs, the determinist value is reduced, as shown in Table 4, where Inconel 718 obtained a value of 0.8402 and AM350 0.8022, lower than the values of more than 0.9 when alloys have a uniform corrosion system.

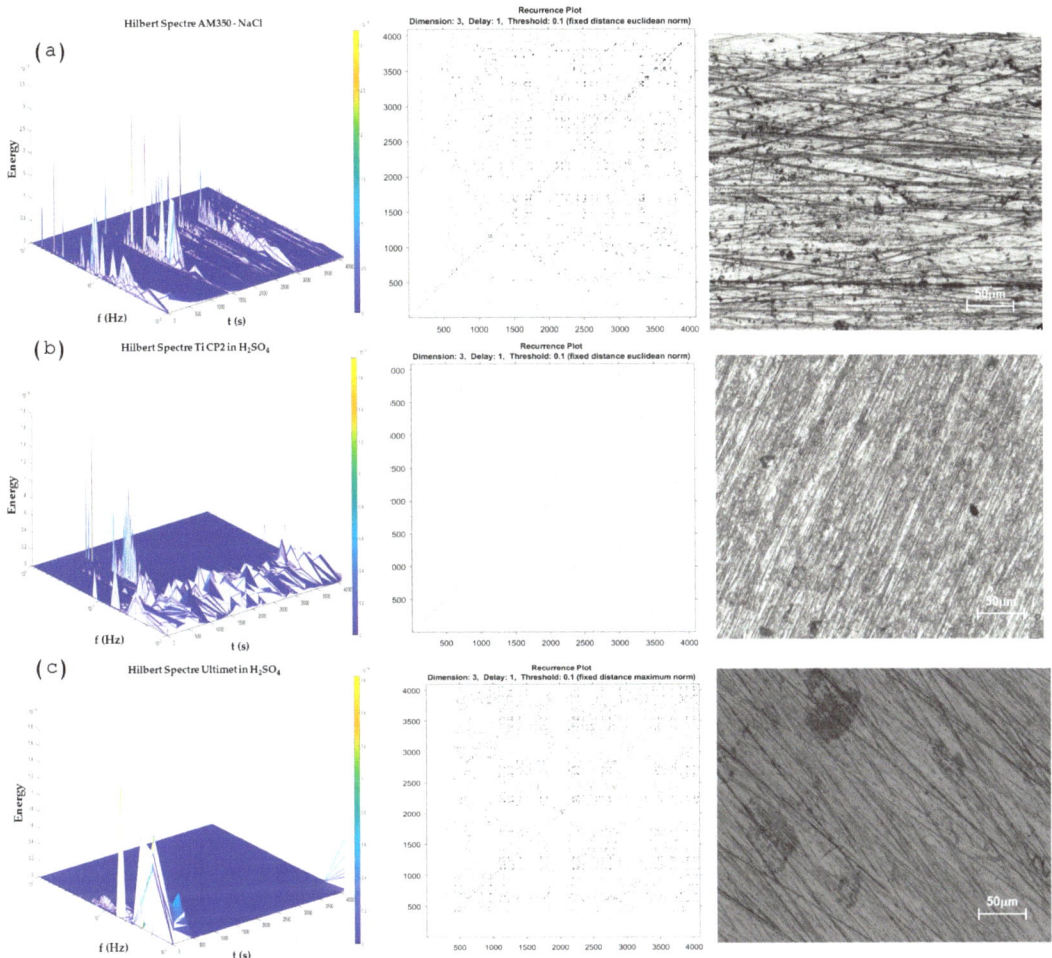

Figure 10. The Hilbert spectra with recurrence plot and morphology by optical microscopy. (**a**) AM350 in NaCl, (**b**) Ti CP2 in H$_2$SO$_4$, and (**c**) Ultimet in H$_2$SO$_4$.

The passive system is present in Figure 10b,c; however, determinism results are lower (0.39 and 0.54 for Ti CP2 and Ultimet exposed to H$_2$SO$_4$). The behavior is sometimes confused with the pitting system due to the nature of how passivation occurs; however, the behavior of RP is more stochastic. The RP can present several differences between a passive and a localized system. The passive system presents dot morphology due to the passive system generating uniform micro-pitting attacks that generate the oxide layer. Therefore, the energy required is very low, and the vertical and horizontal lines are not formed as in the pitting system.

It is important to mention that alloys such as Ti-6Al-4V presented values of 0.7597 when exposed to H$_2$SO$_4$, indicating that the alloy tends to passivate.

Table 4. RP data obtained by quantitative analysis.

	NaCl Solution				H$_2$SO$_4$ Solution		
Alloys	RR	Det	RR/Det	Alloy	RR	Det	RR/Det
304 SS	0.067 ± 0.002	0.985 ± 0.002	0.067 ± 0.002	304 SS	0.028 ± 0.002	0.953 ± 0.001	0.029 ± 0.002
316 SS	0.155 ± 0.03	0.989 ± 0.03	0.156 ± 0.03	316 SS	0.042 ± 0.004	0.970 ± 0.003	0.042 ± 0.003
Inconel 718	0.101 ± 0.07	0.975 ± 0.06	0.103 ± 0.07	Inconel 718	0.056 ± 0.003	0.802 ± 0.004	0.062 ± 0.004
1018 CS	0.042 ± 0.09	0.963 ± 0.07	0.044 ± 0.08	1018 CS	0.011 ± 0.003	0.642 ± 0.002	0.017 ± 0.003
AA 2024	0.017 ± 0.008	0.809 ± 0.007	0.021 ± 0.008	AA 2024	0.172 ± 0.004	0.990 ± 0.004	0.171 ± 0.004
AA 2055	0.171 ± 0.08	0.992 ± 0.06	0.173 ± 0.07	AA 2055	0.026 ± 0.004	0.800 ± 0.003	0.028 ± 0.003
AA 6061	0.001 ± 0.0002	0.269 ± 0.0001	0.004 ± 0.0002	AA 6061	0.025 ± 0.003	0.932 ± 0.003	0.030 ± 0.003
AM350	0.029 ± 0.005	0.802 ± 0.005	0.036 ± 0.005	AM350	0.058 ± 0.003	0.956 ± 0.001	0.060 ± 0.002
CUSTOM450	0.095 ± 0.002	0.990 ± 0.003	0.096 0.003	CUSTOM450	0.048 ± 0.007	0.958 ± 0.008	0.051 ± 0.008
Ti-6Al-4V	0.076 ± 0.001	0.960 ± 0.002	0.079 ± 0.002	Ti-6Al-4V	0.017 ± 0.004	0.759 ± 0.005	0.022 ± 0.005
Ti CP2	0.085 ± 0.007	0.986 ± 0.009	0.086 ± 0.008	Ti CP2	0.002 ± 0.003	0.393 ± 0.002	0.006 ± 0.003
Ultimet	0.039 ± 0.004	0.907 ± 0.003	0.043 ± 0.003	Ultimet	0.013 ± 0.007	0.544 ± 0.008	0.025 ± 0.008
Waspaloy	0.099 ± 0.002	0.967 ± 0.001	0.102 ± 0.001	Waspaloy	0.008 ± 0.001	0.064 ± 0.001	0.123 ± 0.001

4. Discussion

The statistical method showed that the type of corrosion does not correspond in almost all cases. This occurs due to the variability of LI; authors such as Eden and Mansfeld [4,9,71] mentioned that statistical analysis presents limitations, considering that Eden proposed the LI years before to establish the corrosion type. For this reason, LI should be used to determine the corrosion type at your discretion. Furthermore, the signal analyzed by the statistical method must not present a DC signal to reduce the standard deviation and present a more specific result [72]. In this research, the values obtained by LI do not converge with those obtained by skewness, kurtosis, wavelets, slope, HHT, and RPs in almost all results, creating confusion when analyzing the system. Only in steel has LI behavior presented some certainty; however, the limitations of this method are present.

Other statistical methods that are more certain are kurtosis and skewness. For those analyses, some results, such as 304 SS, reflect a result similar to that obtained by the HHT and RP analysie. However, skewness and kurtosis do not present a value for a passive system. The analysis by skewness is more exact than the kurtosis analyses. However, authors such as Cottis, Turgoose, Abella, Jaquez, and Sánchez-Amaya [10,11,72–76] recommend using the skewness with discretion due to statistical analysis' limitations. Something similar occurred with kurtosis and skewness than in LI for this research, where the results between kurtosis and skewness do not match; meanwhile, kurtosis established a localized process, and skewness presented results of uniform corrosion in several alloys. This occurs due to the limitations of statistics and conventional methods for studying complex systems for corrosion. Furthermore, it is important to mention that the skewness results presented more certainty than kurtosis to determine a corrosion process in this research.

The results of PSD show the ψ^0, Z_n, and slope to characterize the corrosion system. The slope analysis to determine the corrosion type presented values related to localized corrosion to 304 SS, 316 SS, Inconel 718, AISI 1018 CS, AA 2024, Ti-6Al-4V, and Ti CP2; the rest of the alloys presented uniform corrosion according to slope parameters in NaCl solution. In H$_2$SO$_4$, only Ultimet presented a uniform corrosion value. The values of the slope parameters used to determine the corrosion type diverge from the results of the statistical method. That divergence is a parameter to consider in the slope analysis with some limitations; however, it is crucial to analyze the change at different PSD frequencies because this indicates changes in the corrosion process [77].

On the other hand, the values of Z_n0 correspond to those obtained by R_n, which shows that both methods are a good option for determining the corrosion resistance. The use of ψ^0 to determine the corrosion kinetics is an option. However, in this case, it only helps to see what system is more active; some authors recommend using that value with a system where the measure of potential is compromised, such as microbial systems [78–80].

The wavelet analysis shows how almost all the samples in NaCl presented energy accumulation in the last crystals, indicating that the alloys presented a long corrosion process. Alloys such as Ti CP2 and Ti-6Al-4V presented low energy, indicating a passive system [81,82]. Several authors [41–44,56,57] mentioned that the wavelet method fits more with studying some dynamic systems. However, the results presented in this research present some limitations due to the necessity of studying a stationary signal; for that reason, more specialized methods must be implemented.

Methods such as HHT and RP present an excellent option. The HHT helps determine the type and process of corrosion that occurs when it is happening. Authors such as Homborg explain that the passive layer's passivation, breaker, and regeneration can be observed in the Hilbert specter. HHT is considered one of the best options for analyzing some signals.

Wavelets and HHT have also been used for economic and mechanical analysis, biomedical industry, and corrosion applications. However, wavelet transform can be affected by the signal type and reconstruction, implying resolution limitations in the time and frequency domains. In contrast, HHT does not present those limitations due to the mathematical adaptation of the transform, allowing the analysis of non-stationary and stationary signals [83–85]. Hence, HHT is more relevant to the analysis of EN signals because it allows the study of many signals, reducing errors. Consequently, this research showed that HHT analysis presented good performance in characterizing the corrosion process in the system and the corrosion type. The results of HHT in this research presented the difference between a pitting and a uniform system. The pitting system presents a Hilbert specter with a high presence of energy at high and middle frequencies, and the uniform system presents energy at low frequencies. As suggested by these research authors, complementing the HHT analysis with the RPs is necessary. However, HHT does not present the limitations of statistical and slope methods.

In previous research, Calabrese et al. [34] concluded that data analysis by HHT is a better option to analyze EN in comparison with wavelets; in this research, that information was confirmed, and HHT is better than statistical PSD analysis to determine the corrosion process and type that occurs.

Several authors have used RP to determine corrosion type [51,74–78,83–85]. Valavanis et al. [86] suggest that RP can be used to determine dynamic states where transitions occur and physiochemically complex processes such as passivity, pitting, and uniform corrosion. Garcia-Ochoa [87] suggests that RP can be used to study non-linear systems and analyze the electrochemical process. Garcia-Ochoa concludes that RP and its quantifications are necessary for analyzing non-linear systems in different scientific disciplines. It is essential to mention that corrosion is a chaotic or non-linear system [88–90]. Due to those, the RP and RQA analyses presented values related to the corrosion process that occurs in the system. For that reason, the results showed that HHT and RP presented more congruence with the corrosion process that occurred on the alloy's surface [91,92]. By this method, the difference between a pitting system (Figure 7a,b) and a passive system (Figure 8b,c) did not occur by conventional statistical methods or wavelets (in more complex systems). However, it is important to make a graphic analysis of RPs due to the values of determinism and recurrence (DET and RR), which presented similar values for a pitting-localized system than for a passive system. It occurred due to the passivation mechanism, propitiated by small pitting processes that generate the oxide layer. In DET and RR values, it cannot be observed and is nerveless; in RPs, a passive system is shown as a dot system, and a localized-pitting system is shown with vertical and horizontal lines that indicate the moment at which pitting occurs.

5. Conclusions

- This study concludes that statistical and slope (PSD) analysis present limitations in determining the corrosion type and some corrosion processes. This can be attributed to the complexity of the EN signal. It is the divergences that determine the type of corrosion that occurs in alloys.

- Statistical analysis presented limitations in all the analyses; localization index, kurtosis, and skewness showed different results in all the analyses. Some authors attribute this to the presence of different corrosion processes in the system; however, it can be confusing and speculative. For that reason, it is important to use a different method to determine the corrosion process in a material.

- The analysis of wavelets presented better results in determining corrosion type compared with statistical and PSD (slope) methods; however, the limitation of analyzing different types of signals limits this method.

- The analysis by HHT and RP presented the best results for determining the corrosion process and type. This is because methods present several advantages to analyzing chaotic signals. Hence, HHT and RP are recommended for analyzing EN signals more than statistical, wavelet, or slope methods. That is because of the nature of the corrosion signal; it is a complex signal, and the EN should be analyzed with the correct tools. The results converged with the results presented by the different authors mentioned in the discussion.

- It is important to complement the results of determinism and RR with the RPs as a visual reference. This is due to the results of DET and RR when a passive and a pitting system are presented. The pitting process presents DET values from 0.5 to 0.8 as well as a passivation system, but the graphic system is different. In passivation, a dot system is present, while in the pitting system, there are horizontal and vertical lines.

- In the HHT and RPs methods, the transition of corrosion processes as the break and regeneration of the passive layer, as well as the pitting generation of the predominance of a uniform process, can be observed. At HHT, the breaking of the passive layer is shown with energy at high frequencies and energy accumulation at low frequencies at subsequent seconds of energy presented at high frequencies.

- The use of R_n and Z_n to determine corrosion resistance can be accepted, and they presented similar results. The higher values were obtained by Ti C2 and Ti-6Al-4V exposed in H_2SO_4 with 271,851 and 325,751 $\Omega \cdot cm^2$ by R_n and 742,824 and 939,575 $\Omega \cdot cm^2$ by Z_n. Although the values are not the same, R_n and Z_n can be considered homologues.

- It is important to define a method for correct EN analysis. If one analysis is correct, it can easily be applied to study neuronal networks or machine learning.

- EN is a powerful technique that can be employed in situ due to its non-destructive properties. Also, using this technique to detect different corrosion systems and some galvanic couples in the alloy phase is helpful for a correct alloy design.

Author Contributions: Conceptualization, J.M.J.-M., C.G.-T., C.M.-R. and F.A.-C.; methodology, J.M.J.-M., F.E.-L., C.M.-R., L.L.-R., C.T.M.-R. and F.A.-C.; data curation, F.E.-L., F.A.-C., M.A.B.-Z., G.S.-H., L.L.-R., D.N.-M., J.M.J.-M. and C.G.-T.; formal analysis, C.G.-T., D.N.-M., M.A.B.-Z., C.M.-R., G.S.-H., C.T.M.-R. and F.A.-C.; writing—review and editing, J.M.J.-M., C.G.-T. and F.A.-C. All authors have read and agreed to the published version of the manuscript.

Funding: This research received no external funding.

Institutional Review Board Statement: Not applicable.

Informed Consent Statement: Not applicable.

Data Availability Statement: The data presented in this study are available on request from the corresponding author.

Acknowledgments: The authors wish to thank the Academic Body UANL—CA-316 "Deterioration and integrity of composite materials". This research is dedicated to the memory of Joan Genescá Llongueras.

Conflicts of Interest: The authors declare no conflict of interest.

References

1. Stern, M.; Geary, A.L. Electrochemical polarization. I. A theoretical analysis of the shape of the polarization curves. *J. Electrochem. Soc.* **1957**, *104*, 56–63. [CrossRef]
2. Pellegrini-Cervantes, M.J.; Almeraya-Calderon, F.; Borunda-Terrazas, A.; Bautista-Margulis, R.G.; Chacón-Nava, J.G.; Fajardo-San-Miguel, G.; Almaral-Sanchez, J.L.; Barrios-Durstewitz, C.; Martinez-Villafañe, A. Corrosion Resistance, Porosity and Strength of lended Portland Cement Mortar Containing Rice Husk Ash and Nano-SiO$_2$. *Int. J. Electrochem. Sci.* **2013**, *8*, 10697–10710. [CrossRef]
3. Volmer, M.; Weber, A. Keimbildung in nbersättigten Gebilden. *Z. Phys. Chem.* **1959**, *119*, 277–3013. [CrossRef]
4. Butler, J.A.V. Studies in heterogeneous equilibria, II. The kinetic interpretation of the Nernst theory of electromotive force. *Trans. Faraday Soc.* **1924**, *19*, 729–733. [CrossRef]
5. Butler, J.A.V. Studies in heterogeneous equilibria, I. Conditions at the boundary surface of crystalline solids and liquids, and the application of statistical mechanics. *Trans. Faraday Soc.* **1924**, *19*, 659–665. [CrossRef]
6. Macdonald, D.D. Review of mechanistic analysis by electrochemical impedance spectroscopy. *Electrochim. Acta* **1990**, *35*, 1509–1525. [CrossRef]
7. Silverman, D.C. Tutorial on Cyclic Potentiodynamic Polarization Technique. In Proceedings of the CORROSION 98, San Diego, CA, USA, 22–27 March 1998.
8. Silverman, D.C. Practical Corrosion Prediction Using Electrochemical Techniques. In *Uhlig's Corrosion Handbook*, 3rd ed.; John Wiley & Sons: Hoboken, NJ, USA, 2011; pp. 1129–1166. [CrossRef]
9. Kearns, J.R.; Scully, J.R.; Roberge, P.R.; Reichert, D.L.; Dawson, J.L. *Electrochemical Noise Measurement for Corrosion Applications*; ASTM International: West Conshohocken, PA, USA, 2015; ISBN 080312032X.
10. Sanchez-Amaya, J.M.; Cottis, R.A.; Botana, F.J. Shot Noise and Statistical Parameters for the Estimation of Corrosion Mechanisms. *Corros. Sci.* **2005**, *47*, 3280–3299. [CrossRef]
11. Cottis, R.A. Interpretation of Electrochemical Noise Data. *Corrosion* **2001**, *57*, 265–285. [CrossRef]
12. Bertocci, U.; Huet, F. Noise Analysis Applied to Electrochemical Systems. *Corrosion* **1995**, *51*, 131–144. [CrossRef]
13. Bertocci, U.; Kruger, J. Studies of Passive Film Breakdown by Detection and Analysis of Electrochemical Noise. *Surf. Sci.* **1980**, *101*, 608–618. [CrossRef]
14. Al-Zanki, I.A.; Gill, J.S.; Dawson, J.L. Electrochemical Noise Measurements on Mild Steel in 0.5 M Sulphuric Acid. *Mater. Sci. Forum* **1986**, *8*, 463–476. [CrossRef]
15. Eden, D.A.; Rothwell, A.N. *Electrochemical Noise Data: Analysis, Interpretation and Presentation*; NACE International: Houston, TX, USA, 1992; pp. 1–12.
16. Bertocci, U.; Gabrielli, C.; Huet, F.; Keddam, M.; Rousseau, P. Noise Resistance Applied to Corrosion Measurements: II. Experimental Tests. *J. Electrochem. Soc.* **1997**, *144*, 37–43. [CrossRef]
17. Eden, D.A.; John, D.G.; Dawson, J.L. Corrosion Monitoring. International Patent Wo 87/0722, 19 November 1997.
18. Mansfeld, F.; Sun, Z.; Hsu, C.H.; Nagiub, A. Concerning Trend Removal in Electrochemical Noise Measurements. *Corros. Sci.* **2001**, *43*, 341–352. [CrossRef]
19. Lee, C.C.; Mansfeld, F. Analysis of Electrochemical Noise Data for a Passive System in the Frequency Domain. *Corros. Sci.* **1998**, *40*, 959–962. [CrossRef]
20. Homborg, A.M.; Tinga, T.; Van Westing, E.P.M.; Zhang, X.; Ferrari, G.M.; De Wit, J.H.W.; Mol, J.M.C. A Critical Appraisal of the Interpretation of Electrochemical Noise for Corrosion Studies. *Corrosion* **2014**, *70*, 971–987. [CrossRef] [PubMed]
21. Chen, A.; Cao, F.; Liao, X.; Liu, W.; Zheng, L.; Zhang, J.; Cao, C. Study of Pitting Corrosion on Mild Steel during Wet–Dry Cycles by Electrochemical Noise Analysis Based on Chaos Theory. *Corros. Sci.* **2013**, *66*, 183–195. [CrossRef]
22. Mansfeld, F.; Sun, Z. Technical Note: Localization Index Obtained from Electrochemical Noise Analysis. *Corrosion* **1999**, *55*, 915–918. [CrossRef]
23. Reid, S.A.; Eden, D.A. Assessment of Corrosion. U.S. Patent US9264824B1, 24 July 2001.
24. Eden, D.A. Electrochemical Noise—The First Two Octaves. In Proceedings of the CORROSION 98, San Diego, CA, USA, 22–27 March 1998; Volume 1998.
25. Villegas-Tovar, J.; Gaona-Tiburcio, C.; Lara-Banda, M.; Maldonado-Bandala, E.; Baltazar-Zamora, M.A.; Cabral-Miramontes, J.; Nieves-Mendoza, D.; Olguin-Coca, J.; Estupiñan-Lopez, F.; Almeraya-Calderón, F. Electrochemical Corrosion Behavior of Passivated Precipitation Hardening Stainless Steels for Aerospace Applications. *Metals* **2023**, *13*, 835. [CrossRef]
26. Almeraya-Calderón, F.; Jáquez-Muñoz, J.M.; Maldonado-Bandala, E.; Cabral-Miramontes, J.; Nieves-Mendoza, D.; Olgui-Coca, J.; Lopez-Leon, L.D.; Estupiñán-López, F.; Lira-Martínez, A.; Gaona Tiburcio, C. Corrosion Resistance of Titanium Alloys Anodized in Alkaline Solutions. *Metals* **2023**, *13*, 1510. [CrossRef]
27. Coakley, J.; Vorontsov, V.A.; Littrell, K.C.; Heenan, R.K.; Ohnuma, M.; Jones, N.G.; Dye, D. Nanoprecipitation in a Beta-Titanium Alloy. *J. Alloys Compd.* **2015**, *623*, 146–156. [CrossRef]
28. Legat, A.; Doleček, V. Corrosion Monitoring System Based on Measurement and Analysis of Electrochemical Noise. *Corrosion* **1995**, *51*, 295–300. [CrossRef]
29. Uruchurtu, J.C.; Dawson, J.L. Noise Analysis of Pure Aluminum under Different Pitting Conditions. *Corrosion* **1987**, *43*, 19–26. [CrossRef]

30. Bertocci, U.; Huet, F.; Nogueira, R.P.; Rousseau, P. Drift Removal Procedures in the Analysis of Electrochemical Noise. *Corrosion* **2002**, *58*, 337–347. [CrossRef]
31. Botona Pedemonte, F.J.; Aballe Villero, A.; Marcos Bárcena, M. *Ruido Electroquímico. Métodos de Análisis*; Septem Ediciones, S.L.: Oviedo, Spain, 2002; ISBN 84-95687-33-X.
32. Mehdipour, M.; Naderi, R.; Markhali, B.P. Electrochemical Study of Effect of the Concentration of Azole Derivatives on Corrosion Behavior of Stainless Steel in H_2SO_4. *Prog. Org. Coatings* **2014**, *77*, 1761–1767. [CrossRef]
33. Cappeln, F.; Bjerrum, N.J.; Petrushina, I.M. Electrochemical Noise Measurements of Steel Corrosion in the Molten $NaCl-K_2SO_4$ System. *J. Electrochem. Soc.* **2005**, *152*, B228. [CrossRef]
34. Calabrese, L.; Galeano, M.; Proverbio, E. Identifying Corrosion Forms on Synthetic Electrochemical Noise Signals by the Hilbert–Huang Transform Method. *Corros. Eng. Sci. Technol.* **2018**, *53*, 492–501. [CrossRef]
35. Mansfeld, F.; Han, L.T.; Lee, C.C.; Zhang, G. Evaluation of Corrosion Protection by Polymer Coatings Using Electrochemical Impedance Spectroscopy and Noise Analysis. *Electrochim. Acta* **1998**, *43*, 2933–2945. [CrossRef]
36. Legat, A.; Doleček, V. Chaotic Analysis of Electrochemical Noise Measured on Stainless Steel. *J. Electrochem. Soc.* **1995**, *142*, 1851–1858. [CrossRef]
37. Martínez-Aparicio, B.; Martínez-Bastidas, D.; Gaona-Tiburcio, C.; Martin, U.; Cabral-Miramontes, J.; Almeraya-Calderón, F. Localized corrosion of 15–5 PH and 17–4 PH stainless steel in NaCl solution. *J. Solid State Electrochem.* **2023**, *27*, 2993–3001. [CrossRef]
38. Lentka, Ł.; Smulko, J. Methods of Trend Removal in Electrochemical Noise Data—Overview. *Measurement* **2019**, *131*, 569–581. [CrossRef]
39. Homborg, A.M.; Cottis, R.A.; Mol, J.M.C. An Integrated Approach in the Time, Frequency and Time-Frequency Domain for the Identification of Corrosion Using Electrochemical Noise. *Electrochim. Acta* **2016**, *222*, 627–640. [CrossRef]
40. Homborg, A.M.; Oonincx, P.J.; Mol, J.M.C. Wavelet Transform Modulus Maxima and Holder Exponents Combined with Transient Detection for the Differentiation of Pitting Corrosion Using Electrochemical Noise. *Corrosion* **2018**, *74*, 1001–1010. [CrossRef] [PubMed]
41. Brockwell, P.J.; Davis, R.A. *Introduction to Time Series and Forecasting*; Springer: Cham, Switzerland, 2016. [CrossRef]
42. Cai, C.; Zhang, Z.; Cao, F.; Gao, Z.; Zhang, J.; Cao, C. Analysis of Pitting Corrosion Behavior of Pure Al in Sodium Chloride Solution with the Wavelet Technique. *J. Electroanal. Chem.* **2005**, *578*, 143–150. [CrossRef]
43. Homborg, A.M.; van Westing, E.P.M.; Tinga, T.; Zhang, X.; Oonincx, P.J.; Ferrari, G.M.; de Wit, J.H.W.; Mol, J.M.C. Novel Time–Frequency Characterization of Electrochemical Noise Data in Corrosion Studies Using Hilbert Spectra. *Corros. Sci.* **2013**, *66*, 97–110. [CrossRef]
44. Huang, N.E.; Shen, Z.; Long, S.R.; Wu, M.C.; Shih, H.H.; Yen, N.; Tung, C.C.; Liu, H.H. The Empirical Mode Decomposition and the Hilbert Spectrum for Non-linear and Non-Stationary Time Series Analysis. *Proc. R. Soc. A* **1996**, *454*, 903–995. [CrossRef]
45. Lafront, A.M.; Safizadeh, F.; Ghali, E.; Houlachi, G. Study of the Copper Anode Passivation by Electrochemical Noise Analysis Using Spectral and Wavelet Transforms. *Electrochim. Acta* **2010**, *55*, 2505–2512. [CrossRef]
46. Hai, L.; Guo-qiang, X.; Pan, Z.; Hua-sen, Z.; Khan, M.Y. The Hilbert–Huang Transform-Based Denoising Method for the TEM Response of a PRBS Source Signal. *Pure Appl. Geophys.* **2016**, *173*, 2777–2789. [CrossRef]
47. Marwan, N.; Carmen Romano, M.; Thiel, M.; Kurths, J. Recurrence Plots for the Analysis of Complex Systems. *Phys. Rep.* **2007**, *438*, 237–329. [CrossRef]
48. Zbilut, J.P.; Webber, C.L. Recurrence Quantification Analysis: Introduction and Historical Context. *Int. J. Bifurcat. Chaos* **2007**, *17*, 3477–3481. [CrossRef]
49. Garcia-Ochoa, E. Recurrence Plots: A New Methodology for Electrochemical Noise Signal Analysis. *J. Electroanal. Chem.* **2020**, *864*, 114092. [CrossRef]
50. Marwan, N.; Kraemer, K.H. Trends in Recurrence Analysis of Dynamical Systems. *Eur. Phys. J. Spec. Top.* **2023**, *232*, 5–27. [CrossRef]
51. Montoya-Rangel, M.; de Oca, N.G.M.; Gaona-Tiburcio, C.; Colás, R.; Cabral-Miramontes, J.; Nieves-Mendoza, D.; Maldonado-Bandala, E.; Chacón-Nava, J.; Almeraya-Calderón, F. Electrochemical Noise Measurements of Advanced High-Strength Steels in Different Solutions. *Metals* **2020**, *10*, 1232. [CrossRef]
52. Arellano-Pérez, J.H.; Escobar-Jiménez, R.F.; Granados-Lieberman, D.; Gómez-Aguilar, J.F.; Uruchurtu-Chavarín, J.; Alvarado-Martínez, V.M. Electrochemical Noise Signals Evaluation to Classify the Type of Corrosion Using Synchrosqueezing Transform. *J. Electroanal. Chem.* **2019**, *848*, 113249. [CrossRef]
53. Ren, Z.; Li, Q.; Yang, X.; Wang, J. A Novel Method for Identifying Corrosion Types and Transitions Based on Adaboost and Electrochemical Noise. *Anti-Corros. Methods Mater.* **2023**, *70*, 78–85. [CrossRef]
54. Montoya-Rangel, M.; Garza-Montes-de-Oca, N.F.; Gaona-Tiburcio, C.; Almeraya-Calderón, F. Corrosion Mechanism of Advanced High Strength Dual-Phase Steels by Electrochemical Noise Analysis in Chloride Solutions. *Mater. Today Commun.* **2023**, *35*, 105663. [CrossRef]
55. Ye, Z.; Guan, L.; Li, Y.; Zhong, J.; Liao, L.; Xia, D.; Huang, J. Understanding the Galvanic Corrosion of Cu-Ni Alloy/2205 DSS Couple Using Electrochemical Noise and Microelectrochemical Studies. *Corros. Sci.* **2023**, *224*, 111512. [CrossRef]
56. Shahri, Z.; Allahkaram, S.R.; Soltani, R.; Jafari, H. Electrochemical Analysis on Localized Corrosion of PEO/Magnesium Oxide Coating. *J. Alloys Compd.* **2024**, *1003*, 175572. [CrossRef]

57. Homborg, A.; Mol, A.; Tinga, T. Corrosion Classification through Deep Learning of Electrochemical Noise Time-Frequency Transient Information. *Eng. Appl. Artif. Intell.* **2024**, *133*, 108044. [CrossRef]
58. Salunkhe, V.G.; Khot, S.M.; Desavale, R.G.; Yelve, N.P. Unbalance Bearing Fault Identification Using Highly Accurate Hilbert–Huang Transform Approach. *J. Nondestruct. Eval. Diagn. Progn. Eng. Syst.* **2023**, *6*, 031005. [CrossRef]
59. Jamali, S.S.; Wu, Y.; Homborg, A.M.; Lemay, S.G.; Gooding, J.J. Interpretation of Stochastic Electrochemical Data. *Curr. Opin. Electrochem.* **2024**, *46*, 101505. [CrossRef]
60. Lee, Y.N.; Jung, M.J.; Jo, S.W.; Rahim, G.; Lee, S.G.; Choi, K.S. Fast-Settling Onboard Electrochemical Impedance Spectroscopy System Adopting Two-Stage Hilbert Transform. In Proceedings of the IECON 2023—49th Annual Conference of the IEEE Industrial Electronics Society, Singapore, 16–19 October 2023. [CrossRef]
61. Ortíz-Corona, J.; Uruchurtu-Chavarin, J.; García-Ochoa, E.M.; González-Sánchez, J.A.; Larios-Duran, E.R.; Rodríguez-Gómez, F.J. Monitoring of Silver Alloy Tarnishing in Sulphides by Electrochemical Noise Measurements: Application of Statistical and Recurrence Plot Analysis. *Electrochim. Acta* **2024**, *495*, 144388. [CrossRef]
62. Luo, B.; Wei, Q.; Hu, S.; Manoach, E.; Deng, T.; Cao, M. A Novel Cross-Domain Identification Method for Bridge Damage Based on Recurrence Plot and Convolutional Neural Networks. *J. Vibroeng.* **2024**, *26*, 1–22. [CrossRef]
63. Lin, R.L.; Li, W.Y.; Li, L.; Liao, C.J. High-Dissolution Mechanism of Ti–48Al–2Cr–2Nb Alloy in Electrochemical Machining from a Non-linear Dynamic Perspective. *Adv. Eng. Mater.* **2024**, *26*, 2302083. [CrossRef]
64. Martínez-Villafañe, A.; Almeraya-Calderón, M.F.; Gaona-Tiburcio, C.; Gonzalez-Rodriguez, J.G.; Porcayo-Calderón, J. High-Temperature Degradation and Protection of Ferritic and Austenitic Steels in Steam Generators. *J. Mater. Eng. Perform.* **1997**, *7*, 108–113. [CrossRef]
65. *ASTM E3-95*; Standard Practice for Preparation of Metallographic Specimens. ASTM International: West Conshohocken, PA, USA, 2007.
66. *ASTM G199*; Standard Guide for Electrochemical Noise Measurement. ASTM International: West Conshohocken, PA, USA, 2020.
67. Jaquez-Muñoz, J.; Gaona-Tiburcio, C.; Lira-Martinez, A.; Zambrano-Robledo, P.; Maldonado-Bandala, E.; Samaniego-Gamez, O.; Nieves-Mendoza, D.; Olguin-Coca, J.; Estupiñan-Lopez, F.; Almeraya-Calderon, F. Susceptibility to Pitting Corrosion of Ti CP2, Ti-6Al-2Sn-4Zr-2Mo, and Ti-6Al-4V Alloys for Aeronautical Applications. *Metals* **2021**, *11*, 1002. [CrossRef]
68. Jáquez-Muñoz, J.M.; Gaona-Tiburcio, C.; Méndez-Ramírez, C.T.; Baltazar-Zamora, M.Á.; Estupinán-López, F.; Bautista-Margulis, R.G.; Cuevas-Rodríguez, J.; Flores-De los Rios, J.P.; Almeraya-Calderón, F. Corrosion of Titanium Alloys Anodized Using Electrochemical Techniques. *Metals* **2023**, *13*, 476. [CrossRef]
69. Almeraya-Calderon, F.; Villegas-Tovar, M.; Maldonado-Bandala, E.; Lara-Banda, M.; Baltazar-Zamora, M.A.; Santiago-Hurtado, G.; Nieves-Mendoza, D.; Lopez-Leon, L.D.; Jaquez-Muñoz, J.M.; Estupiñán-López, F.; et al. Use of Electrochemical Noise for the Study of Corrosion by Passivated CUSTOM 450 and AM 350 Stainless Steels. *Metals* **2024**, *14*, 341. [CrossRef]
70. Coakley, J.; Isheim, D.; Radecka, A.; Dye, D.; Stone, H.J.; Seidman, D.N. Microstructural Evolution in a Superelastic Metastable Beta-Ti Alloy. *Scr. Mater.* **2017**, *128*, 87–90. [CrossRef]
71. Nazarnezhad-Bajestani, M.; Neshati, J.; Siadati, M.H. Determination of SS321 Pitting Stage in FeCl3 Solution Based on Electrochemical Noise Measurement Data Using Artificial Neural Network. *J. Electroanal. Chem.* **2019**, *845*, 31–38. [CrossRef]
72. Aballe, A.; Bethencourt, M.; Botana, F.J.; Marcos, M.; Sánchez-Amaya, J.M. Use of Wavelets to Study Electrochemical Noise Transients. *Electrochim. Acta* **2001**, *46*, 2353–2361. [CrossRef]
73. Cottis, R.A.; Homborg, A.M.; Mol, J.M.C. The Relationship between Spectral and Wavelet Techniques for Noise Analysis. *Electrochim. Acta* **2016**, *202*, 277–287. [CrossRef]
74. Wang, Z.; Zeng, X.; Hu, X.; Hu, J. The Multi-Disturbance Complex Power Quality Signal HHT Detection Technique. In Proceedings of the 2012 IEEE Innovative Smart Grid Technologies—ISGT Asia, Tianjin, China, 21–24 May 2012. [CrossRef]
75. Jáquez-Muñoz, J.M.; Gaona-Tiburcio, C.; Chacón-Nava, J.; Cabral-Miramontes, J.; Nieves-Mendoza, D.; Maldonado-Bandala, E.M.; Delgado, A.D.; Flores-De Los Rios, J.P.; Bocchetta, P.; Almeraya-Calderón, F. Electrochemical Corrosion of Titanium and Titanium Alloys Anodized in H2SO4 and H3PO4 Solutions. *Coatings* **2022**, *12*, 325. [CrossRef]
76. Lara-Banda, M.; Gaona-Tiburcio, C.; Zambrano-Robledo, P.; Delgado, E.M.; Cabral-Miramontes, J.A.; Nieves-Mendoza, D.; Maldonado-Bandala, E.; Estupiñan-López, F.; Chacón-Nava, J.G.; Almeraya-Calderón, F. Alternative to Nitric Acid Passivation of 15-5 and 17-4PH Stainless Steel Using Electrochemical Techniques. *Materials* **2020**, *13*, 2836. [CrossRef] [PubMed]
77. Bocchetta, P.; Chen, L.Y.; Tardelli, J.D.C.; Dos Reis, A.C.; Almeraya-Calderón, F.; Leo, P. Passive Layers and Corrosion Resistance of Biomedical Ti-6Al-4V and β-Ti Alloys. *Coatings* **2021**, *11*, 487. [CrossRef]
78. Addison, P.S. *The Illustrated Wavelet Transform Handbook: Introductory Theory and Applications in Science, Engineering, Medicine and Finance*, 2nd ed.; CRC Press: Boca Raton, FL, USA, 2017; pp. 1–446. [CrossRef]
79. Liu, W.; Wang, D.; Chen, X.; Wang, C.; Liu, H. Recurrence Plot-Based Dynamic Analysis on Electrochemical Noise of the Evolutive Corrosion Process. *Corros. Sci.* **2017**, *124*, 93–102. [CrossRef]
80. Acuña-González, N.; García-Ochoa, E.; González-Sánchez, J. Assessment of the Dynamics of Corrosion Fatigue Crack Initiation Applying Recurrence Plots to the Analysis of Electrochemical Noise Data. *Int. J. Fatigue* **2008**, *30*, 1211–1219. [CrossRef]
81. Li, J.; Du, C.W.; Liu, Z.Y.; Li, X.G.; Liu, M. Effect of Microstructure on the Corrosion Resistance of 2205 Duplex Stainless Steel. Part 2: Electrochemical Noise Analysis of Corrosion Behaviors of Different Microstructures Based on Wavelet Transform. *Constr. Build. Mater.* **2018**, *189*, 1294–1302. [CrossRef]

82. Cui, J.; Yu, D.; Long, Z.; Xi, B.; He, X.; Pei, Y. Application of Electrochemical Noise (EN) Technology to Evaluate the Passivation Performances of Adsorption and Film-Forming Type Corrosion Inhibitors. *J. Electroanal. Chem.* **2019**, *855*, 113584. [CrossRef]
83. Gaona-Tiburcio, C.; Montoya, R.M.; Cabral, M.J.A.; Estupiñan, L.F.; Zambrano, R.P.; Orozco, C.R.; Chacon-Nava, J.G.; Baltazar, Z.M.A.; Almeraya-Caldero, F. Corrosion Resistance of Multilayer Coatings Deposited by PVD on Inconel 718 Using Electrochemical Impedance Spectroscopy Technique. *Coatings* **2020**, *10*, 521. [CrossRef]
84. Abdulmutaali, A.; Hou, Y.; Aldrich, C.; Lepkova, K. An Online Monitoring Approach of Carbon Steel Corrosion via the Use of Electrochemical Noise and Wavelet Analysis. *Metals* **2024**, *14*, 66. [CrossRef]
85. Hou, Y.; Aldrich, C.; Lepkova, K.; Machuca, L.L.; Kinsella, B. Monitoring of Carbon Steel Corrosion by Use of Electrochemical Noise and Recurrence Quantification Analysis. *Corros. Sci.* **2016**, *112*, 63–72. [CrossRef]
86. Valavanis, D.; Spanoudaki, D.; Gkili, C.; Sazou, D. Using Recurrence Plots for the Analysis of the Nonlinear Dynamical Response of Iron Passivation-Corrosion Processes. *Chaos* **2018**, *28*, 085708. [CrossRef] [PubMed]
87. Garcia-Ochoa, E.; Maldonado, P.; Corvo, F. Non-linear Dynamics of Potassium Iodide Adsorption on the Interface of Carbon Steel in Acidic Medium. *Mater. Corros.* **2020**, *71*, 1152–1159. [CrossRef]
88. Marwan, N.; Webber, C.L.; Macau, E.E.N.; Viana, R.L. Introduction to Focus Issue: Recurrence Quantification Analysis for Understanding Complex Systems. *Chaos* **2018**, *28*, 85601. [CrossRef]
89. Martínez-Ramos, C.; Olguin-Coca, J.; Lopez-Leon, L.D.; Gaona-Tiburcio, C.; Lara-Banda, M.; Maldonado-Bandala, E.; Castañeda-Robles, I.; Jaquez-Muñoz, J.M.; Cabral-Miramontes, J.; Nieves-Mendoza, D.; et al. Electrochemical Noise Analysis Using Experimental Chaos Theory, Power Spectral Density and Hilbert–Huang Transform in Anodized Aluminum Alloys in Tartaric–Phosphoric–Sulfuric Acid Solutions. *Metals* **2023**, *13*, 1850. [CrossRef]
90. Gaona-Tiburcio, C.; Almeraya-Calderón, F.; Chacon-Nava, J.G.; Matutes-Aquino, J.A.; Martinez-Villafañe, A. Electrochemical response of permanent magnets in different solutions. *J. Alloys Compd.* **2004**, *369*, 78–80. [CrossRef]
91. Gaona-Tiburcio, C.; Jáquez-Muñoz, J.M.; Nieves-Mendoza, D.; Maldonado-Bandala, E.; Lara-Banda, M.; Lira-Martinez, M.A.; Reyes-Blas, H.; Baltazar-Zamora, M.Á.; Landa-Ruiz, L.; Lopez-Leon, L.D.; et al. Corrosion Behavior of Titanium Alloys (Ti CP2, Ti-6Al-2Sn-4Zr-2Mo, Ti-6Al-4V and Ti Beta-C) with Anodized and Exposed in NaCl and H_2SO_4 Solutions. *Metals* **2024**, *14*, 160. [CrossRef]
92. Cerezo, H.R.; Tiburcio, C.G.; Miramontes, J.A.C.; Almeraya-Calderón, F. Electrochemical characterization of Al–Li alloys AA2099 and AA2055 for aeronautical applications: Effect of thermomechanical treatments. *J. Solid State Electrochem.* **2023**, *27*, 3101–3117. [CrossRef]

Article

Effect of Single Particle High-Speed Impingement on the Electrochemical Step Characteristics of a Stainless-Steel Surface

Meihong Liu [1], Long Chai [2], Min Yang [2] and Jiarui Cheng [3,*]

[1] School-Run Factory (Engineering Training Center), Xi'an Aeronautical Institute, Xi'an 710077, China; liumeihong1223@163.com
[2] CCDC Changqing Downhole Technology Company, China National Petroleum Corporation, Xi'an 712042, China; cqjx_chail@cnpc.com.cn (L.C.); yang-min521@163.com (M.Y.)
[3] Xi'an Key Laboratory of Wellbore Integrity Evaluation, Xi'an Shiyou University, Xi'an 710065, China
* Correspondence: cjr88112@163.com; Tel.: +86-18092492490

Abstract: In the process of particle erosion and electrochemical corrosion interaction, the electrolyte flow state change, product film destruction, and matrix structure change caused by particle impact affect the electrochemical corrosion process. Such transient, complex physical and electrochemical changes are difficult to capture because of the short duration of action and the small collision area. The peak, step time, and recovery time in this transient step cycle can indirectly reflect the smoothness and reaction rate of the electrochemical reaction system, and thus characterize the resistance to scouring corrosion coupling damage of metals in liquid–solid two-phase flow. In this study, in order to obtain the electrochemical response at the moment of particle impact, electrochemical monitoring experiments using a specially designed miniature three-electrode system were used to test step-critical values, including step potential, current, and resistance, among others. Meanwhile, an electrochemical step model under particle impact considering boundary layer perturbation was developed. The experimental results reflect the effect law of particle impact velocity and particle size on the peak step and recovery period. Meanwhile, the effect of particle impingement on the electrochemical step of stainless steel in different electrolyte solutions was obtained by comparing the step curves in distilled water and Cl-containing water. The connection between the parameters in the electrochemical step model and in the particle impact, as well as the effect of the variation of these parameters on the surface repassivation process are discussed in this paper. By fitting and modeling the test curves, a new mathematical model of electrochemical step-decay under single-particle impact was obtained, which can be used to characterize the change pattern of electrochemical parameters on the metal surface before and after the impingement.

Keywords: erosion–corrosion; particle disturbance; electrochemical step; stainless steel repassivation

Citation: Liu, M.; Chai, L.; Yang, M.; Cheng, J. Effect of Single Particle High-Speed Impingement on the Electrochemical Step Characteristics of a Stainless-Steel Surface. *Materials* 2024, 17, 3043. https://doi.org/10.3390/ma17123043

Academic Editor: Jordi Sort

Received: 19 May 2024
Revised: 5 June 2024
Accepted: 17 June 2024
Published: 20 June 2024

1. Introduction

Erosion–corrosion damage is influenced by multiple factors, including material properties, various hydrodynamic conditions, and the surrounding environment, making its study relatively complex [1]. The mechanism of damage caused by single-particle impacts in liquid–solid two-phase flow is a critical aspect of damage analysis. In practical conditions, erosion and corrosion occur simultaneously. The mechanical processes induced by single-particle erosion can lead to material deformation or detachment, while corrosion typically manifests as chemical or electrochemical reactions [2]. Additionally, the loss of pipe wall material occurs not only through the detachment of solid particles from the metal surface but also through dissolution in ionic form [3]. Current research on particle erosion and its impact on electrochemical reactions on metal surfaces encompasses many aspects, including corrosion current density, the influence of passive films, and impact damage models. These areas have received significant attention and in-depth study from numerous scholars.

Through experiments exploring the corrosion rate of electroplated copper–manganese materials in different media, Supriyatna et al. [4] concluded that higher applied current densities on steel significantly reduce its corrosion rate, with the corresponding electroplated layer thickness increasing with the current density. Additionally, Jan Mayén [5] discovered that the corrosion current density is also influenced by the tensile strength of the material, indicating a close relationship with its microstructural characteristics. Wang et al. [6] developed a novel non-destructive method for predicting corrosion current density in the presence of stray currents by integrating electrochemical laboratory measurements with data-driven techniques. Alhumade et al. [7] proposed an accurate corrosion current density model based on the adaptive neuro-fuzzy inference system (ANFIS) and experimentally validated the precision of the ANFIS model for corrosion current density.

In studies on the impact of passivation films, Li et al. [8] found that both repassivation and depassivation processes are related to the kinetic energy of solid particles, with depassivation delaying repassivation. They further theoretically predicted and experimentally validated the dependence of the critical flow velocity (CFV) for erosion–corrosion on solid particle concentration and diameter. Xu et al. [9] explored the fine structure of passivation films at the atomic scale, demonstrating the growth mode of the crystals. Ion channels within the passivation film explain the emergence of defect regions and the localized attack of chloride ions during corrosion. Hou et al. [10] discovered that introducing defects into the passivation film structure promotes corrosion degradation, experimentally demonstrating and revealing the fundamental deterioration process of passivation films. Wu et al. [11] studied the effect of high temperatures on the corrosion resistance of Fe-based amorphous coatings, finding that high temperatures accelerate the corrosion process and promote the thickening of the passivation film. Wang et al. [12] proposed a new method for studying repassivation time, which allows for comparison of the repassivation performance of different materials without the need for curve fitting or consideration of film growth mechanisms.

In the context of particle impact models, Khalifa [13] proposed a novel modeling approach for the agglomerate impact damage in wall-bounded particle turbulence based on artificial neural networks (ANN). This model incorporates the effects of shear factors, thereby enhancing its applicability. Mohammad et al. [14] investigated the formation of splat fragmentation during the flattening and solidification of droplets under plasma particle spraying conditions by establishing a numerical model. This study provided a kinetic perspective to explain the mechanisms affecting droplet flattening on solid surfaces. To predict the deformation of individual particles at different impact velocities, Schreiber [15] validated and optimized the Preston–Tonks–Wallace (PTW) model, which is effective in predicting the deformation of cold-sprayed particles. Ren [16] developed a semi-analytical model to estimate rock-breaking depth under stable damage conditions in particle jet impact, conducting numerical analyses on influencing factors. This model offers theoretical support for parameter optimization and field applications of particle jet impact drilling technology. In corrosive solutions, particle impacts on metal walls cause changes in near-wall fluid flow, leading to transient variations in electrochemical reaction parameters [17]. Changes in particle impact velocity, particle diameter, and particle count result in corresponding transient variations. Therefore, further research is needed to quantify the electrochemical responses induced by variations in particle parameters.

Based on the considerations above, this study investigates the effects of corrosive fluid flow rate and particle geometric parameters. The electrochemical response of pipe wall materials under single particle impact in chloride-containing solutions is examined, analyzing the interaction between mechanical and electrochemical damage. The results are used to explore the mechanisms by which particle impact influences electrochemical corrosion.

2. Experiment

In the pipe flow particle impact experiments, the low particle concentration in the liquid, combined with the small effective area of the working electrode in the micro three-electrode system, made sustained effective collisions difficult. Increasing the particle concentration further proved impractical for pumping and caused significant damage to the screw blades, pipes, and nozzles. Therefore, jet experiments were employed to study the electrochemical response of metal wall surfaces to particle impact.

The particle impact experiments were conducted on the multiphase flow erosion–corrosion test rig [18], including the screw pump, liquid flowmeter, test chamber, sample holder, stirred tank in the liquid flow system and air compressor, buffer tank, filter drier, gas flowmeter, sand storage tank, and PLC feeder in the gas flow system. During the experiments, particles were added to the pipeline from the sand storage tank and carried by the liquid to the nozzle for ejection, impacting the target material surface (as shown in Figure 1). Once the particles collided with the sample surface, they were collected in the test section and the stirred tank because the sand would break after impact and affect the experimental results.

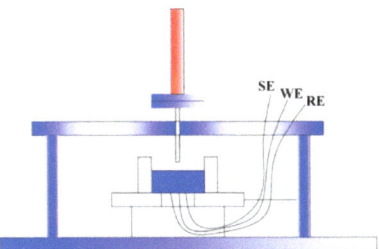

Figure 1. Diagram of particle impact experiment.

The electrochemical experimental setup for particle impact is illustrated in Figure 2. An electrochemical workstation PARSTAT MC-1000 was used to test current changes (Ametek. Co., Ltd., Berwyn, PA, USA). An AgCl-saturated calomel electrode was employed as the reference electrode, and graphite served as the auxiliary electrode. The graphite electrode area was 20 times larger than the working electrode. The 304 stainless steel was used as the working electrode because its surface can be easily passivated. The saturated calomel reference electrode, which was placed in the test chamber, was connected to the standard three-electrode system. A long platinum wire was used as the counter electrode. For electrochemical monitoring, polarization curves were recorded by changing the electrode potential at a sweep rate of 0.2 mV/s.

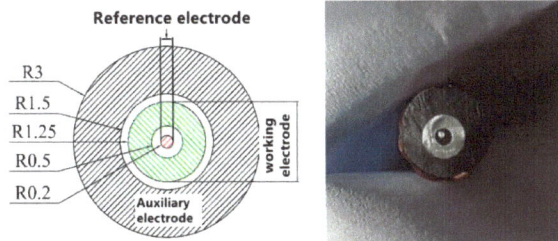

Figure 2. Diagram of three-electrode setup for particle impact electrochemical experiment.

Two media were utilized in the experiments: H_2O and H_2O + 3.5 wt% NaCl. The jet nozzles had diameters of 10 mm and 15 mm, respectively. The particle impact angle was set to 90° to maximize the acceleration of reactant mass transfer. Single particle impact experiments utilized large-diameter spherical plastic particles made of methyl methacrylate (PMMA), characterized by their large diameter and low density. The large diameter maximizes disturbance of the mass transfer boundary layer and disrupts corrosion product films. The diameters used were 4 mm, 6 mm, 8 mm, 10 mm, and 12 mm, with a density of approximately 1200 kg/m^3, as depicted in Figure 3.

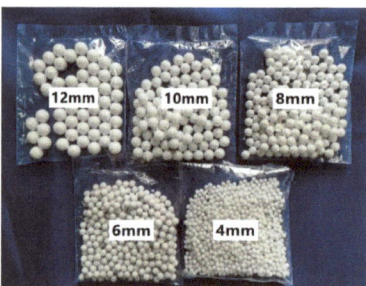

Figure 3. Single-particle impact experiment using plastic particles.

Before initiating the experimental apparatus, the specimen was secured within the testing chamber, as depicted in Figure 1. The angle between the nozzle and the specimen was adjusted to 90°, with the nozzle aligned with the surface of the working electrode. The preparation of the solution in the mixing tank followed the same procedure as that for the fluid-induced corrosion experiment. Once the solution was prepared, mixing was initiated, the liquid circuit valve was opened, and the gas circuit valve was closed. The single-screw pump was then activated, and the flow rate was adjusted to the desired value. After stabilizing the flow rate for 10 min, the temperature and pressure inside the pipe were measured and recorded. Simultaneously, the electrochemical workstation was turned on for machine preheating. Once the flow rate in the experimental circuit stabilized, the three test channels of the electrochemical workstation were sequentially connected to the three electrodes for open-circuit potential and open-circuit current monitoring. The valve of the sand storage tank was then opened, allowing the liquid to carry the particles to impact the surface of the working electrode. The trend of potential and current changes during particle impact was observed. Upon completion of the experiment, the particles were collected within the testing chamber, the electrochemical workstation and pipeline instruments were shut off, and the connections to the three electrodes were disconnected.

3. Results

Due to the considerable size of the erosion–corrosion experimental rig, significant vibrations occur in the pipeline system at high flow rates, which can affect the monitoring results of potential and current. To minimize experimental errors, medium flow rates of 1 m/s, 2 m/s, 3 m/s, 4 m/s, and 5 m/s were selected for the single-particle impact experiments. For ease of analysis, the initial values of current and potential were uniformly set to 0 mV.

3.1. Results of Open-Circuit Potential Response

Differing from the current density exhibiting a peak value, the open-circuit potential (OCP) of the 304 stainless steel's surface underwent a sudden negative shift after particle impact, displaying a minimum peak value. Similar to the trend observed in current density variations, the absolute value of the OCP peak increased with increasing flow velocity and particle diameter. Moreover, compared to the current density, the time required for the OCP to return to its initial value was longer, and the deviation between the recovery value and the initial value was greater, resulting in a negative shift in the final potential. In Figure 4, under different particle diameters, the OCP peak gradually increases with increasing flow velocity. For particle diameters of 4, 6, 8, and 10 mm, the absolute value of the potential exhibited relatively uniform increments. However, when the particle diameter increased to the maximum value of 12 mm, a sudden change in potential value occurred from 4 m/s (−0.00552 V) to 5 m/s (−0.01534 V), with the largest absolute value increment reaching 117.9%. Additionally, the test results indicate that with an increase in particle diameter, the absolute value of the peak potential exhibited an order of magnitude increase. The maximum potential peak values corresponding to particle diameters of 4 to 12 mm are -8.1×10^{-4} V, −0.00122 V, −0.00184 V, −0.00307 V, and −0.01534 V, respectively. In Figure 5, it is evident that the recovery time for the 12 mm diameter particles at various flow velocities is longer, with the longest recovery time observed at a flow velocity of 5 m/s and a particle diameter of 12 mm, reaching 20 s, indicating that the potential is more difficult to recover to its initial value compared to the current density.

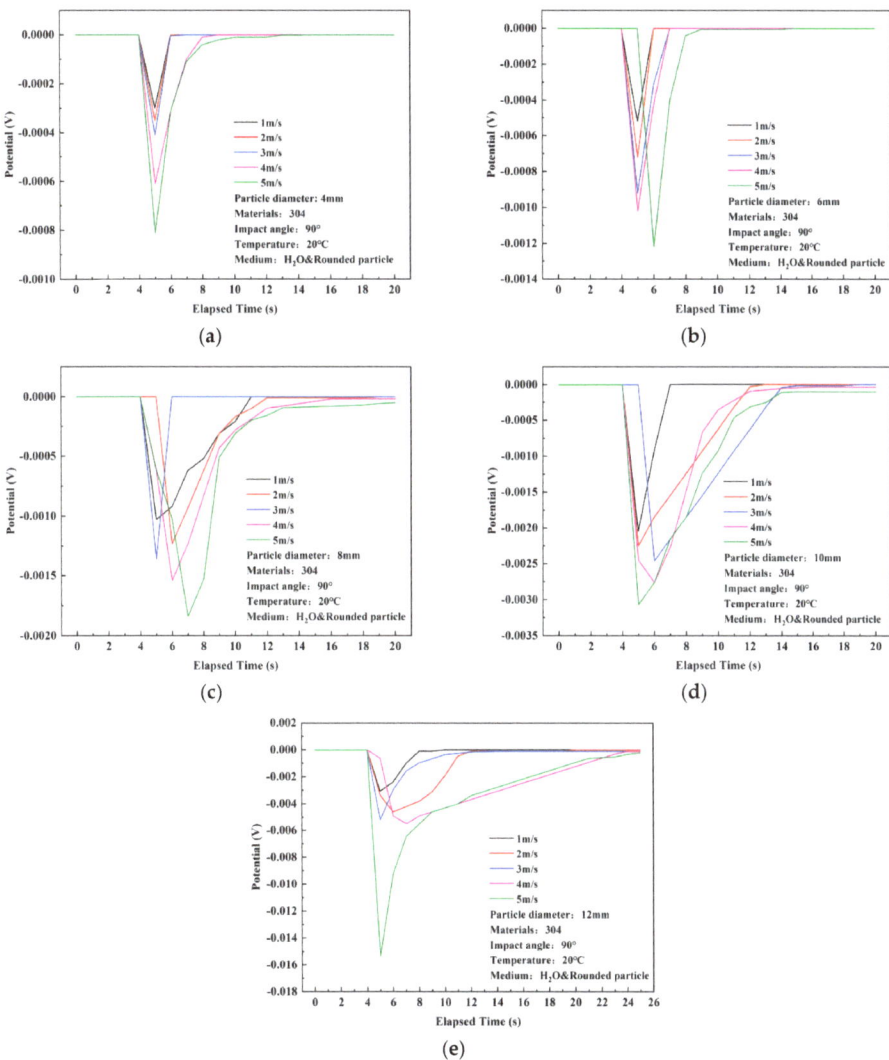

Figure 4. Experimental results of potential variation of 304 stainless steel under single particle impact in distilled water (flow rate comparison). (**a**) 4 mm, (**b**) 6 mm, (**c**) 8 mm, (**d**) 10 mm, (**e**) 12 mm.

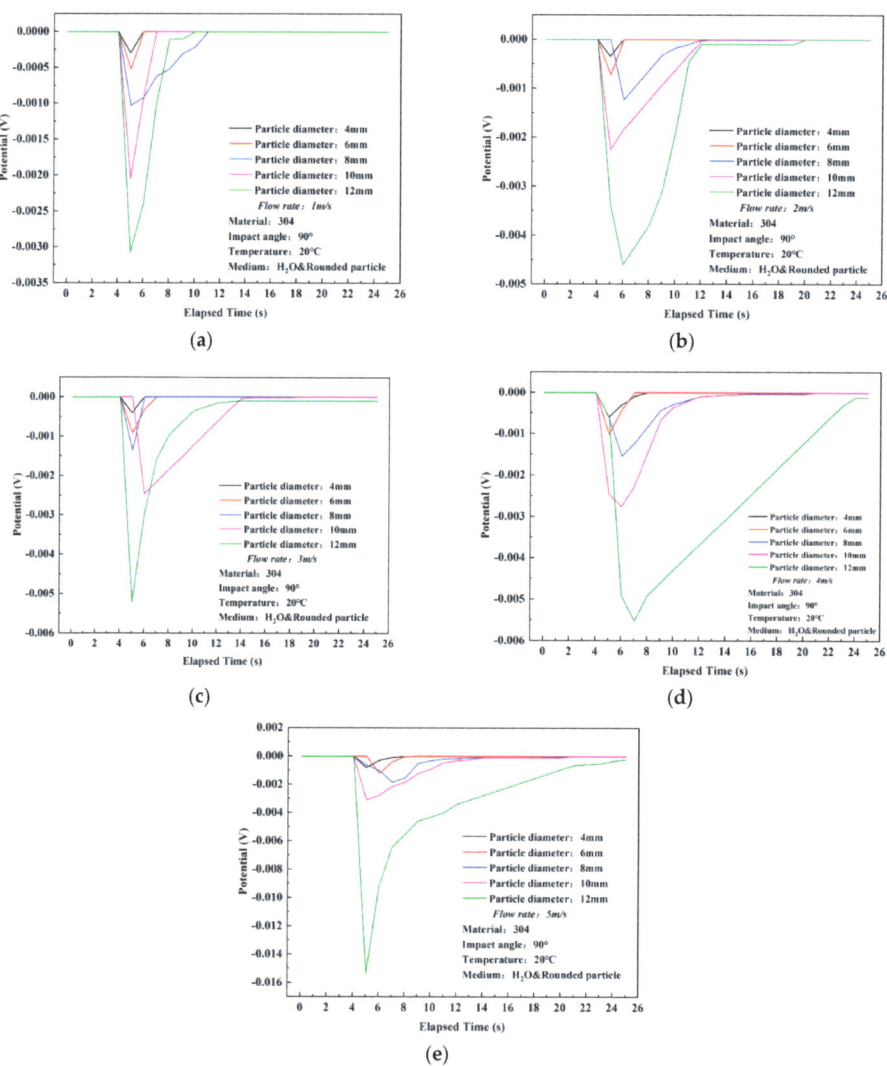

Figure 5. Experimental results of potential variation of 304 stainless steel under single particle impact in distilled water (comparison by particle diameter). (**a**) 1 m/s, (**b**) 2 m/s, (**c**) 3 m/s, (**d**) 4 m/s, (**e**) 5 m/s.

3.2. Results of Surface Current Density Response

The experimental results of single particle impact on 304 stainless steel in water are shown in Figure 6. The transient peak value of surface current density of 304 stainless steel increased with the increase in flow velocity. Taking the example of the surface results under 4 mm particle diameter impact, the current density peak values at 1–5 m/s are 1.27×10^{-7} A/cm^2, 1.54×10^{-7} A/cm^2, 1.90×10^{-7} A/cm^2, 2.04×10^{-7} A/cm^2, and 2.15×10^{-7} A/cm^2, respectively. With each 1 m/s increase in flow velocity, the percentage increase in current density peak values is 21.26%, 23.38%, 7.37%, and 5.39%. The maximum percentage increase in current density peak values occurs when the flow velocity increases from 2 m/s to 3 m/s, reaching 23.38%. This indicates that the increase in flow velocity significantly increases the peak value of surface current density caused by particle impact. Similar trends in changes in current density peak values under impact of particles of other diameters indicate a positive correlation between flow velocity and changes in current density peak values. Additionally, as shown in Figure 7, after being impacted by particles, the current density starts to recover from the peak value, but there is a deviation between the final recovery value and the initial value. This is because the particles cause damage to the surface of the material, resulting in deviation from the initial morphology after impact, and the material cannot self-repair to its original state. However, the deviation is small, indicating the good repassivation ability of 304 stainless steel after being impacted by particles in flowing water. At the same time, the step time of current density is short, while the recovery time of current density is longer, remaining within 6 s, which also reflects the good repassivation ability of 304 stainless steel.

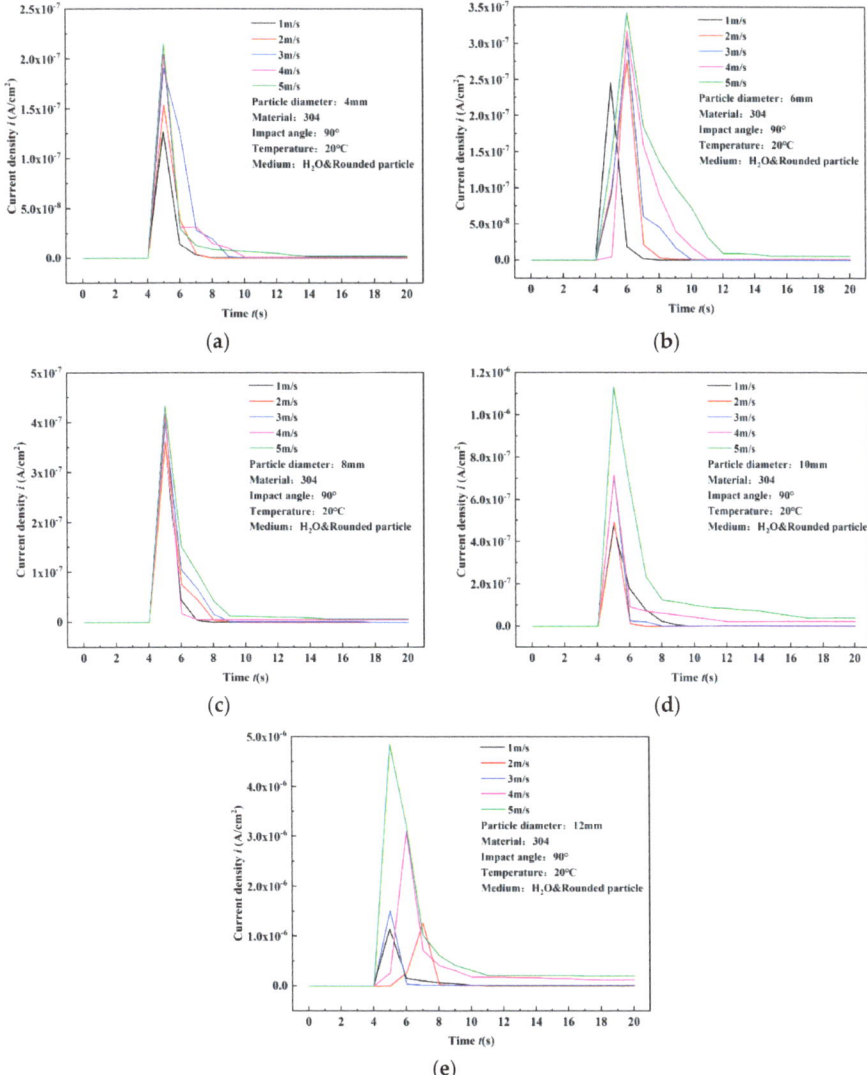

Figure 6. Experimental results of surface current density variation of 304 stainless steel subjected to single particle impact in water (comparison of flow velocities). (**a**) 4 mm, (**b**) 6 mm, (**c**) 8 mm, (**d**) 10 mm, (**e**) 12 mm.

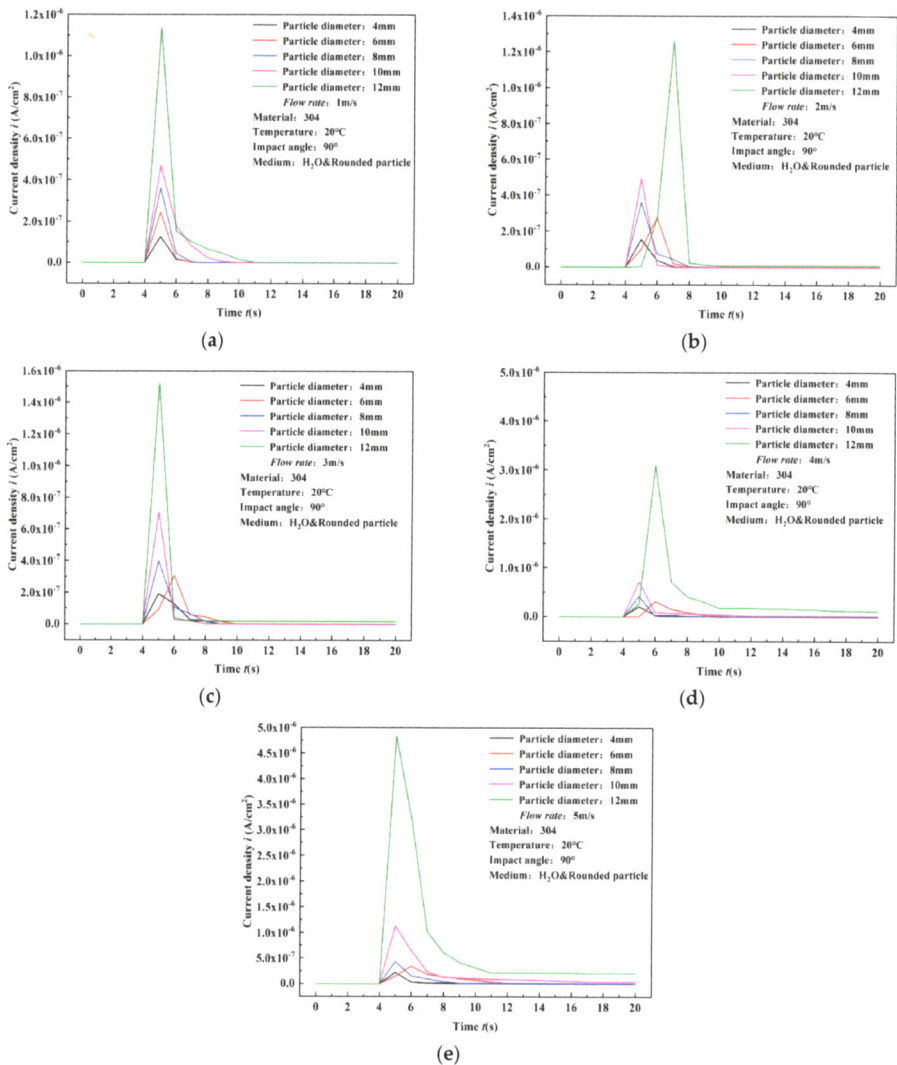

Figure 7. Experimental results of surface current density variation of 304 stainless steel subjected to single particle impact in water (comparison of particle diameters). (**a**) 1 m/s, (**b**) 2 m/s, (**c**) 3 m/s, (**d**) 4 m/s, (**e**) 5 m/s.

Figure 7 illustrates the peak current density under different particle diameters at the same flow velocity. Comparative results show a noticeable increase in current density peak values with the enlargement of particle diameter at each flow velocity. When the flow velocity is 1 m/s, the current density peak values for particle diameters of 4, 6, 8, 10, and 12 mm are, respectively, 1.27×10^{-7} A/cm^2, 2.45×10^{-7} A/cm^2, 3.62×10^{-7} A/cm^2, 4.71×10^{-7} A/cm^2, and 1.14×10^{-6} A/cm^2. With every 2 mm increase in particle diameter, the percentage increase in current density peak values is 92.91%, 47.76%, 30.11%, and 142.04%, respectively. The largest increase occurs between diameters 10 mm and 12 mm, representing an order of magnitude increase, indicating a nonlinear growth relationship between current density peak values and particle diameter. Similar trends in variations are observed under different flow conditions. This phenomenon is attributed to the cubic relationship between the diameter of spherical particles and their volume, which is proportional to the volume. Consequently, with constant particle density, the volume undergoes nonlinear changes, resulting in nonlinear changes in particle mass, and, subsequently, the impact energy of particles exhibits nonlinear amplification. Furthermore, with the increase in flow velocity, the time required for the surface current density peak value of 304 stainless steel to recover to its initial value also increases. Therefore, the variation in current density peak values with particle diameter becomes more significant. The highest increase occurs at a flow velocity of 5 m/s, where the current density corresponding to a diameter of 10 mm increases from 1.13×10^{-6} A/cm^2 to 4.85×10^{-6} A/cm^2 for a diameter of 12 mm, representing a percentage increase of 329.12%.

When 3.5 wt% NaCl is added to the experimental medium, the peak current density also increases with the increase in flow rate and particle diameter. Both the flow rate and particle diameter are positively correlated with the peak current density. Compared to the pure water medium, under the same flow rate and particle diameter conditions, the peak current density increases by orders of magnitude. In the pure water medium, the peak current density is in the range of 10^{-6} to 10^{-7}, while in the 3.5 wt% NaCl medium, it is in the range of 10^{-3}. Therefore, after particle impact, the deviation between the recovery value and the initial value of the current density is larger.

Due to the larger peak current density, the recovery time of the current density in the 3.5 wt% NaCl medium is longer. As shown in Figure 8e, when the 12 mm diameter particles impact the material surface at 4 m/s, the recovery time of the current density is 8 s, while at 5 m/s impact velocity, the recovery time reaches 12 s. It can be observed that the rising phase curve of the current density after particle impact also has a slope, rather than being vertical to the time axis, as shown in the typical response curve of Figure 9e corresponding to the 12 mm diameter particles.

For stainless steel, the transient increase in current density is attributed to passivation occurring on its surface. Due to the presence of a passivation film on the surface, the initial current density value is relatively low as the film impedes reaction processes. When large-diameter particles impact the surface of stainless steel, the passivation film ruptures, allowing active reactants to contact the metal substrate, intensifying the reaction process momentarily, thereby increasing the current density value. After reaching its peak, the current density diminishes with the repassivation process, gradually approaching the pre-impact initial value.

With increasing particle impact velocity, the peak current density also increases. Additionally, particles with greater mass (larger diameter) generate larger peak current density values. Since both impact velocity and particle diameter significantly influence the peak current density, this phenomenon is associated with particle kinetic energy. As reducing impact velocity or decreasing mass can diminish particle kinetic energy, there exists a kinetic energy threshold below which transient currents cannot be detected. The existence of this threshold is because there are no transient currents below the critical kinetic energy threshold. Upon impact on passivated surfaces, rapid transient current spikes do not occur because particle impacts below the threshold kinetic energy do not rupture the passivating oxide film on the metal surface.

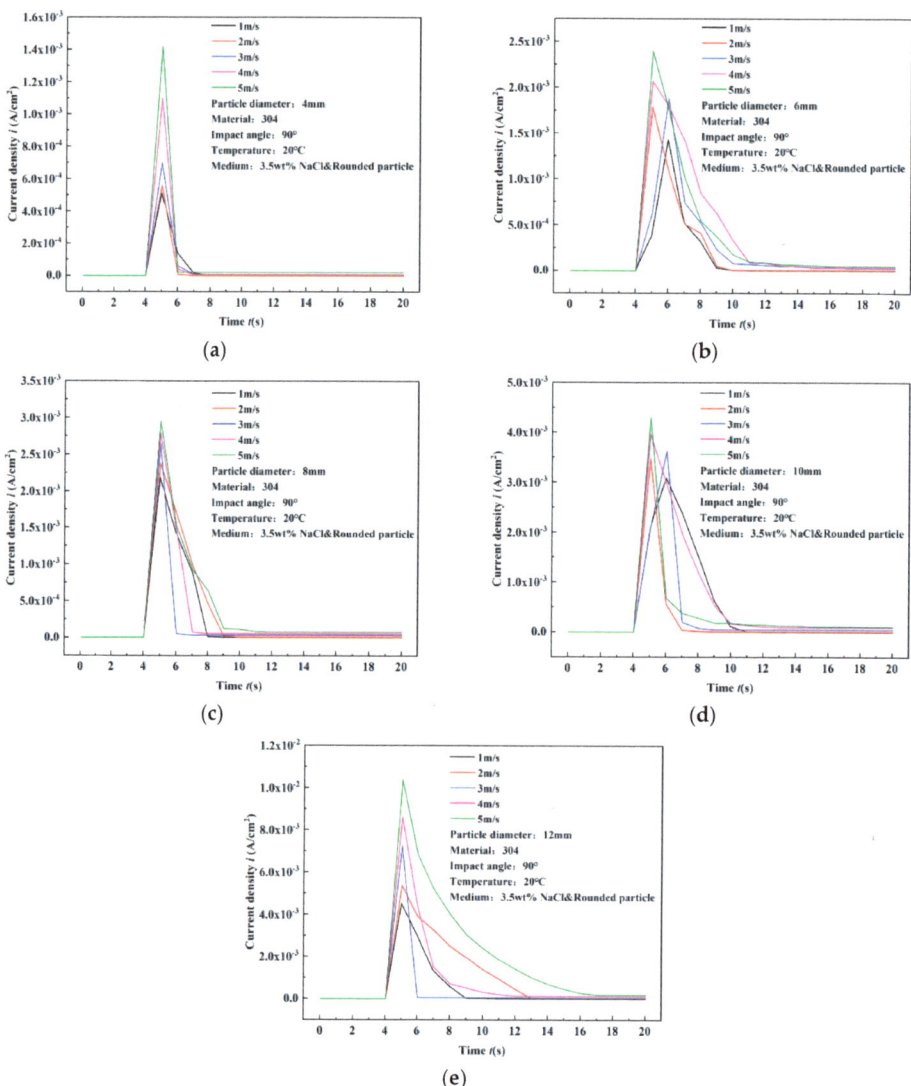

Figure 8. Experimental results of surface current density variation of 304 stainless steel under single particle impact in 3.5 wt% NaCl solution (comparison of flow rates). (**a**) 4 mm, (**b**) 6 mm, (**c**) 8 mm, (**d**) 10 mm, (**e**) 12 mm.

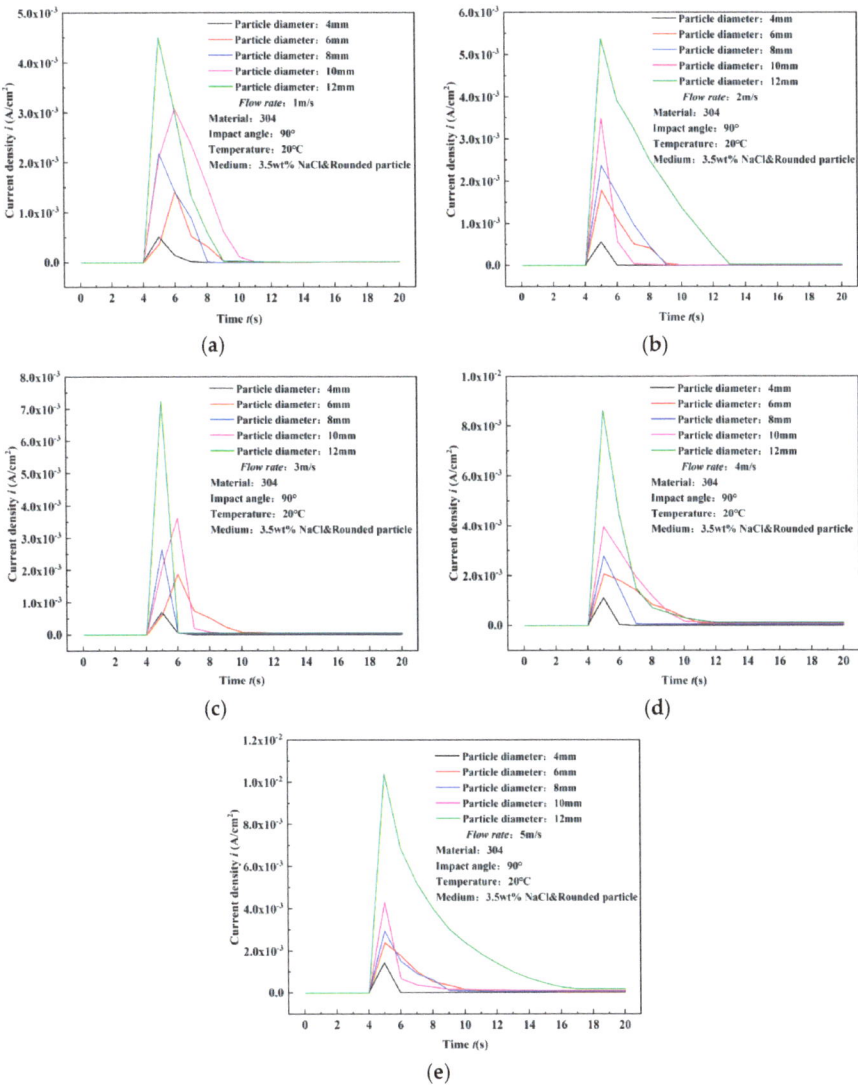

Figure 9. Experimental results of surface current density variation of 304 stainless steel under single particle impact in 3.5 wt% NaCl solution (comparison of particle diameters). (**a**) 1 m/s, (**b**) 2 m/s, (**c**) 3 m/s, (**d**) 4 m/s, (**e**) 5 m/s.

4. Discussion

The impact of particles and electrochemical reaction corrosion exhibit a synergistic relationship with mutual influence. Particles disturb the mass transfer boundary layer, altering the mass transfer rate of reactants. Simultaneously, the impact dislodges the oxide film, carrying away reactants. Moreover, electrochemical reactions alter the surface structure of the metal, weakening local hardness. Concurrently, localized pitting occurs, disrupting the geometric organization structure.

In the experiment investigating the material's current density response under single-particle impact, a positive step change was observed on the surface of the 304 stainless-steel electrode following particle impact [19]. As the diameter and impact velocity of the individual particles increased, the peak current density on the surface of the 304 stainless-steel electrode also increased. Concurrently, the time required for the current density to recover from its peak to near-initial values increased, indicating greater difficulty in maintaining system stability. The relatively short recovery time for 304 stainless steel suggests its robust self-repair capability. The results indicate a positive correlation between the particle impact-induced electrochemical corrosion rate and particle diameter, as well as impact velocity.

The surface open-circuit potential of the electrode undergoes a negative step change following particle impact, reaching a peak and gradually recovering with a nonlinear passivation trend. The open-circuit potential value after passivation recovery is lower than the potential value before particle impact, indicating that particle impact weakens the corrosion resistance of stainless steel in Cl^- medium. When particles impact the metal surface, they squeeze the near-wall flow mass transfer boundary layer, accelerating the reaction process, leading to a sudden positive step change in the current density and the appearance of a current density peak. As the particles move away from the surface, the current density gradually returns to near-initial values. Thus, based on the different response processes of current density, a model for the variation of current density on the electrode surface due to particle impact can be established, thereby predicting the material surface corrosion rate [20].

After metal is impacted by particles in a liquid–solid two-phase flow, the metal surface is exposed to the corrosive medium. The electrochemical reaction system on the surface experiences transient disturbances but eventually stabilizes. The electrochemical reaction current density is positively correlated with the corrosion rate. Following particle impact, the trend of current density variation is depicted in Figure 10.

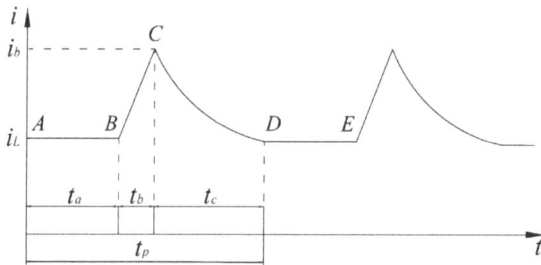

Figure 10. The schematic diagram illustrates the trend of current density variation caused by particle impact on the metal surface.

Sections A–B represent pure liquid flow, where the surface undergoes a reaction equilibrium process, resulting in a stabilized current density. In sections B–C, the process involves the response of current due to particle impact. As a result of film rupture or sediment dispersion, the surface reaction current density reaches its maximum value, considered as the limit current density at point C. In sections C–D, following the departure of particles from the surface after impact, a transient process of recovery occurs, leading to the re-establishment of the surface electrochemical equilibrium system, ultimately reaching stability. Sections D–E are analogous to sections A–B, representing a stable process. For sections B–C, due to the lag in mass transfer and reaction, the actual time of current response lags behind the time of impact. However, because of the high impact velocity and short contact time with the surface, the lag in current change is neglected. After the disappearance of the impact effect in sections B–C, the reaction system re-establishes, initiating a repassivation process. The entire process of single particle impact on surface current density change can be segmented for calculation, thus further establishing a model for the current density response of single particle impact on the surface.

When particles enter the mass transfer boundary layer, they compress the boundary layer thickness, increase reactant concentration, and accelerate the mass transfer rate of the system. The near-wall viscous sublayer is characterized by laminar flow. After particles impact the viscous sublayer, the flow velocity changes from a laminar to a turbulent state. Therefore, Equation (1) is utilized to calculate the change in current density.

$$i_L = \frac{0.0189 n_e F C^a \mathrm{Re}^{1/5} u^a}{\left[1 + 0.6203 \mathrm{Re}^{-1/10}(Sc - 1)\right] L_x} \tag{1}$$

By altering the flow velocity within the conduit, the variation in the average flow velocity near the wall boundary layer at different velocities is computed. This provides a fitting function for the change in flow velocity and conduit diameter. By integrating the numerical changes in conduit diameter, the effect of particle impact on the convective growth current density is determined.

$$i_{bl}(t) = A_b \cdot \lambda_u = i_L \cdot (0.0118 u^a + 0.1967 d_l + 0.2288) \tag{2}$$

By analyzing the convective, diffusive, and migratory current densities, as they are connected in parallel, the increase in current density due to particle impact on the metal wall is determined as follows:

$$i_b(t) = i_{b1}(t) + i_{b2}(t) = i_L \cdot (0.0118 u^a + 0.1967 d_l + 0.2288) + \left(\sqrt{t_b} - 1/\sqrt{t_b}\right) n_e F A_e D_O^{1/2} C^a \tag{3}$$

Upon completion of particle impact on the surface, firstly, there is a transition of the metal surface concentration from a small gradient to a large gradient, leading to a rapid decrease in the diffusion rate and a nonlinear decrease in the current density. Secondly, for stainless-steel surfaces, influenced by the passivation film coverage, the polarization resistance increases, resulting in a decrease in exchange current density. Through calculation, the time function of material diffusion current density after particle impact can be obtained:

$$i_{c1}(t) = \frac{n_e F A_e D_O^{1/2} C_O^a}{\pi^{1/2} t_c^{1/2} (1 + \kappa\varphi)} \tag{4}$$

For the single-particle impact process, the metal current density undergoes a full cycle variation, namely, impact–transient–recovery stages. Therefore, in calculating the corrosion rate, the current density–time function can be integrated to obtain the average value.

5. Conclusions

This paper begins with an experimental study on liquid–solid jet flow particle impact on a wall surface, focusing on the surface electrochemical response to individual particle impacts on 304 stainless steel. Subsequently, the process of a single-particle impact on the wall surface current density variation was calculated in stages, and a model for the response of the wall surface current density to single-particle impact was established. Based on the experimental and computational research conducted above, the following conclusions were drawn:

(1) Particle impact experiments indicate that there is a positive directional surge in current density on the surface of 304 stainless-steel electrodes after particle impact. With the gradual increase in the single particle's diameter and impact velocity, the peak current density on the surface of 304 stainless-steel electrodes increases accordingly. Simultaneously, it takes longer for the current density to recover from its peak value to near its initial value, indicating that the system finds it increasingly difficult to maintain a stable state. The results demonstrate a positive correlation between the electrochemical corrosion rate of particle impact on the wall surface and the particle diameter and impact velocity.

(2) The open-circuit potential on the electrode surface undergoes a sudden negative deviation after being impacted by particles, reaching a peak value, and gradually recovers with a nonlinear passivation trend. The open-circuit potential value after passivation recovery is lower than the potential value before particle impact, indicating that particle impact weakens the corrosion resistance of stainless steel in Cl^- medium.

(3) When particles impact the metal surface, they squeeze the near-wall flow boundary layer, accelerating the reaction process, causing a sudden positive deviation in current density, leading to the appearance of a current density peak. As the particles move away from the wall, the current density gradually returns to near its initial value. Based on the different response processes of current density, a model for the variation of current density on the electrode surface due to particle impact was established to predict the corrosion rate of the material surface.

Author Contributions: J.C. and M.L. carried out modeling and data calculations; J.C. and M.L. wrote the program; M.Y. and L.C. verified and changed the manuscript. All authors have read and agreed to the published version of the manuscript.

Funding: This research was funded by National Natural Science Foundation of China (grant no. 52105209). It was also supported by Natural Science Basic Research Program of Shaanxi Province [No.2023-JC-YB-407], Science and Technology Plan Project of Xi'an [No.2022JH-RYFW-0183].

Informed Consent Statement: Informed consent was obtained from all subjects involved in the study.

Data Availability Statement: The original contributions presented in the study are included in the article, further inquiries can be directed to the corresponding author.

Conflicts of Interest: Authors Long Chai and Min Yang were employed by the CCDC Changqing Downhole Technology Company. The remaining authors declare that the research was conducted in the absence of any commercial or financial relationships that could be construed as a potential conflict of interest.

References

1. Xu, Y.; Zhang, Q.; Chen, H.; Zhao, Y.; Huang, Y. Experimental study on erosion-corrosion of carbon steel in flowing NaCl solution of different pH. *J. Mater. Res. Technol.* **2022**, *20*, 4432–4451. [CrossRef]
2. Zhao, Y.L.; Ye, F.X.; Zhang, G.; Yao, J.; Liu, Y.F.; Dong, S.G. Investigation of erosion-corrosion behavior of Q235B steel in liquid-solid flows. *Pet. Sci.* **2022**, *19*, 2358–2373. [CrossRef]
3. Du, Y.; Yang, G.; Chen, S.; Ren, Y. Research on the erosion-corrosion mechanism of 304 stainless steel pipeline of mine water in falling film flow. *Corros. Sci.* **2022**, *206*, 110531. [CrossRef]
4. Supriyatna, Y.I.; Noviyana, R.; Suka, E.G.; Kambuna, B.N.; Sumardi, S. Influence of current density in Cu-Mn electroplating of AISI 1020 steel corrosion rate. *Mater. Today Proc.* **2021**, *44*, 3289–3295. [CrossRef]
5. Mayén, J.; Hernández-Hernández, M.; Del Carmen Gallegos-Melgar, A.; Pereyra, I.; Barredo, E.; Abundez-Pliego, A.; Porcayo-Calderón, J. Statistical Analysis of Corrosion Current Density and Tensile Strength of Al-6061 Alloy by Ageing and Retrogression Heat Treatments. *Trans. Indian Inst. Met.* **2022**, *75*, 2725–2735. [CrossRef]
6. Wang, C.; Li, W.; Xin, G.; Wang, Y.; Xu, S.; Fan, M. Novel method for prediction of corrosion current density of gas pipeline steel under stray current interference based on hybrid LWQPSO-NN model. *Measurement* **2022**, *200*, 111592. [CrossRef]
7. Alhumade, H.; Rezk, H. An Accurate Model of the Corrosion Current Density of Coatings Using an Adaptive Network-Based Fuzzy Inference System. *Metals* **2022**, *12*, 392. [CrossRef]
8. Li, L.L.; Wang, Z.B.; He, S.Y.; Zheng, Y.G. Correlation between depassivation and repassivation processes determined by single particle impingement: Its crucial role in the phenomenon of critical flow velocity for erosion-corrosion. *J. Mater. Sci. Technol.* **2021**, *89*, 158–166. [CrossRef]
9. Xu, L.; Wu, P.; Zhu, X.; Zhao, G.; Ren, X.; Wei, Q.; Xie, L. Structural characteristics and chloride intrusion mechanism of passive film. *Corros. Sci.* **2022**, *207*, 110563. [CrossRef]
10. Hou, D.; Zhang, K.; Hong, F.; Wu, S.; Wang, Z.; Li, M.; Wang, M. The corrosion deterioration of reinforced passivation Film: The impact of defects. *Appl. Surf. Sci.* **2022**, *582*, 152408. [CrossRef]
11. Wu, L.; Zhou, Z.; Zhang, K.; Zhang, X.; Wang, G. Electrochemical and passive film evaluation on the corrosion resistance variation of Fe-based amorphous coating affected by high temperature. *J. Non-Cryst. Solids* **2022**, *597*, 121892. [CrossRef]
12. Wang, Z.; Sun, C.; Li, L.; Roostaei, M.; Fattahpour, V.; Mahmoudi, M.; Zeng, H.; Zheng, Y.; Luo, J.L. A new method to obtain the repassivation time of passive materials based on the single particle impingement. *Corros. Sci.* **2020**, *170*, 108717. [CrossRef]

13. Khalifa, A.; Breuer, M.; Gollwitzer, J. Neural-network based approach for modeling wall-impact breakage of agglomerates in particle-laden flows applied in Euler–Lagrange LES. *Int. J. Heat Fluid Flow* **2022**, *94*, 108897. [CrossRef]
14. Nasiri, M.M.; Dolatabadi, A.; Moreau, C. Modeling of liquid detachment and fragmentation during the impact of plasma spray particles on a cold substrate. *Int. J. Heat Mass Transf.* **2022**, *189*, 122718. [CrossRef]
15. Schreiber, J.M.; Smid, I.; Eden, T.J.; Koudela, K.; Cote, D.; Champagne, V. Cold spray particle impact simulation using the Preston-Tonks-Wallace plasticity model. *Finite Elem. Anal. Des.* **2021**, *191*, 103557. [CrossRef]
16. Ren, F.; Fang, T.; Cheng, X. Theoretical modeling and experimental study of rock-breaking depth in particle jet impact drilling process. *J. Pet. Sci. Eng.* **2019**, *183*, 106419. [CrossRef]
17. Birkin, P.R.; Lear, R.; Webster, L.; Powell, L.; Martin, H.L. In-situ detection of single particle impact, erosion/corrosion and surface roughening. *Wear* **2021**, *464*, 203527. [CrossRef]
18. Cheng, J.; Yan, Q.; Pan, Z.; Wei, W. On-Line Measurement and Characterization of Electrochemical Corrosion of 304L Stainless Steel Pipe Wall in High-Speed Cl-Containing Solution. *Metal* **2022**, *12*, 1324. [CrossRef]
19. Szklarz, Z.; Kołczyk-Siedlecka, K.; Vereshchagina, E.; Herbjørnrød, A.; Wittendorp, P.; Jain, S.; Wójcik, P.J. A Study of the Long-Term Electrochemical Stability of Thin-Film Titanium–Platinum Microelectrodes and Their Comparison to Classic, Wire-Based Platinum Microelectrodes in Selected Inorganic Electrolytes. *Materials* **2024**, *17*, 1352. [CrossRef] [PubMed]
20. Xiang, Y.H.; Han, F.X.; Liu, Q. Effect of Salt Solution Erosion on Mechanical Properties and Micropore Structure of Recycled Fine Aggregate ECC. *Materials* **2024**, *17*, 2498. [CrossRef] [PubMed]

Article

Study on Shear Performance of Corroded Steel Fiber Reinforced Concrete Beams under Impact Load

Jianxiao Gu [1], Liancheng Li [2], Xin Huang [3] and Hui Chen [1,*]

1   Wenzhou Key Laboratory of Intelligent Lifeline Protection and Emergency Technology for Resilient City, College of Architecture and Energy Engineering, Wenzhou University of Technology, Wenzhou 325035, China; 102015295@glut.edu.cn
2   College of Architecture and Civil Engineering, Wenzhou University, Wenzhou 325035, China; 21461544036@stu.wzu.edu.cn
3   Wenzhou Traffic Engineering Testing and Testing Co., Ltd., Wenzhou 325035, China; 20461542031@stu.wzu.edu.cn
*   Correspondence: chenhui0306@wzu.edu.cn

Abstract: With the growing use of steel-fiber-reinforced-concrete (SFRC) beams in environmentally friendly and rapid construction, it is essential to assess their impact performance. These beams may encounter unexpected impact loadings from accidents or terrorist attacks during service life. This study explored the impact of steel fiber content and drop hammer height on the impact load testing of corrosion-treated SFRC beams. Experiments were conducted with varying steel fiber contents (0%, 0.25%, 0.5%, 0.75%, and 1.0%), and drop hammer height (1 m, 2 m, and 3 m). The corrosion test demonstrates that SFRC beams supplemented with steel fibers showcase a diminished surface rust spot area in comparison to those lacking fibers. This improvement is ascribed to the bonding between fibers and the concrete matrix, along with their current-sharing properties. SFRC beams, subjected to impact testing, exhibit concrete crushing at the top without spalling, showcasing improved impact resistance due to increased fiber content, which reduces crack formation. Additionally, different fiber contents yield varied responses to impact loads, with higher fiber content notably enhancing overall beam performance and energy dissipation capacity. Energy dissipation analysis shows a moderate increase with higher fiber contents, and impulse impact force generally rises with fiber content, indicating improved impact resistance.

Keywords: impact performance; SFRC beams; failure mode; dynamic response; energy dissipation

Citation: Gu, J.; Li, L.; Huang, X.; Chen, H. Study on Shear Performance of Corroded Steel Fiber Reinforced Concrete Beams under Impact Load. *Materials* **2024**, *17*, 2566. https://doi.org/10.3390/ma17112566

Academic Editor: Yuri Ribakov

Received: 15 April 2024
Revised: 3 May 2024
Accepted: 15 May 2024
Published: 27 May 2024

1. Introduction

Worldwide, the extensive utilization of reinforced concrete (RC) structures can be primarily attributed to the favorable properties of concrete. These encompass high ductility, considerable stiffness, minimal maintenance requirements, enhanced fire resistance, and significant compressive strength despite its inherent deficiency in tensile strength [1–3]. However, corrosion poses a significant threat to the durability of RC structures, impacting both mechanical strength and long-term performance [4]. Normally, rust formation weakens the bond between steel and concrete, causing delamination and spalling. Moreover, corrosion-induced volumetric expansion imposes splitting stresses on surrounding concrete, resulting in cracking and eventual cover spalling. Increased exposure exacerbates corrosion rates, accelerating structural deterioration. To mitigate the impact of corrosion on durability and ensure the structure's long-term service life, significant investments are necessary for maintenance and repair [5].

In recent years, the world has experienced an increase in accidents, earthquakes, heavy objects, or rockfalls impacting on concrete slabs, as well as terrorist attacks, landslides, floods, and incidents involving the release of corrosive gases and liquids, all of which have caused damage to concrete structures [6–10]. Moreover, rust and corrosion phenomena typically manifest in moist environments with the presence of oxygen, along with salts in the soil or groundwater [11,12]. Understanding the response of corroded reinforced concrete (RC) beams to impact loads is essential in assessing their ability to withstand unexpected forces, such as those encountered during accidents or deliberate acts. In light of the escalating frequency of unpredictable and extreme events, enhancing the resilience and safety of structures necessitates thorough research into the behavior of concrete beams under impact conditions [13–18]. The above-mentioned research has paid little attention to the mechanical behavior of concrete beams under impact loads in the presence of rebar corrosion [19]. The weakening of the bond between steel and concrete due to rust formation leads to delamination and spalling. Therefore, the objective of this study is to examine the mechanical response of RC beams affected by corrosion when subjected to impact loading.

Recognizing the potential severe issues arising from the corrosion of stirrups, numerous studies have been conducted to examine the performance of RC beams containing corroded stirrups [2,20–26]. The presence of corroded stirrups compromises the shear capacity and ductility of RC beams, with the shear span ratio, stirrup spacing, and concrete strength being significant influencing factors. Research conducted by Fernandez et al. [27] and Al-Sibahy et al. [28] using electrically accelerated corrosion has revealed a correlation between the reduction in load-carrying capacity and the corrosion-induced loss of cross-sectional areas in steel reinforcement. Moreover, the corrosion-induced structural deterioration and defects in steel reinforcement have led to a significant need for maintenance in numerous concrete highway bridges. Extending the service life of corroded RC structures involves repairing them to enhance both their load-bearing capacity and deformation resistance [23,26]. Several strengthening technologies, including fiber-reinforced polymer (FRP), steel plates, and minor concrete repair, are available for restoring degraded RC structures. For instance, experiments by Chen et al. [29] on RC specimens immersed for 60 days in a 10% Na_2SO_4 solution showed a loss of load-bearing capacity of 17.8%, which rose to 34.8% under coupled cyclic loading. Lu et al. [30] also observed alterations in crack pattern, ductility, and stiffness evolution in beams experiencing corrosion-induced mass losses ranging from 4% to 18%. Nevertheless, there has been relatively limited investigation into the impact loading behavior resulting from dynamic loads in corroded RC beams [20,31,32]. This study conducted impact load tests, considering the variables of steel fiber content, corrosion conditions, and impact load level, to investigate the mechanical response and analyze the test results.

Moreover, extensive research has explored detailed experimental, numerical, and analytical methods to investigate the dynamic response of fiber-reinforced-concrete (FRC) beams [1,16,33–38]. For instance, considerable attention has been devoted to investigating the performance of concrete beams reinforced with Glass FRP (GFRP) bars, particularly under impact loading conditions [39–41]. The results indicated that these beams exhibited an average 15% increase in moment capacity under impact loading (using a drop mass of 110 kg at a height of 1.2 m) compared to static loading conditions. While this research has shed light on the mechanical response of FRC beams while considering various variables, it has not been comprehensive in its approach. However, there has been limited research in the last decade concerning the impact of drop hammer height, corrosion condition, and fiber content on concrete structures. Examining these factors is vital for assessing concrete structural safety. This study investigates the mechanical response and strength parameters of corroded steel-fiber-reinforced-concrete (SFRC) beams under different impact energy and fiber content in an impact load test.

This study explores the mechanical behavior of SFRC beams, examining the influence of corrosion condition, steel fibers content, and impact energy on their properties through drop hammer impact load tests. The study includes fifteen specimens for drop hammer impact tests. It monitors and compares failure modes, midspan deflection, impact force, and absorbed energy. Additionally, impact mechanical behavior responses based on test data are analyzed to assess the beam's damage level under loadings. The failure modes of SFRC beams are investigated using high-speed camera images, and energy dissipation is calculated. Overall, the inclusion of steel fibers in SFRC beams improves impact resistance and deformation capacity, enhancing structural performance under impact loading conditions. Conclusions drawn from this paper will enhance understanding of the differential impact response of SFRC beams under varying steel fiber contents and impact energies.

2. Experimental Program

2.1. Materials

The experiment employed ordinary Portland cement, and the results of compressive strength tests for concrete cube specimens are detailed in Table 1 (following the testing method outlined in GB/T 50081-2002 [42]). This in turn gives a concrete mixture with a mix proportion of 1:1.5:2.8:4.7 (water–cement–sand–gravel) by weight. The composition of each part is controlled proportionally based on the total mass of the RC beams. Natural river sand was used as the fine aggregate, while the coarse aggregate was crushed limestone with a maximum size of 20 mm. The study utilized high-performance cold-drawn hooked end steel-fibers with a length of 30 mm, a diameter of 0.5 mm, a density of 7800 kg/m^3, and an ultimate tensile strength of 1200 MPa. Five dosages of steel fibers were adopted, at 0%, 0.25%, 0.5%, 0.75% and 1% by weight of cement. The pictures of the steel fibers are presented in Figure 1, while Table 2 shows the results of compressive strength tests for steel-fiber-reinforced concrete cube specimens. For the steel bars, longitudinal steel bars (HRB400, the product comes from HBIS, Tangshan, Hebei, China) with diameters of 12 mm and 10 mm were used, displaying average yield strengths of 465 MPa and 534 MPa and ultimate yield strengths of 625 MPa and 674 MPa, respectively. Additionally, 8 mm diameter stirrup steel bars (HPB300, the product comes from HBIS, Tangshan, Hebei, China) were incorporated, exhibiting average yield strengths of 363 MPa and ultimate yield strengths of 576 MPa.

Table 1. Basic parameters of concrete materials (results of compressive strength tests for concrete cube specimens).

Curing Period (d)	Average Cubic Compressive Strength (MPa)	Axial Compressive Strength (MPa)
14	32.1	24.4
21	34.2	26.0
28	37.6	28.6

Figure 1. Materials and steel-fiber-reinforced concrete cube specimens, (**a**) steel fiber (SF) material, (**b**) steel-fiber-reinforced-concrete cube specimens, and (**c**) compressive strength tests device diagram.

Table 2. Basic parameters of steel-fiber-reinforced concrete materials (results of compressive strength tests for steel-fiber-reinforced concrete cube specimens).

Type	Steel Fiber Dosage (%)	Curing Period (d)	Average Cubic Compressive Strength (MPa)	Axial Compressive Strength (MPa)
DL-0	0		33.00	25.08
DL-0.25	0.25		41.85	31.81
DL-0.50	0.50	28	42.97	32.66
DL-0.75	0.75		45.63	34.68
DL-1.00	1.0		45.27	34.41

2.2. Details of Beam Specimens

This section presents tests conducted on a total of fifteen full-scale steel-fiber-reinforced concrete (SFRC) beams which comprise three specimens for drop hammer impact tests without steel fiber reinforcement. The dimensions of the beams, including width (b), height (h), total length (L), and clear-span length (Lc), were 100 mm, 150 mm, 1400 mm, and 1160 mm, respectively, as illustrated in Figure 2. The cover thickness for longitudinal reinforcement is 15 mm. To ease the transportation of specimens, 8 mm diameter lifting loops are installed on both sides of the SFRC beam. The stirrup reinforcement ratio (e.g., 0.670%) is designed to study the mechanical response of SFRC beams under impact loads. Additionally, one set of SFRC beams were tested with a curing period of 28 days.

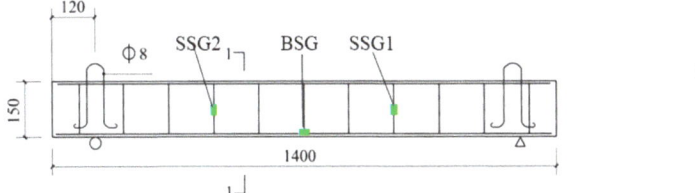
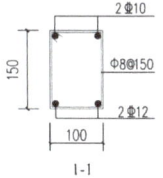

Figure 2. Dimensions and reinforcement configuration of SFRC beams (unit: mm).

3. Accelerated Corrosion and Test Setup

For measuring the mechanical characteristics of SFRC beams under impact loads, strain gauges (SG) are positioned at the bottom longitudinal reinforcement near the mid-span of the beam and at 1/3 span locations on the stirrup reinforcement, as illustrated in Figure 2. In the SFRC beams, three strain gauges (SGs) are affixed to the reinforcements: the bottom strain gauge (BSG) and two stirrup strain gauges (SSG1 and SSG2), as depicted in Figure 2. These gauges record strain values for analysis. Impact results, such as failure mode, peak load, load–mid-span deflection curve, and the load–strain curve of reinforcements, are subsequently examined. Two sets of strain values are measured, and their averages are calculated to enhance the reliability of the strain data. The process of installing the strain gauges involves the following steps: (1) the surface of the steel bars is polished smoothly using a grinding machine. (2) After polishing, the SGs are attached in the same direction as the steel bars using a strong adhesive, and a single-component room temperature vulcanizing silicone rubber is applied to waterproof the strain gauges. (3) Once the adhesive dries and cures, the area where the strain gauge is attached is covered with yellow waxed paper to reduce the likelihood of strain gauge failure during loading. The strain gauges (SGs) utilized are of the model BE-120-3AA (The product comes from Jiangsu Donghua Testing Technology Co., Ltd., Jingjiang, China), with a sensitivity of 2.08 ± 1%, and the materials for the SGs and attachment are depicted in Figure 3a. Additionally, after casting the specimens, they are wrapped with plastic film, and water is periodically applied after the concrete has been set to maintain humidity during the curing process. The casting and curing process of the specimens is portrayed in Figure 3b.

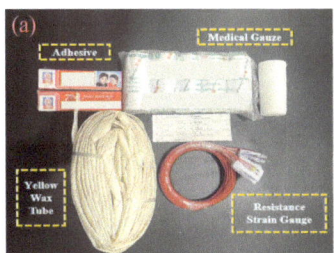

Figure 3. Materials and SFRC beams, (**a**) strain gauge, paste auxiliary material, (**b**) SFRC beams casting, and curing condition.

3.1. Corrosion Method

A composite solution containing chloride and sulfate salts was formulated to simulate the attack environment for the SFRC beams. The concentrations of these salts mirror those found in the soil and groundwater of southern Wenzhou, China. The primary constituent of the aforementioned salts was NaCl. The corrosion process was stimulated electrochemically by chloride ingress [43]. The chloride ion concentration for this class was set at 5% by mass. In accordance with the methodology outlined in this study, all SFRC beams underwent an initial curing period of 28 days. Subsequently, those SFRC beams slated for exposure to aggressive environments were subjected to a 3-day immersion in NaCl solution, as illustrated in Figure 4. This step aimed to replicate real-world conditions, allowing for a specified period before salt solutions could initiate concrete deterioration. Additionally, a copper plate was positioned beneath the SFRC beams, as depicted in Figure 4a. The corrosion process employed the copper plate as the cathode and the steel bar as the anode. Both were connected to the power supply equipment, delivering the predetermined current density as specified earlier (200 μA/cm²). Various SFRC beams were interconnected in a series to ensure uniform intensity, as depicted in Figure 4b. It is important to note that this study exclusively applies electrochemical acceleration corrosion to the stirrups. To facilitate this, a 4 mm diameter steel bar was employed to connect and extend the stirrups externally from the beam. The section of the steel bar in contact with the stirrups was insulated with epoxy resin and wrapped with cloth, as illustrated in Figure 4c.

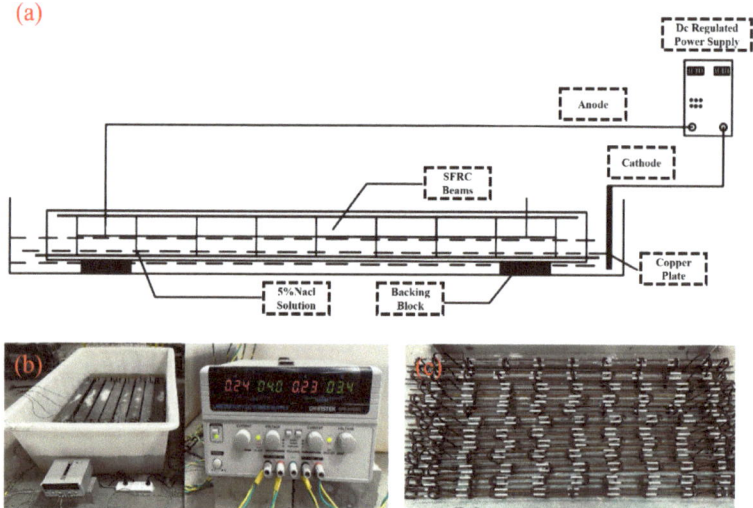

Figure 4. Corrosion method of SFRC beams, (**a**) test setup induced corrosion, (**b**) laboratory accelerated corrosion device, (**c**) a voltmeter device for power supply.

The initiation of steel reinforcement corrosion was induced through the passage of an electrical current. Utilizing Faraday's law (Equation (1)), it becomes possible to calculate the steel's weight loss over time due to corrosion. This calculation hinges on the applied electrical current intensity, represented as I(t), along with the diameter and exposed length of the steel bar.

$$E = \frac{m_{Fe} \int I \cdot dt}{V \cdot F} \tag{1}$$

Here, m_{Fe} represents the atomic mass, V denotes the steel valency (set to two), and F stands for Faraday's constant. Given that the applied intensity remained consistent throughout the entire test period, Faraday's law can be restated as Equation (2) to determine the weight loss of steel.

$$\Delta m = \frac{m_{Fe} I \cdot t}{V \cdot F} \tag{2}$$

Multiple researchers have posited that utilizing current densities below 200 μA/cm² in accelerated corrosion tests yields steel weight-loss estimates comparable to those predicted by Faraday's law, with a discrepancy of 5–10% [44,45]. The SFRC beams corrosion test in this study was conducted under electric current for 48 days. According to theoretical calculations, the corrosion rate of the stirrups was determined to be 12.3%.

3.2. Impact Test

The design and experimental procedure of the impact load test refer to GB/T 2423.5-2019 (Mechanical Shock Testing Methods and Standards) [46]. The impact tests were conducted using a drop hammer test system comprising control devices and a loading device, as depicted in Figures 5 and 6. The beams, with a clear span of 1160 mm, were secured at both ends using clamps to prevent detachment from the supports during testing. A loading apparatus consisting of a cylindrical drop hammer and counterweight was utilized, with the counterweight mass being adjusted to control the impact energy. During the tests, both the hammer and counterweight were released from a predetermined height, achieving the desired speed and energy upon free-fall impact onto the beams. Additionally, the SFRC beams used in the experimental program for impact load testing are presented in Table 3. In the experiment, three identical beams were tested for each loading condition to ensure the reliability of the experiment. In the analysis, the parameters were averaged across the three beams for analysis. The beams are categorized in a DL-A-B format, wherein 'DL' represents the beam group, 'A' denotes the steel fiber content ranging from 0 to 1.0%, and 'B' signifies the falling height utilized in the impact load test. Data during the impact testing process are collected using the DH-5960 (The product comes from Jiangsu Donghua Testing Technology Co., Ltd., Jingjiang, China) for super-dynamic signal testing and analysis (Figure 6a,b). The DH-5960 features a total of 12 channels, each equipped with an 8 M point buffer. It boasts a maximum transient sampling rate of 20 MHz and a frequency response of up to 1 MHz. Simultaneous acquisition of acceleration, displacement, and strain data are supported. In this study, the sampling frequency is set at 100 kHz. Two channels are employed to collect acceleration data during the experiment, while the remaining five channels are dedicated to collecting displacement and strain data. The sampling frequency remains consistent across all channels.

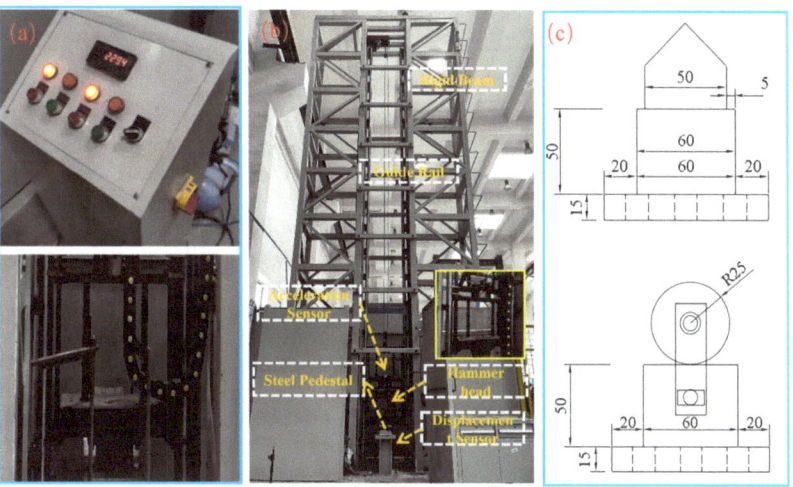

Figure 5. Loading device diagram, (**a**) control device, (**b**) impact load test device diagram, and (**c**) support detail (unit: mm).

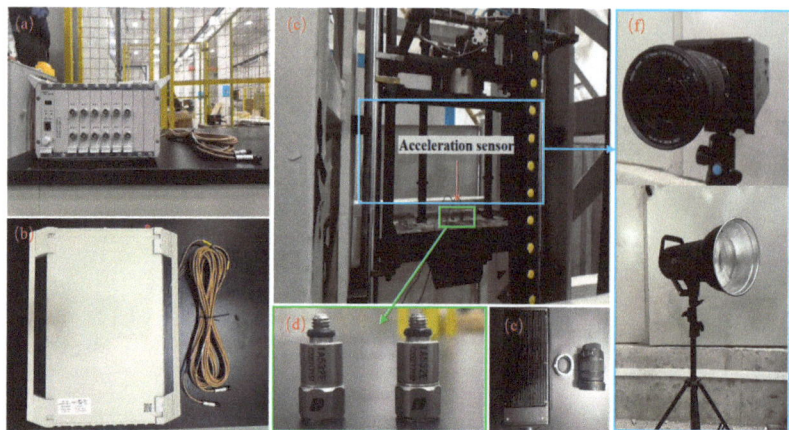

Figure 6. Physical diagram of the impact load test, (**a**,**b**) hyperdynamic signal test and analysis system, (**c**) drop weight impact test setup, (**d**) acceleration sensor and installation location, (**e**) high-speed camera parts, and (**f**) high-speed camera and lighting equipment.

Table 3. Experimental program of impact load testing of SFRC beams.

Types	Hammer Mass (kg)	Falling Height (m)	Impact Velocity (m/s)	Corrosion Time (d)	Steel Fibers Content (%)
DL-0-1m		1	4.43		0
DL-0-2m		2	6.26		0
DL-0-3m		3	7.67		0
DL-0.25-1m		1	4.43		0.25
DL-0.25-2m		2	6.26		0.25
DL-0.25-3m		3	7.67		0.25
DL-0.5-1m		1	4.43		0.5
DL-0.5-2m	100	2	6.26	48	0.5
DL-0.5-3m		3	7.67		0.5
DL-0.75-1m		1	4.43		0.75
DL-0.75-2m		2	6.26		0.75
DL-0.75-3m		3	7.67		0.75
DL-1.0-1m		1	4.43		1.0
DL-1.0-2m		2	6.26		1.0
DL-1.0-3m		3	7.67		1.0

Displacement data are gathered using the Panasonic standard-type HL-G125-A-C5 laser displacement sensor (The product comes from Jiangsu Donghua Testing Technology Co., Ltd., Jingjiang, China), featuring a measuring distance of 250 mm and a measurement range of ±150 mm. The laser displacement sensor is positioned directly beneath the center of the support steel plate. To facilitate this alignment, a circular hole is pre-drilled to enable the infrared laser emitted by the displacement sensor to pass through and align with the mid-span position of the specimen. Moreover, acceleration signals are acquired using IEPE piezoelectric accelerometer sensors with the model number 1A532E and a measurement range of $\pm10,000$ g (The product comes from Jiangsu Donghua Testing Technology Co., Ltd., Jingjiang, China). Two identical accelerometers of the same model are simultaneously utilized during the experiment to measure the instantaneous acceleration of the hammer. This approach is adopted to reduce the risk of accelerometer failure during the test caused by vibrations induced by impacts. The acceleration sensors are depicted in Figure 6d, and their location is at the mid-span of the RC beams.

Furthermore, to document the failure characteristics of the beams during the experiment, an AE120M high-speed camera (The product comes from Jiangsu Donghua Testing Technology Co., Ltd., Jingjiang, China) is employed during the impact load testing process to capture the deformation and crack propagation of the specimens. The instrument features a full-frame resolution of 1280 × 1024 and can achieve a high frame rate of 2000 FPS. When operating in a 1024 × 128 frame, it can attain a frame rate of up to 16,000 FPS. In this experiment, the high-speed camera has a data acquisition cycle of 1000 microseconds and operates at a resolution of 1280 × 1024, as illustrated in Figure 6f.

4. Test Results

This section entails impact testing on SFRC beams treated for corrosion to assess the impact of stirrup ratio, drop height, and steel fiber content on shear performance. The experiments entailed gathering data on impact force, mid-span deflection, and steel strain. High-speed cameras were utilized to capture crack development in the SFRC beams, followed by analysis of corresponding time–history curves and failure modes.

4.1. Failure Modes

Figure 7 illustrates the outcomes of accelerated corrosion tests conducted on SFRC beams with different steel fiber contents. The data show that SFRC beams incorporating steel fibers exhibit reduced rust spot areas on their surfaces compared to those without steel fibers (DL-0). This phenomenon is credited to the steel fibers bonding with the concrete matrix and their capacity to share the corrosion current, resulting in fewer pronounced rust marks on the concrete surface. Moreover, no expansion cracks due to rust have emerged, and the integrity of the beam specimens remains unaffected. In order to better observe the failure mode of the SFRC beams under impact load, the SFRC beams after accelerated corrosion testing were painted with white paint. The results of the impact load test are shown in Figures 8–10.

Figure 8 illustrates the failure modes of SFRC beams subjected to 1 m drop hammer impact loads with varying steel fiber contents. Concrete crushing was observed at the top of both SFRC beams, while no concrete spalling occurred at the bottom. Flexural cracks initiated at the mid-span, with subsequent development of flexure-shear cracks in the shear span region. Furthermore, the SFRC beams DL-0-1m and DL-0.25-1m experienced failure in a combined flexure-shear mode, exhibiting both flexural and flexure-shear cracks. However, an increase in steel fiber content led SFRC beams DL-0.5-1m and DL-1.0-1m to fail in a flexure-shear mode. Compared to RC beams DL-0-1m, DL-0.25-1m, DL-0.5-1m, and DL-0.5-1m, there was a reduction in the number of flexure and flexure-shear cracks. Additionally, under the 28-day curing period, most of the SFRC beams experienced diagonal shear failure, manifested by a diagonal crack on the right side of the beam extending from the point of impact to the supports. As the steel fiber content increased, the occurrence of shear cracks and flexural-shear cracks in SFRC beams during impact load tests decreased or, in some cases, vanished (as shown in Figures 8 and 9).

To investigate the impact of energy levels on SFRC beams, three different impact energy levels were employed, as depicted in Figures 8–10. SFRC beams DL-0-2m and DL-0-3m experienced diagonal shear failure marked by diagonal cracks on the left side of the mid-span and accompanied by concrete spalling on the bottom. SFRC beams of DL-0.25-2m, DL-0.5-2m, DL-0.75-2m, and DL-1.0-2m exhibited a flexure-shear combined failure mode, characterized by broader shear cracks at the mid-span and increased post-impact residual deformation. Furthermore, SFRC beams of DL-0.25-3m, DL-0.5-3m, DL-0.75-3m, and DL-1.0-3m exhibited a larger damaged area and more severe local damage. Specifically, SFRC beam DL-0.5-3m suffered significant splitting damage to the compression steel bars and displayed greater post-impact residual deformation at mid-span compared to SFRC Beam DL-0.5-2m, as illustrated in Figures 9 and 10.

Figure 10 illustrates a transition in failure mode from flexure-shear dominance under a 2 m drop hammer height to shear dominance under higher impact loads, accompanied by more severe localized damage. The observed damage comprises concrete crushing at the top and spalling at the bottom of the SFRC beams. With higher impact loads, the beams experienced failure in diagonal shear, leading to more severe concrete crushing, additional diagonal shear cracks on both sides of the beams, and a broader distribution area of cracks. This stands in contrast to the outcomes depicted in Figures 8 and 9 for SFRC beams subjected to lower impact loads. Consequently, the test results underscore the beneficial impact of higher steel fiber content on the impact performance of SFRC beams, resulting in reduced values for both maximum and residual deflection.

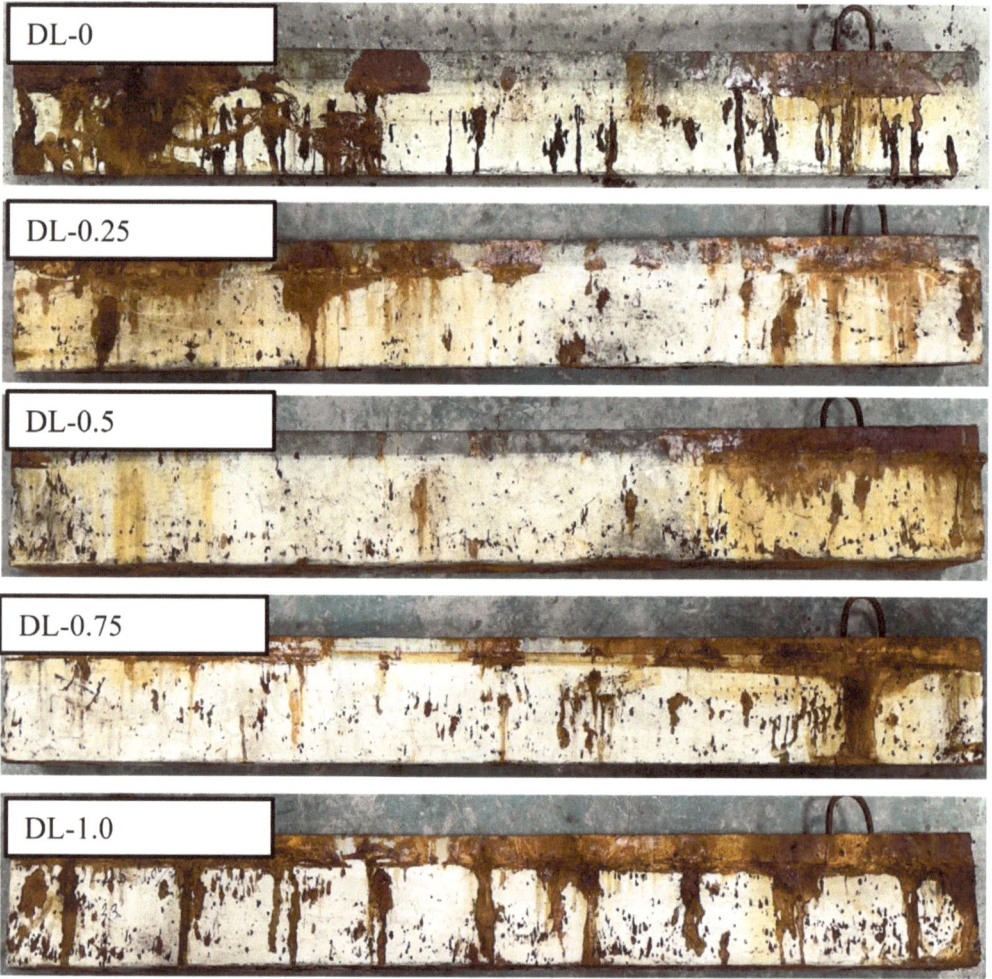

Figure 7. The accelerated corrosion test results of the SFRC beams. The area of rust spots on the surface of SFRC beams is less compared to the DL-0 RC beam without steel fibers.

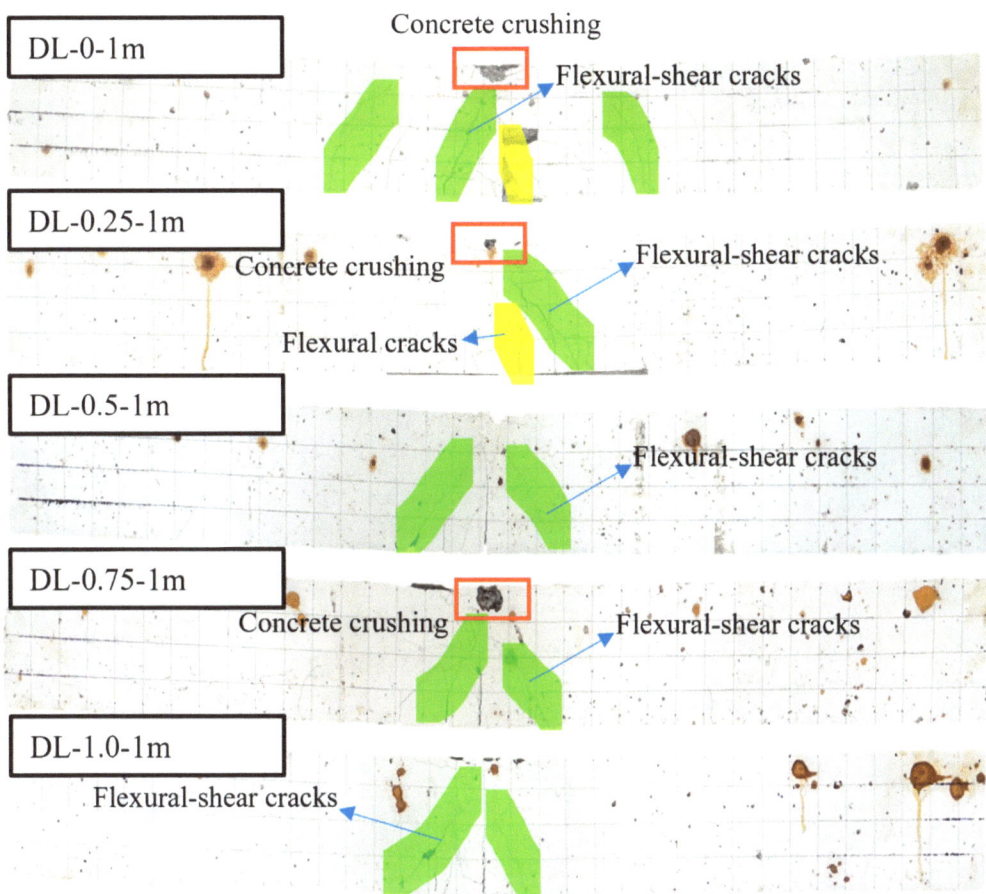

Figure 8. Failure modes of the corrosion-treated SFRC beams under impact loads. The failure modes of SFRC beams transition from a combined flexure failure mode to a flexural-shear mode as the steel fiber content increases from 0% to 1.0%.

Due to the extensive high-speed camera data, only a subset is presented here, focusing on the DL-0-1m SFRC beam (hammer drop height: 3 m) and the DL-1.0-3m SFRC beam (hammer drop height: 3 m). Following the initial impact and the subsequent rise in mid-span bending cracks, inclined cracks penetrated the bottom of the beam. Figure 11 illustrates the failure progression of SFRC beams captured by a high-speed camera, with the contact of the hammer with the SFRC beams marked at 1 ms. Initially, during the impact load's onset, the segment of the SFRC beam in contact with the hammer exhibited no discernible damage. For SFRC beam DL-0-3m, three flexural cracks emerged at the top of the mid-span at 1 ms. Between 5 ms and 10 ms, as mid-span flexural cracks expanded, inclined cracks propagated through the bottom of the SFRC beam, with further elongation and widening occurring. Furthermore, a fresh crack emerged on the right side of the mid-span at 10 ms. Between 10 ms and 15 ms, inclined cracks started forming from the mid-span towards the support. In the case of the DL-1.0-3m SFRC beam, one flexural crack and a lengthy longitudinal crack at the bottom were noticeable at 1 ms. Between 5 ms and 10 ms, the beam displayed a mix of flexure-shear, shear, and longitudinal cracks. By 15 ms, notably wide shear cracks appeared, leading to subsequent concrete spalling at the bottom.

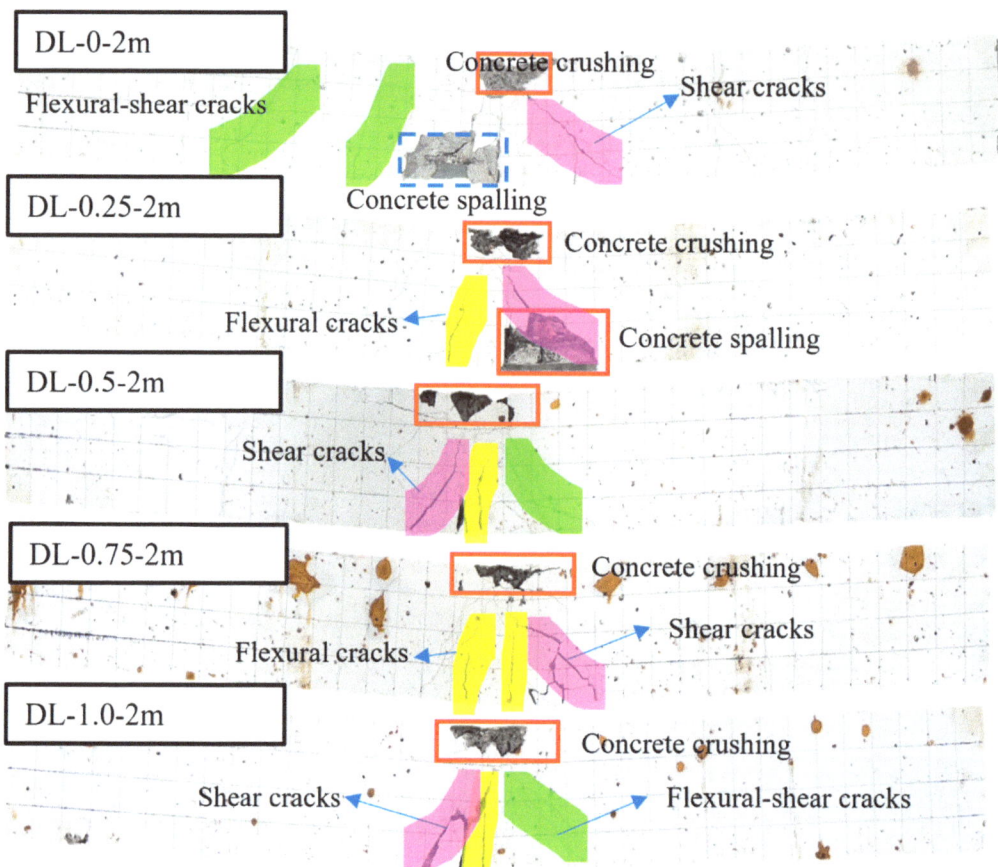

Figure 9. Failure modes of the corrosion-treated SFRC beams under impact loads (drop hammer height: 2 m). The SFRC beams exhibited a flexure-shear combined failure mode, characterized by broader flexural cracks at mid-span and increased post-impact residual deformation.

Particularly noteworthy, at a 3 m hammer drop height, both SFRC Beams exhibited prominent concrete damage at 4 ms in the impact zone where the hammer contacted the beam. The pre-existing cracks gradually extended and widened, accompanied by the appearance of shorter secondary cracks. Additionally, as the impact continued, the mid-span cracks displayed a triangular distribution pattern. Furthermore, for the SFRC beam without steel fiber reinforcement subjected to impact from a 3 m drop hammer height, the extent of damage was significantly greater than that of the SFRC beam with a 1.0% steel fiber content. Taking DL-0-3m as an example, the beam exhibits pronounced diagonal cracks, with cracks intersecting to form a network-like distribution. Concrete spalling occurs at the impact location, with longitudinal reinforcement being exposed at the bottom of the beam. In contrast, for the DL-1.0-3m beam, the development of diagonal cracks is inhibited, and two main bending cracks converge to form a triangular area at the mid-span impact point where a vertical main crack is observed. There is no extensive concrete spalling in the tension zone. The mid-span deflection of the specimens without steel fibers is larger and closer to the bottom of the camera frame, indicating that the addition of steel fibers can effectively enhance the impact resistance and deformation capacity of reinforced concrete, improve its stiffness, and alter the failure mode of the concrete beam specimens, concentrating cracks at the impact location.

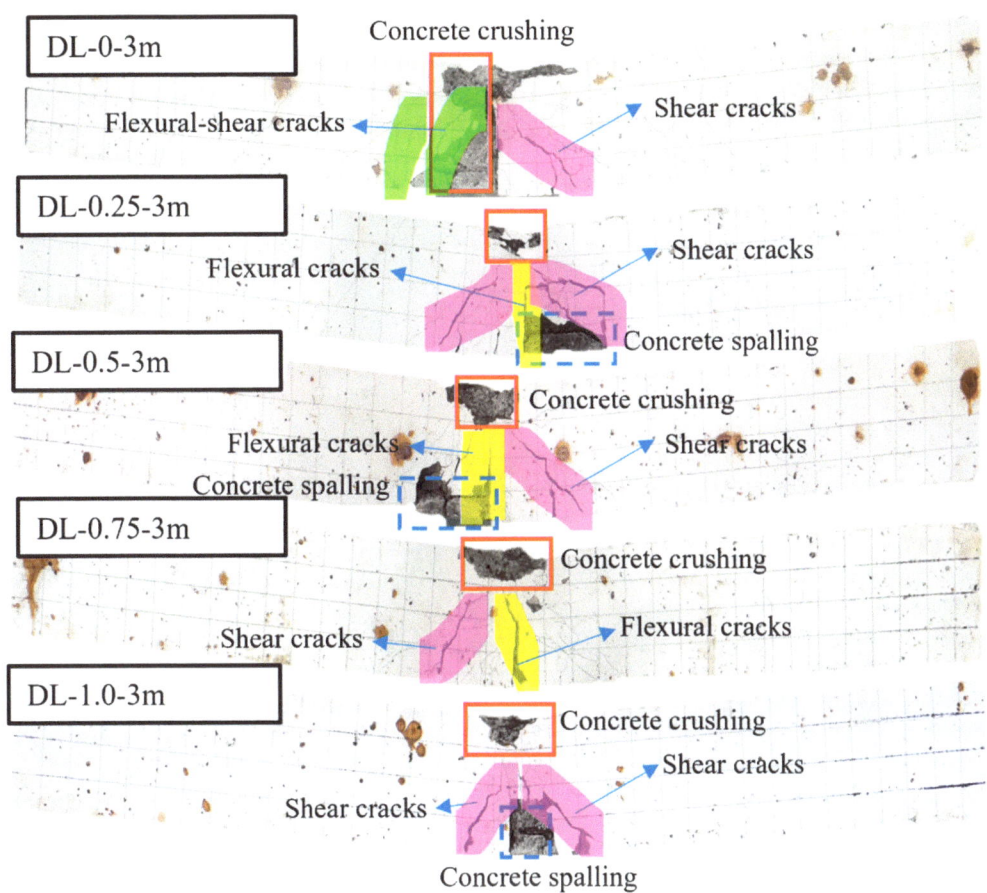

Figure 10. Failure modes of the corrosion-treated SFRC beams under impact loads (drop hammer height: 3 m). The failure mode transitioned from flexure-shear dominated under 2 m drop hammer height to shear dominated under impact loads, accompanied by more severe local damage, including concrete crushing on the top and spalling at the bottom.

4.2. Dynamic Responses

Each SFRC beam, subjected to different steel fiber contents, underwent a singular impact from the drop hammer at heights of 1 m, 2 m, and 3 m, respectively. Table 4 presents the typical impact testing results, including parameters such as impact speed (corresponding to dropping heights of 1 m, 2 m, 3 m), peak impact load, peak mid-span deflection, and residual mid-span deflection. Figure 12 displays the temporal evolution of impact loads for SFRC beams DL-0-1m, DL-0.25-1m, DL-0.5-1m, DL-0.75-1m, and DL-1.0-1m. After 48 days of accelerated corrosion under electric current, the overall stiffness of the concrete beams decreased. As depicted, the impact loads exhibit a two-impulse zone profile. Figure 12 illustrates that the SFRC beams (DL-0-1m, DL-0.25-1m, DL-0.5-1m, DL-0.75-1m, and DL-1.0-1m) underwent an initial impulse with peak values of 266.92 kN, 280.62 kN, 336.47 kN, 387.53 kN, and 413.86 kN, respectively, lasting approximately 5–6 ms. Compared to the SFRC beams of DL-0.25-1m, DL-0.5-1m, DL-0.75-1m, and DL-1.0-1m, the peak impact force increased by 5.1%, 19.9%, 15.2%, and 6.8%, respectively. A suitable amount of steel fibers can significantly enhance the overall performance of the specimen beams. Furthermore, as seen in Figure 12, this was followed by an impulse oscillation with an impact load ranging from 0 to 100 kN, lasting about 6–30 ms. This phenomenon contrasts with the findings of Huang et al. [16], who observed only two impulses in SFRC beams under impact loads. This can be explained by the splitting damage of the compression basalt-fiber-reinforced-polymer bars, which could reduce the stiffness of the beam, resulting in a lower second-peak impulse. In contrast, the present study reveals the presence of an impulse oscillation zone. Additionally, Li et al. [25] summarized the reasons in the profiles of impact loads and identified the factors influencing these profiles; however, this process was not explained in detail. Impact loading is generally characterized by the application of a force of high intensity over a brief period. Structural components under impact loading may exhibit the following two response phases: Phase one (first impulse) involves the local response to the stress wave occurring at the loading point immediately after impact. Phase two (impulse oscillation zone) involves the response, including free vibration effects due to elastic–plastic deformation occurring throughout the structural member over a prolonged period after impact. The response in the impulse oscillation zone primarily depends on the loading rate effect and the dynamic behavior of the structural component.

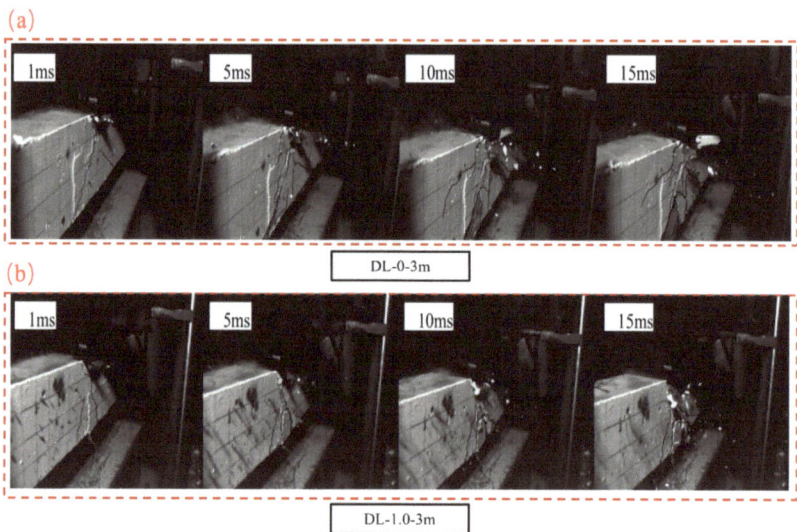

Figure 11. Failure process of SFRC beams recorded by a high-speed camera. The contact of the hammer on the SFRC beams is defined as 1 ms. (**a**) Failure process of SFRC beams DL-0-3m recorded. (**b**) Failure process of SFRC beams DL-1.0-3m recorded.

Table 4. Impact test results under varying conditions of steel fiber content.

Type	Steel Fibers Content (%)	Impact Speed (m/s)	Peak Impact Load (kN)	Peak Mid-Span Deflection (mm)	Residual Mid-Span Deflection(mm)
DL-0-1m	0	4.43	266.92	18.16	12.34
DL-0.25-1m	0.25	4.43	280.62	16.26	9.66
DL-0.5-1m	0.5	4.43	336.47	16.24	9.15
DL-0.75-1m	0.75	4.43	387.53	16.03	8.55
DL-1.0-1m	1.0	4.43	413.86	15.26	6.75
DL-0-2m	0	6.26	541.79	38.45	30.28
DL-0.25-2m	0.25	6.26	556.92	31.92	24.83
DL-0.5-2m	0.5	6.26	610.48	30.65	24.68
DL-0.75-2m	0.75	6.26	665.20	28.07	20.10
DL-1.0-2m	1.0	6.26	720.88	26.87	19.83
DL-0-3m	0	7.67	653.70	55.33	50.69
DL-0.25-3m	0.25	7.67	715.81	45.20	41.22
DL-0.5-3m	0.5	7.67	837.91	41.40	37.29
DL-0.75-3m	0.75	7.67	920.13	38.29	33.81
DL-1.0-3m	1.0	7.67	1051.27	35.63	29.58

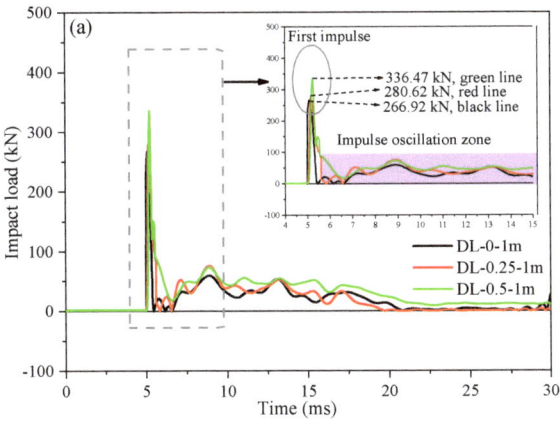

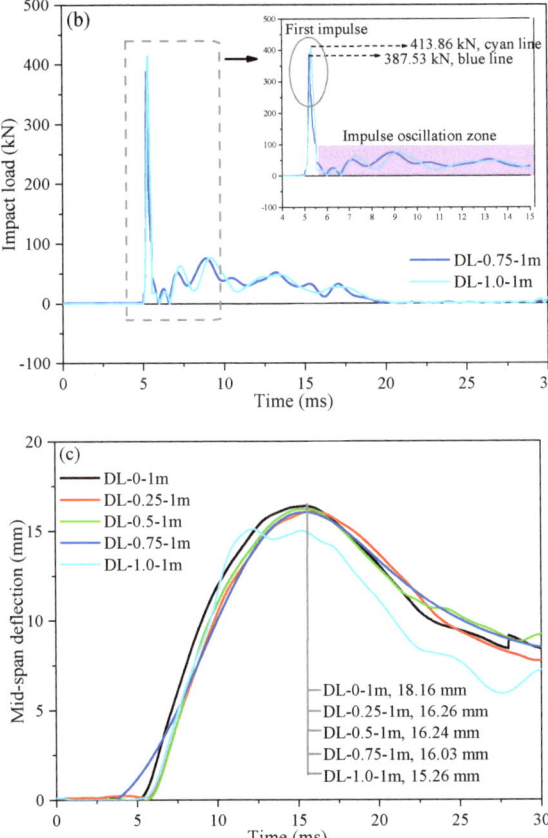

Figure 12. Impact force and mid-span-deflection time histories of the tested SFRC beams under the conditions of 28 d curing period and 1 m drop hammer height. (**a**) Impact force histories of SFRC beams DL-0-1m, DL-0.25-1m, and DL-0.5-1m. (**b**) Impact force histories of SFRC beams DL-0.75-1m, and DL-1.0-1m. (**c**) Mid-span-deflection time histories.

Moreover, upon examination of Figure 12a,b and comparison of the peak impact load of SFRC beams, it was noted that the peak load exhibited an insignificant effect on the stiffness of SFRC beams as the steel fiber content increased during the impact test, while the peak deflection at mid-span showed a decreasing trend. This indicated that the addition of steel fibers can enhance the stiffness of the specimen beams, thereby demonstrating improved impact resistance and deformation resistance. Additionally, Figure 13 depicts how SFRC beams (DL-0-2m, DL-0.25-2m, DL-0.5-2m, DL-0.75-2m, and DL-1.0-2m) experienced an initial impulse with peak values of 514.79 kN, 556.92 kN, 610.48 kN, 665.20 kN, and 720.88 kN, respectively, lasting approximately 4–6 ms. The impulse oscillation zone, as observed in Figures 12–14, is consistent, characterized by an impact load ranging from 0 to 100 kN and lasting approximately 6–30 ms. When comparing two sets of SFRC beams (one comprising DL-0.75-1m and DL-1.0-1m, and the other DL-0.75-2m and DL-1.0-2m), the response time of the peak impact load and the impulse oscillation zone ranged from approximately 4 to 6 ms and 6 to 30 ms, respectively. This variation depended on the steel fiber content and corrosion degree, influenced by the overall integrity of the SFRC beam strength and pouring process. Additionally, from Figure 14a, SFRC beams (DL-0-2m, DL-0.5-2m, DL-1.0-2m, DL-0.5-3m, and DL-1.0-3m) experienced an initial impulse with peak values of 541.79 kN, 610.48 kN, 720.88 kN, 837.91 kN, and 1051.27 kN, respectively, lasting approximately 5–6 ms. The impulse oscillation zone was more pronounced compared to the 1 m drop hammer impact test, as the hammer had to dissipate a greater amount of energy through the beam. In Figure 14, there is an increase in the peak impact load of SFRC beams with the rise in drop hammer height. For instance, the peak impact load for SFRC beam DL-0.5-3m was 837.91 kN, which exceeded that of SFRC beam DL-0.5-2m by approximately 1.37 times. Moreover, the response of the peak impact load intensified with an increase in steel fiber content. For example, the peak impact load for SFRC beam DL-1.0-3m was 1051.27 kN, which surpassed that of SFRC beam DL-0.5-3m by approximately 1.25 times.

Figures 12c, 13c and 14b, accompanied by Table 4, depict the mid-span deflection time histories for the tested SFRC beams. It is evident that the peak mid-span deflection decreases with an increase in steel fiber content during the impact load test, although the data for the peak mid-span deflection of the 1 m drop hammer impact test are not very clear. For instance, the peak mid-span deflections of SFRC beams DL-0-2m, DL-0.25-2m, DL-0.5-2m, DL-0.75-2m, and DL-1.0-2m are 38.45 mm, 31.92 mm, 30.65 mm, 28.07 mm, and 26.87 mm, respectively. Additionally, it is noted that the peak mid-span deflection increases with an increase in drop hammer height during the impact load test during a 28-day curing period. For example, SFRC beams DL-1.0-2m and DL-1.0-3m experience peak mid-span deflections of 26.87 mm and 35.63 mm and residual deflections of 19.83 mm and 29.58 mm, respectively. It is crucial to emphasize that the shear resistance of beams relies on various factors, including the dowel action of tension reinforcements, stirrups, interlocking aggregates, and the shear resistance of concrete in the compression zone [47]. Interestingly, SFRC beams DL-0.5-3m and DL-1.0-3m, subjected to higher drop hammer heights, exhibit larger peak and residual mid-span deflections compared to SFRC beams DL-0.5-1m and DL-1.0-1m, which experienced lower drop hammer heights. With the increase in the steel fiber content, the reduction in SFRC beam stiffness caused by the electric corrosion is mitigated. The peak impact force has been enhanced, and the increase in beam stiffness results in decreased peak deflection and residual displacement at mid-span, with rebound values fluctuating within the range of 6 to 8 mm (Figure 13c and Table 4). The addition of steel fibers to concrete can improve its impact resistance.

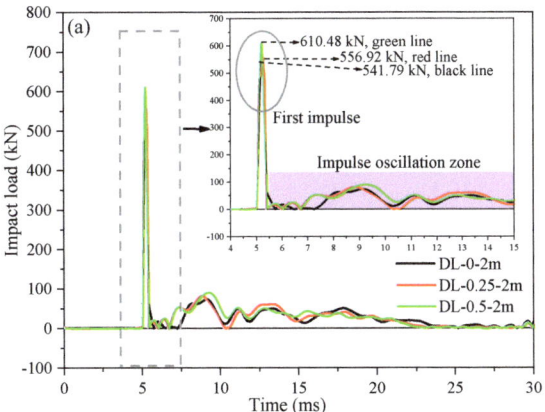

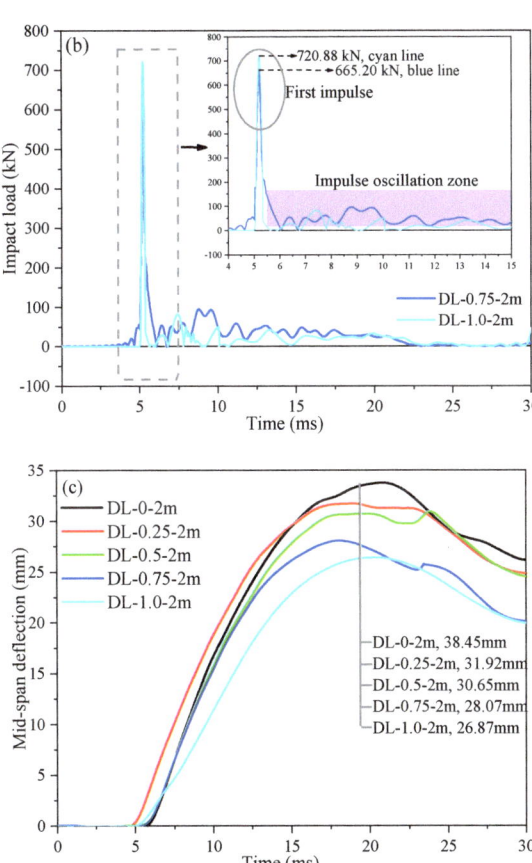

Figure 13. Impact force and mid-span-deflection time histories of the tested SFRC beams under the conditions of 28 d curing period and 2 m drop hammer height. (**a**) Impact force histories of SFRC beams DL-0-2m, DL-0.25-2m, and DL-0.5-2m. (**b**) Impact force histories of SFRC beams DL-0.75-2m, and DL-1.0-2m. (**c**) Mid–span–deflection time histories.

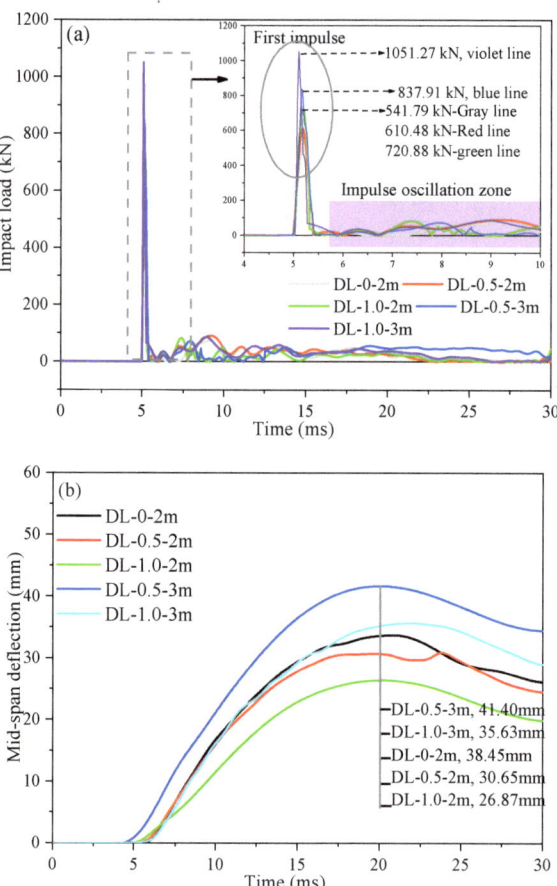

Figure 14. Impact force and mid-span-deflection time histories of the tested SFRC beams under the conditions of 28 d curing period, 2 m, and 3 m drop hammer height. (**a**) Impact force histories of SFRC beams. (**b**) Mid-span-deflection time histories.

Figure 15 presents the temporal evolution of stirrup reinforcement strain for the SFRC beams subjected to impact loading. The data indicate a progressive rise in stirrup strain corresponding to the increasing drop height. Moreover, as the steel fiber content increases, the change in stirrup strain gradually diminishes, albeit with some irregularities in the data. This indicates that, under the impact load, the stress state of SFRC beams gradually shifts towards bending, resulting in a gradual decrease in the load carried by stirrups. Additionally, as shown in Figure 15, the stirrup strain at the mid-span of SFRC beams initially exhibits negative values (compressive) within the first 5 ms, attributed to stress wave propagation. Subsequently, a shift to positive values (tensile) occurs from 5 ms to 30 ms, attributed to the displacement of the neutral axis above the position of the compression steel bars. Starting from 9 ms, the concrete cover commences crushing, as illustrated in Figure 11, and is incapable of withstanding compressive stress, resulting in a downward movement of the compression zone and neutral axis. This phenomenon has been previously observed and elucidated in a study by Tran et al. [48].

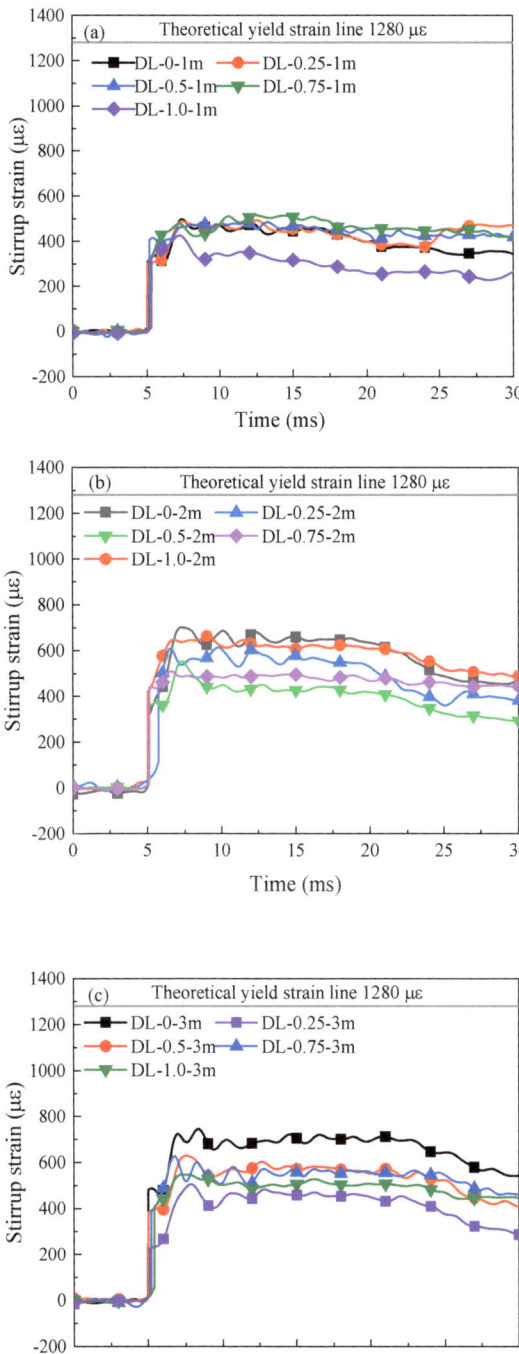

Figure 15. Impact load-strain curves of SFRC beams. (**a**–**c**) SFRC beams treated with varying steel fiber content reinforcement under drop hammer height 1 m, 2 m, and 3 m, respectively.

At the beginning of impact load (around 5 ms), the strain curve of all SFRC beams undergoes a transformation, reaching a peak value within the next 4 ms and gradually decreasing during the subsequent impact process. Throughout the entire impact process, the stirrup reinforcement remains under tension. Regarding the longitudinal strain time–history curves of the SFRC beams, let us consider the example of the longitudinal strain for SFRC beams with different steel fiber contents at dropping heights of 1 m, 2 m, and 3 m, respectively, as depicted in Figure 16. The response of longitudinal strain in the SFRC beams is slightly faster than that of stirrup strain. At 5 ms, following the impact of the drop hammerhead, the longitudinal strain begins to rise. In the case of all the SFRC beams, the longitudinal strain value starts to decline after reaching the initial peak, possibly due to bond failure between the strain gauge and the longitudinal reinforcement under the intense impact load, resulting in inaccurate strain gauge data. Additionally, the longitudinal strain values sharply increase after reaching the initial peak, indicating that the longitudinal reinforcement exceeds the maximum strain value of the strain gauge post-yielding, leading to strain gauge overload. At approximately 7 ms, the longitudinal strain gauges on the SFRC beams fail due to severe tensile damage to the longitudinal reinforcement.

4.3. Analysis of Energy Dissipation

The impact resistance of the SFRC beams is further investigated in this study by integrating the impact load–deflection curves to calculate the total deformation energy, as depicted in Figure 17. The impulse experienced during the impact process is determined by integrating the impact load over the duration of the impact. The average impact load is then computed by dividing the impulse by the duration of the impact. Furthermore, the parameters depicted in Figure 17 can be represented as follows:

$$I_{\mathrm{P}} = \int_0^{T_{\mathrm{d}}} P dt \tag{3}$$

$$E_{\mathrm{P}} = \int_0^{D_{\max}} P d\delta \tag{4}$$

$$P_{\mathrm{m}} = \frac{I_{\mathrm{P}}}{T_{\mathrm{d}}} \tag{5}$$

where I_{P} is the impulse, E_{P} is the energy dissipation, $D_{\max}$ is the peak mid-span deflection, P_{m} is average impact load, and T_d is the impact time.

Tables 4 and 5 list the parameters of energy dissipation, impulse, average impact force, and peak mid-span deflection for all SFRC beams. With an increase in drop hammer height, the overall energy dissipation of the SFRC beams also increases, as depicted in Figure 18. It is observed that the overall energy dissipation ranges from approximately 600 J to 700 J, 1200 J to 1400 J, and 2500 J to 2900 J with 1 m, 2 m, and 3 m drop hammer heights in the impact load test, respectively. Furthermore, an increase in steel fiber content is associated with a certain degree of increase in overall energy dissipation. For example, the energy dissipation of SFRC beam DL-1-3m is 2916.4 J greater than that of SFRC beam DL-0.25-3m, which is 2612 J. Additionally, Table 5 indicates a noticeable improvement in energy dissipation for SFRC beams with an increase in drop hammer height.

In Figure 19, the correlation between the impulse, average impact force, and steel fiber content in SFRC beams is depicted across different drop hammer heights during the impact load test. As shown in Figure 19a, the impulse I_p of the SFRC beams demonstrated an increase with higher steel fiber content during both the 1 m and 3 m drop hammer impact tests, with the effect being more pronounced. However, the results of the 2 m drop hammer impact test exhibited inconsistencies compared to the 1 m and 3 m tests, possibly due to incomplete sensor adherence to the steel bars, leading to data deviation. Additionally, the average impact force P_m of the SFRC beams increased with higher steel fiber content across all drop hammer heights, with the effect being more prominent.

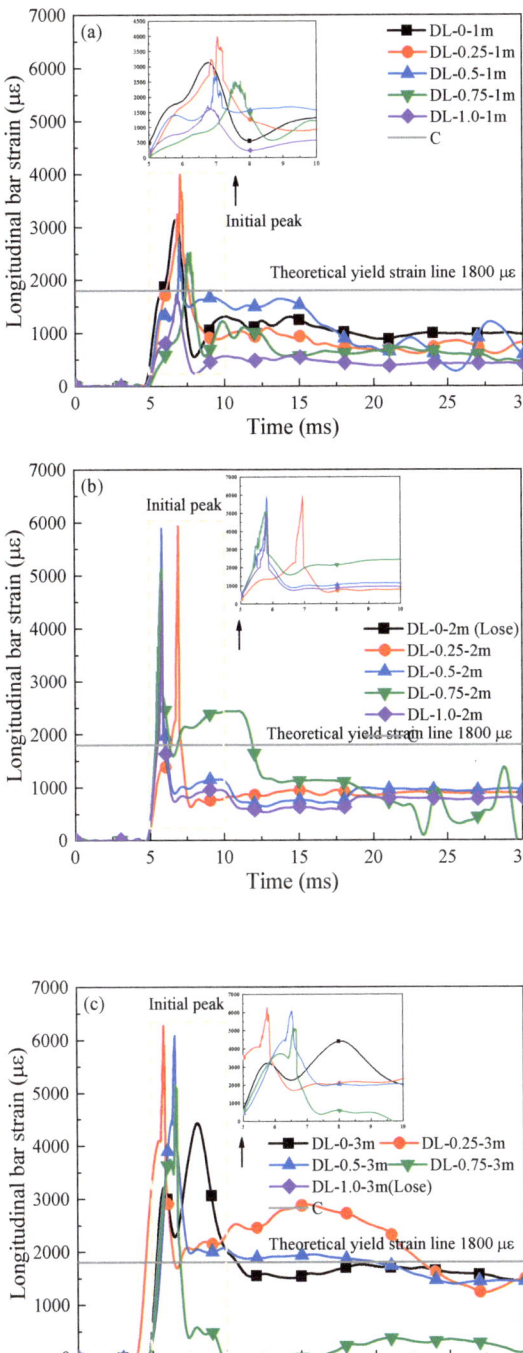

Figure 16. Impact load-strain curves of SFRC beams. (**a–c**) SFRC beams treated with varying steel fiber content reinforcement under drop hammer heights of 1 m, 2 m, and 3 m, respectively.

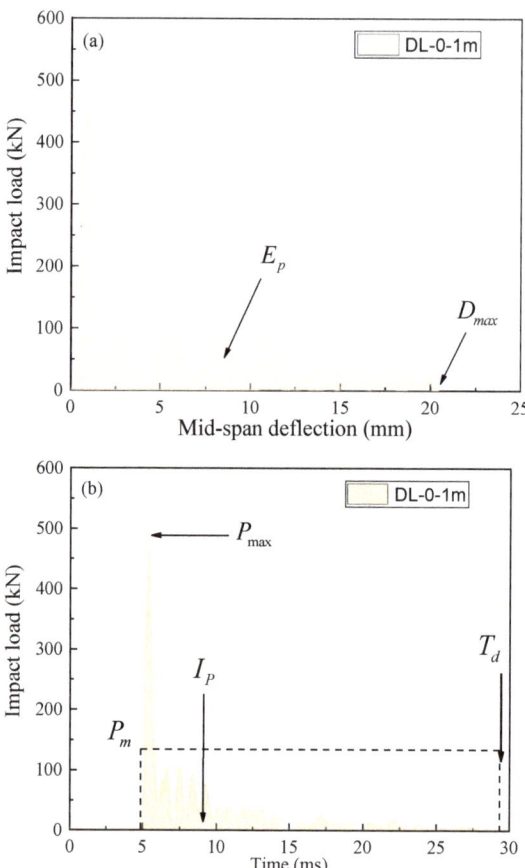

Figure 17. Energy dissipation analysis diagram of SFRC beams in the impact load test. (**a**) Curves of the mid-span deflection and impact load, and (**b**) curves of the time and impact load.

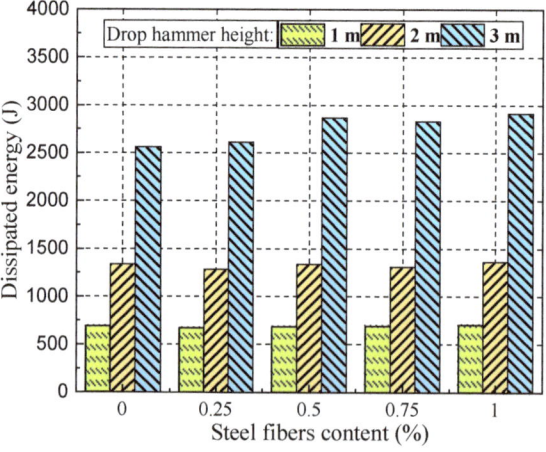

Figure 18. Energy dissipation of SFRC beams with different drop hammer heights in the impact load test.

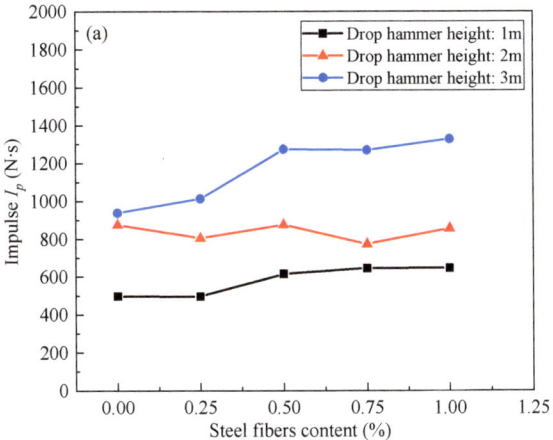

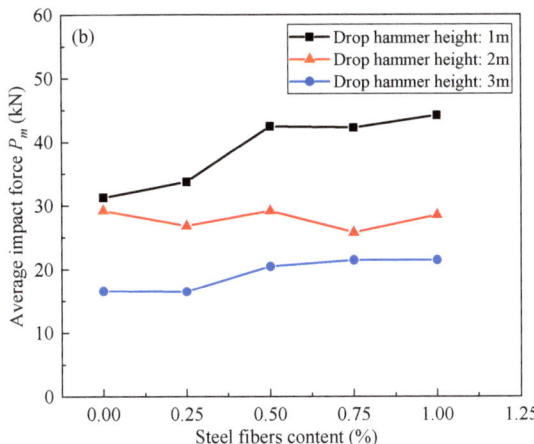

Figure 19. The curves depict the relationship between impulse, average impact force, and steel fiber content for SFRC beams under various drop hammer heights in the impact load test. (**a**) The curves of impulse and steel fibers content. (**b**) The curves of average impact force.

Table 5. The overall energy dissipation results of SFRC beams in the impact load test with varying factors (drop hammer height and steel fiber content).

Type	Energy Dissipation E_p (J)	Impulse I_p (N·s)	Average Impact Force P_m (kN)
DL-0-1m	691.2	497.7	16.59
DL-0-2m	1336.7	876.5	29.22
DL-0-3m	2562.4	938.5	31.28
DL-0.25-1m	669.4	495.8	16.53
DL-0.25-2m	1281.2	805..2	26.84
DL-0.25-3m	2612.0	1012.5	33.75
DL-0.5-1m	684.7	614.7	20.49
DL-0.5-2m	1335.6	876.2	29.21
DL-0.5-3m	2868.7	1273.6	42.45
DL-0.75-1m	692.4	644.3	21.47
DL-0.75-2m	1307.6	773.8	25.80
DL-0.75-3m	2831.7	1268.8	42.29
DL-1.0-1m	703.1	645.1	21.50
DL-1.0-2m	1365.7	856.3	28.54
DL-1.0-3m	2916.4	1326.8	44.23

5. Conclusions

This study examined how the impact load test of corrosion-treated SFRC beams is affected by variations in steel fiber content and drop hammer height. Experiments were conducted with different steel fiber contents (ranging from 0% to 1.0%) and curing periods of 28 days, along with varying drop hammer heights (1 m, 2 m, and 3 m). The failure modes of the SFRC beams were assessed through examining steel fiber content and drop hammer heights, revealing insights into their mechanical behaviors and load resistance capacities. Key strength parameters discussed included peak load, stirrup strain, longitudinal bar strain, impulse, and energy dissipation. The study's findings are outlined as follows:

1. From the corrosion test results, SFRC beams with added steel fibers exhibit fewer surface rust spot areas compared to those without steel fibers. This is attributed to the bonding of steel fibers with the concrete matrix and their ability to share part of the corrosion current, resulting in fewer severe rust marks on the concrete surface.
2. The impact tests reveal that SFRC beams exhibit concrete crushing at the top without spalling, with flexural and flexure-shear cracks developing at mid-span. Higher steel fiber content correlates with enhanced impact resistance, resulting in fewer cracks. With increased impact loads, a shift towards shear-dominated failure occurs, causing more severe localized damage. High-speed camera analysis confirms that steel fiber inclusion boosts impact resistance by reducing crack spread, enhancing stiffness, modifying failure modes, and mitigating concrete spalling.
3. SFRC beams with different steel fiber contents display varied responses to impact loads at drop heights ranging from 1 m to 3 m. Enhanced steel fiber content correlates with improved overall beam performance, as evidenced by higher peak impact load values and distinctive mid-span deflection characteristics. Moreover, a consistent impulse oscillation zone is noted across various drop heights, indicating an enhanced energy dissipation capacity. Stirrup reinforcement strain gradually increases with drop height, while longitudinal strain response is marginally faster, suggesting strain gauge failure under intense impact loads. In summary, integrating steel fibers into SFRC beams enhances their impact resistance and deformation capacity, thereby improving structural performance under impact loading conditions.
4. The energy dissipation analysis indicates a moderate increase in energy dissipation with higher steel fiber content. Impulse and average impact force generally increase with higher steel fiber content, although some inconsistency is noted in the 2 m drop hammer tests, possibly due to sensor adherence issues. Nonetheless, the trend indicates that higher steel fiber content enhances impulse and average impact force.

Author Contributions: Conceptualization, J.G.; Data curation, L.L. and X.H.; Formal analysis, J.G., L.L. and X.H.; Funding acquisition, H.C.; Investigation, L.L.; Methodology, J.G. and X.H.; Resources, H.C.; Software, X.H.; Supervision, H.C.; Writing—original draft, J.G.; Writing—review and editing, H.C. All authors have read and agreed to the published version of the manuscript.

Funding: This research was funded by Natural Science Foundation of Zhejiang Province (No. LY21E080002), the Natural Science Foundation of Wenzhou Zhejiang Province (No. S20220003), the Science Foundation of the Wenzhou University of Technology (No. ky202212, ky202202), and the National Natural Science Foundation of China (No. 51608393).

Institutional Review Board Statement: Not applicable.

Informed Consent Statement: Not applicable.

Data Availability Statement: The original contributions presented in the study are included in the article, further inquiries can be directed to the corresponding author.

Conflicts of Interest: Author Xin Huang was employed by the company Wenzhou Traffic Engineering Testing and Testing Co., Ltd. The remaining authors declare that the research was conducted in the absence of any commercial or financial relationships that could be construed as a potential conflict of interest.

References

1. Ahmed, E.R.; Wael, A.; Abathar, A.H.; Sarah, H. Shear performance of basalt fiber-reinforced concrete beams reinforced with BFRP bars. *Compos. Struct.* **2022**, *288*, 115443. [CrossRef]
2. Abdullah, M.J.; Beddu, S.; Manan, T.S.A.; Syamsir, A.; Naganathan, S.; Kamal, N.L.M.; Mohamad, D.; Itam, Z.; Yee, H.M.; Yapandi, M.F.K.M.; et al. The strength and thermal properties of concrete containing water absorptive aggregate from well-graded bottom ash (BA) as partial sand replacement. *Constr. Build. Mater.* **2022**, *339*, 127658. [CrossRef]
3. Biswas, R.K.; Iwanami, M.; Chijiwa, N.; Uno, K. Effect of non-uniform rebar corrosion on structural performance of RC structures: A numerical and experimental investigation. *Constr. Build. Mater.* **2020**, *230*, 116908. [CrossRef]
4. Otieno, M.; Beushausen, H.; Alexander, M. Chloride-induced corrosion of steel in cracked concrete—Part I: Experimental studies under accelerated and natural marine environments. *Cem. Concr. Res.* **2016**, *79*, 373–385. [CrossRef]
5. Almusallam, A.A. Effect of degree of corrosion on the properties of reinforcing steel bars. *Constr. Build. Mater.* **2001**, *15*, 361–368. [CrossRef]
6. Auyeung, S.; Alipour, A.; Saini, D. Performance-based design of bridge piers under vehicle collision. *Eng. Struct.* **2019**, *191*, 752–765. [CrossRef]
7. Hao, H.; Hao, Y.; Li, J.; Chen, W. Review of the current practices in blast-resistant analysis and design of concrete structures. *Adv. Struct. Eng.* **2016**, *19*, 1193–1223. [CrossRef]
8. Oliveira, M.C.; Teles, D.; Amorim, D. Shear behaviour of reinforced concrete beams under impact loads by the Lumped Damage framework. *Frat. Ed Integrità Strutt.* **2020**, *14*, 13–25. [CrossRef]
9. Reba, T.C.; Aure, T. WInfluence of aggregate size on shear mechanism of reinforced concrete beam subjected to impact load. *SN Appl. Sci.* **2022**, *4*, 224. [CrossRef]
10. Remennikov, A.M.; Kaewunruen, S. Impact resistance of reinforced concrete columns: Experimental studies and design considerations. In *Progress in Mechanics of Structures and Materials*; CRC Press: Boca Raton, FL, USA, 2020; pp. 817–823. Available online: https://ro.uow.edu.au/engpapers/379/ (accessed on 1 January 2020).
11. Mindess, S.; Young, J.F.; Darwin, D. *Concrete*; Prentice-Hall: Upper Saddle River, NJ, USA, 2002.
12. Wang, K.; Jansen, D.; Shah, S.; Karr, A. Permeability study of cracked concrete. *Cem. Concr. Res.* **1997**, *27*, 381–393. [CrossRef]
13. Adhikary, S.D.; Li, B.; Fujikake, K. Strength and behavior in shear of reinforced concrete deep beams under dynamic loading conditions. *Nucl. Eng. Des.* **2013**, *259*, 14–28. [CrossRef]
14. Fu, Y.; Yu, X.; Dong, X.; Zhou, F.; Ning, J.; Li, P.; Zheng, Y.X. Investigating the failure behaviors of RC beams without stirrups under impact loading. *Int. J. Impact Eng.* **2020**, *137*, 103432. [CrossRef]
15. Liao, W.Z.; Liu, K.X.; Ma, C.; Liang, J.C. Experimental study on consistency of the impact performance between composite beams and reinforced concrete beams. *Compos. Struct.* **2023**, *308*, 116677. [CrossRef]
16. Huang, Z.; Chen, W.; Tran, T.T.; Pham, T.; Hao, H.; Chen, Z.; Elchalakani, M. Experimental and Numerical Study on Concrete Beams Reinforced with Basalt FRP Bars under Static and Impact Loads. *Compos. Struct.* **2021**, *263*, 113648. [CrossRef]
17. Pham, T.M.; Hao, H. Impact Behavior of FRP-Strengthened RC Beams without Stirrups. *J. Compos. Constr.* **2016**, *20*, 04016011.
18. Saatci, S.; Vecchio, F.J. Effects of shear mechanisms on impact behavior of reinforced concrete beams. *ACI Struct. J.* **2009**, *106*, 78–86. [CrossRef]
19. Fu, C.Q.; Huang, J.; Dong, Z.; Song, C.; Zhang, Y. Shear behavior of reinforced concrete beams subjected to accelerated non-uniform corrosion. *Eng. Struct.* **2023**, *286*, 116081. [CrossRef]
20. Anas, S.M.; Alam, M.; Shariq, M. Damage response of conventionally reinforced two-way spanning concrete slab under eccentric impacting drop weight loading. *Def. Technol.* **2023**, *19*, 12–34. [CrossRef]
21. Chung, L.; Najm, H.; Balaguru, P. Flexural behavior of concrete slabs with corroded bars. *Cem. Concr. Compos.* **2008**, *30*, 184–193. [CrossRef]
22. Francois, R.; Khan, I.; Dang, V.H. Impact of corrosion on mechanical properties of steel embedded in 27-year-old corroded reinforced concrete beams. *Mater. Struct.* **2013**, *46*, 899–910. [CrossRef]
23. Khan, I.; François, R.; Castel, A. Structural performance of a 26-year-old corroded reinforced concrete beam. *Eur. J. Environ. Civ. Eng.* **2012**, *16*, 440–449. [CrossRef]
24. Liao, W.; Li, M.; Zhang, W.; Tian, Z. Experimental studies and numerical simulation of behavior of RC beams retrofitted with HSSWM-HPM under impact loading. *Eng. Struct.* **2017**, *149*, 131–146. [CrossRef]
25. Li, H.; Chen, W.; Hao, H. Factors influencing impact force profile and measurement accuracy in drop weight impact tests. *Int. J. Impact Eng.* **2020**, *145*, 103688. [CrossRef]
26. Zhu, W.; Francois, R. Corrosion of the reinforcement and its influence on the residual structural performance of a 26-year-old corroded RC beam. *Constr. Build. Mater.* **2014**, *51*, 461–472. [CrossRef]
27. Fernandez, I.; Etxeberria, M.; Marí, A.R. Ultimate bond strength assessment of uncorroded and corroded reinforced recycled aggregate concretes. *Constr. Build. Mater.* **2016**, *111*, 543–555. [CrossRef]
28. Al-Sibahy, A.; Sabhan, M. Corrosion effects on the bond behaviour of steel bars in self-compacting concrete. *Constr. Build. Mater.* **2020**, *250*, 118568. [CrossRef]
29. Chen, S.F.; Zheng, M.L.; Wand, B.G. Study of High-Performance Concrete Subjected to Coupled Action from Sodium Sulfate Solution and Alternating Stresses. *J. Mater. Civ. Eng.* **2009**, *21*, 148–153. [CrossRef]

30. Lu, Y.; Tang, W.; Li, S.; Tang, M. Effects of simultaneous fatigue loading and corrosion on the behavior of reinforced beams. *Constr. Build. Mater.* **2018**, *181*, 85–93. [CrossRef]
31. Muda, Z.C.; Usman, F.; Syamsir, A.; Chen, S.Y.; Mustapha, K.N.; Beddu, S.; Thiruchelvam, S.; Kamal, N.L.M.; Alam, M.A.; Birima, A.H.; et al. Effect of Thickness and Fibre Volume Fractionon Impact Resistance of Steel Fibre Reinforced Concrete (SFRC). *IOP Conf. Ser. Earth Environ. Sci.* **2016**, *32*, 012026. [CrossRef]
32. Micallef, K.; Sagaseta, J.; Ruiz, M.F.; Muttoni, A. Assessingpunching shear failure in reinforced concrete flat slabs subjected to localised impact loading. *Int. J. Impact Eng.* **2014**, *71*, 17–33. [CrossRef]
33. Huang, Y.Q.; Huang, J.S.; Zhang, W.; Liu, X. Experimental and numerical study of hooked-end steel fiber-reinforced concrete based on the meso- and macro-models. *Compos. Struct.* **2023**, *309*, 116750. [CrossRef]
34. Kim, J.; Kwon, M.; Seo, H.; Lim, J. Experimental study of torsional strength of RC beams constructed with HPFRC composite mortar. *Constr. Build. Mater.* **2015**, *91*, 9–16. [CrossRef]
35. Martinola, G.; Meda, A.; Plizzari, G.A.; Rinaldi, Z. Strengthening and repair of RC beams with fiber reinforced concrete. *Cem. Concr. Compos.* **2010**, *32*, 731–739. [CrossRef]
36. Qin, F.; Zhang, Z.; Yin, Z.; Di, J.; Xu, L.; Xu, X. Use of high strength, high ductility engineered cementitious composites (ECC) to enhance the flexural performance of reinforced concrete beams. *J. Build. Eng.* **2020**, *32*, 101746. [CrossRef]
37. Tran, D.; Pham, T.M.; Hao, H.; Ha, N.S.; Vo, N.; Chen, W.S. Precast segmental beams made of fiber-reinforced geopolymer concrete and FRP tendons against impact loads. *Eng. Struct.* **2023**, *295*, 116862. [CrossRef]
38. Zhang, H.; Huang, Y.; Yang, Z.; Guo, F.; Shen, L. 3D meso-scale investigation of ultra high-performance fibre reinforced concrete (UHPFRC) using cohesive crack model and Weibull random field. *Constr. Build. Mater.* **2022**, *327*, 127013. [CrossRef]
39. Goldston, M.W.; Remennikov, A.; Saleh, Z.; Sheikh, M.N. Experimental investigations on the behavior of GFRP bar reinforced HSC and UHSC beams under static and impact loading. *Structures* **2019**, *22*, 109–123. [CrossRef]
40. Saleh, Z.; Sheikh, M.N.; Remennikov, A.; Basu, A. Damage assessment of GFRP bar reinforced ultra-high-strength concrete beams under overloading impact conditions. *Eng. Struct.* **2020**, *213*, 110581. [CrossRef]
41. Yu, F.; Fang, Y.; Feng, C.C.; Tan, S.; Wang, Y. Shear capacity of PVC-CFRP confined concrete column-RC beam interior joint strengthened with core steel tube. *Thin-Walled Struct.* **2023**, *193*, 111213. [CrossRef]
42. GB/T 50081-2002; Standard for Test Method of Mechanical Properties on Ordinary Concrete. Ministry of Housing and Urban-Rural Development of the People's Republic of China: Beijing, China, 2003.
43. Stein, K.J.; Graeff, Â.G.; Garcez, M.R. Structural performance of reinforced concrete beams subjected to combined effects of corrosion and cyclic loading. *J. Build. Pathol. Rehabil.* **2023**, *8*, 15. [CrossRef]
44. Badawi, M.; Soudki, K. Control of corrosion-induced damage in reinforced concrete beams using carbon fiber-reinforced polymer laminates. *J. Compos. Constr.* **2005**, *9*, 195–201. [CrossRef]
45. El Maaddawy, T.E.A.; Soudki, K.K.A. Effectiveness of impressed current technique to simulate corrosion of steel reinforcement in concrete. *J. Mater. Civ. Eng.* **2003**, *15*, 41–47. [CrossRef]
46. GB/T 2423.5-2019; Environmental Testing for Electric and Electronic Products—Part 2: Test methods—Test Ea and Guidance: Shock. State Bureau of Technical Supervision: Beijing, China, 2019.
47. Dancygier, A.N.; Yankelevsky, D.Z.; Jaegermann, C. Response of high-performance concrete plates to impact of non-deforming projectiles. *Int. J. Impact Eng.* **2007**, *34*, 1768–1779. [CrossRef]
48. Tran, T.T.; Pham, T.M.; Huang, Z.; Chen, W.; Hao, H.; Elchalakani, M. Impact Response of Fibre Reinforced Geopolymer Concrete Beams with BFRP Bars and Stirrups. *Eng. Struct.* **2020**, *231*, 111785. [CrossRef]

Article

Comparative Study on Passive Film Formation Mechanism of Cast and PBF-LB/M-TC4 in Simulated Physiological Solution

Ming Liu [1,2,*,†] , Zhang Liu [3,†] , Jie Wang [4], Yongqiang Zhang [5,6] and Xin Gao [7,*]

1 Department of Materials Science and Engineering, Xi'an University of Technology, Xi'an 710048, China
2 Shaanxi Province Key Laboratory of Corrosion and Protection, Xi'an University of Technology, Xi'an 710048, China
3 Center for Advancing Materials Performance from the Nanoscale (CAMP-Nano), State Key Laboratory for Mechanical Behavior of Materials, Xi'an Jiaotong University, Xi'an 710049, China; zhangliu@stu.xjtu.edu.cn
4 Shaanxi Zhou Doctor Dental Medical Co., Ltd., Xi'an 710086, China; 13379211721@163.com
5 Xi'an QinTi Intelligent Manufacturing Technologies Co., Ltd., Xi'an 710061, China; zhyq613@mail.nwpu.edu.cn
6 State Key Laboratory of Solidification Processing, Northwestern Polytechnical University, Xi'an 710072, China
7 Research and Development Department, Beijing Med-Zenith Medical Scientific Corporation Limited, Beijing 101316, China
* Correspondence: liuming87@xaut.edu.cn (M.L.); 15901462422@163.com (X.G.)
† Those authors contribute equally to this work.

Abstract: Personalized laser powder bed fusion (PBF-LB/M) Ti-6Al-4V (TC4) has a broader application prospect than that of traditional casting. In this paper, the composition and corrosion resistance of the passive film formation mechanism of TC4 prepared by optimization of PBF-LB/M techniques and traditional casting were systematically studied in 0.9 wt.% NaCl at 37 °C by electrochemical technique and surface analysis. The rates of the passive film formation process, corrosion resistance and composition of TC4 show different characteristics for the different preparation processes. Although the rate of passive film formation of cast-TC4 was higher at the initial immersion, the open circuit potential was more positive, and the film thickness was larger after stabilization, those facts show no positive correlation with corrosion resistance. On the contrary, with no obvious defects on the optimized PBF-LB/M-TC4, the passive film resistance is 2.5 times more, the defect concentration is reduced by 30%, and the TiO_2 content is higher than that of the cast-TC4, making the martensitic-based PBF-LB/M-TC4 exhibit excellent corrosion resistance. This also provides good technical support for the further clinical application of PBF-LB/M-TC4.

Keywords: laser powder bed fusion; TC4; passive film; X-ray photoelectron spectroscopy; Auger electron spectroscopy

Citation: Liu, M.; Liu, Z.; Wang, J.; Zhang, Y.; Gao, X. Comparative Study on Passive Film Formation Mechanism of Cast and PBF-LB/M-TC4 in Simulated Physiological Solution. *Materials* **2024**, *17*, 2583. https://doi.org/10.3390/ma17112583

Academic Editors: Reza Hashemi and Thomas Niendorf

Received: 11 April 2024
Revised: 22 May 2024
Accepted: 24 May 2024
Published: 27 May 2024

1. Introduction

Titanium (Ti) and its alloys have been widely employed in the medicine field. The studies on Ti and Ti alloys have already become the focus of study and the front field as biomedical implants [1–3], and they occupy "half of the country" in the field of medical metals. TC4 (Ti-6Al-4V) has a series of advantages, such as small density, high specific strength, excellent corrosion resistance, good toughness and weld ability, and has been resoundingly applied as artificial joints, vascular stents and orthopedic instruments [4–6].

As a kind of easy passivation alloy, the surface of Ti and Ti alloys can form a TiO_2-containing passive film in the air protecting it from corrosion [7]. Nevertheless, the corrosion environment of Ti implants in human body is relatively severe, and the local acidizing and mechanical loading on implants may reduce the stability of its passive films, which may lead to further environmental pollution and damage of the formed film, affecting the durability of the implant. It has been confirmed that the corrosion resistance of TC4 in vivo or in vitro will be affected by many factors, such as the microstructure of materials, the coatings and the corrosive environment [8–11].

At present, the manufacturing process of TC4 includes casting, forging and machining [12,13], which are time-consuming and costly. With its complex operation, casting is at a high cost and causes environmental pollution, but it still is the conventional process for manufacturing oral prostheses. Hence, the design and manufacture of biomedical Ti alloys urgently require a fast and economical manufacturing method. In recent years, PBF-LB/M technology (also known as additive manufacturing) has unique advantages compared with traditional equal or reduced manufacturing. PBF-LB/M is a rapid prototyping technology for the direct manufacturing of terminal and near-terminal TC4 products [14,15]. The phase corrosion resistance in PBF-LB/M-TC4 is $\beta > \alpha > \alpha'$, and the V element in the β phase can stabilize the lattice structure, which is conducive to corrosion resistance; however, the V element in the α' phase is relatively easy to dissolve, resulting in poor corrosion resistance [16–19]. There have also been plenty of studies on the passive film corrosion resistance of other Ti alloys [20–23]; a systematic and comprehensive evaluation of the corrosion behavior of Ti alloys is given, but it is worth noting that the good corrosion resistance of Ti and Ti alloys derives from the compact corrosion-resistant passive films.

In this study, the formation mechanism of the passive film and corrosion resistance of the optimized PBF-LB/M-TC4 prepared by PBF-LB/M and the traditional cast TC4 (cast-TC4) were systematically investigated in a simulated physiological solution (0.9 wt.% NaCl). Electrochemical techniques were applied to evaluate the passive film formation mechanism and corrosion resistance, and the composition and thickness of the passive film were further analyzed by surface technology.

2. Experimental Section

2.1. Material and Sample Preparation

The YLM-120 selective laser melting equipment produced by Jiangsu Yongnian laser forming Technology Co., Ltd. (Wuxi, Jiangsu, China) with high-precision circular working cylinder and multi-level precision guidance-sealing system was applied to fabricate the PBF-LB/M-TC4, which can effectively reduce the loss of metal powder and pollution. The raw material is 0.02 mm of spherical powder. In the preparation process, argon gas was selected as the protection gas. The following parameters could be selected based on our optimization: the hatch distance is 0.12 mm, the laser power is 275 W, the scanning velocity is 1100 mm/s, and the layer thickness is 30 μm. The comparison cast-TC4 sample was fabricated with lost wax.

The PBF-LB/M- and cast-TC4 cube samples (10 × 10 × 4 mm) were chosen for microstructure analysis, electrochemical testing and surface detection, respectively. For electrochemical testing samples, the one side (10 × 10 mm) was spliced with copper wire by conductive adhesive, the exposed testing area is 1 cm², and the rest of the sample was sealed with epoxy resin. All samples were polished step by step with SiC sandpaper from 200–2000 # before the experiment, then washed with alcohol and deionized water and dried.

2.2. Electrochemical Tests

Electrochemical tests were performed on a VMP3 multi-channel electrochemical workstation (Biologic, Seyssinet-Pariset, France) with a standard three-electrode system. The reference electrode is SCE (the saturated calomel reference electrode), the platinum sheet (12 cm²) is the counter electrode, and the PBF-LB/M- and cast-TC4 sample is the working electrode. The simulated physiological solution with a mass fraction of 0.9 wt.% NaCl was applied in the test. Firstly, a long open circuit potential (OCP) was continuously monitored for 168 h. The polarization curves were measured after immersion for 0.5 h and 120 h, respectively. The scanning of the potentiodynamic polarization curve was from cathodic −250 mV vs. OCP to the anode with a 1 mV/s scanning rate and stopped when the anode current density exceeded 100 μA/cm². Four potential ranges were applied for the cyclic voltammetry (CV) test: -2 V$_{SCE}$ to 2.5 V$_{SCE}$, -2 V$_{SCE}$ to 2 V$_{SCE}$, -1.5 V$_{SCE}$ to 1.5 V$_{SCE}$ and -1 V$_{SCE}$ to 1 V$_{SCE}$. The scanning started from cathode to anode and then back to the cathode, with five scanning cycles and a fixed scanning rate of 100 mV/s. The linear polarization was performed from cathode to anode with a scanning potential range of ±20 mV$_{SCE}$ and a scanning rate of 0.2 mV/s. Electrochemical impedance spectroscopy (EIS) measurements were performed under OCP with a signal of 10 mV sine wave and a test frequency range of 100 kHz–10 mHz. ZsimpWin 3.5 software was applied to analyze the test results. The Mott–Schottky was tested with a fixed frequency and scanning rate of 1 kHz and 50 mV/s, and the potential range of scanning was -1.0 V$_{SCE}$ to 3 V$_{SCE}$. All electrochemical measurements were tested three times, and the representative results were given. The temperature for all electrochemical tests was controlled at 37 °C by a thermostat water bath.

2.3. X-ray Photoelectron Spectroscopy Analysis

The X-ray photoelectron spectroscopy (XPS) was applied (Thermo Scientific, Oxford, UK) to identify the passive film composition of PBF-LB/M- and cast-TC4 samples' immersion in physiological solution for 168 h. The monochromator was Al Kα, the sensitivity was 100 kcps, the spectrum scanning range was 0–1350 eV, the wide scanning interval was 1 eV, the narrow scanning interval was 0.1 eV, and the spectrum was calibrated with C1s (285.0 eV). The composition of two TC4 passive films was analyzed by Xpspeak 4.1 software using the Gauss–Newton fitting mode.

2.4. Auger Electron Spectroscopy Analysis

Auger electron spectroscopy (AES) was performed on a PHI-700 (ULVAC-PHI, Chigasaki, Japan) equipped to analyze the passive films' thickness variation of a two TC4 sample's immersion in physiological solution for 168 h. A coaxial electron gun and CMA energy analyzer were adopted. Auger spectra were taken at 5 keV with an energy resolution of 0.1%, the incidence angle was 30°, and the vacuum of the analysis chamber was $<3.9 \times 10^{-9}$ Torr. The depth profile was obtained by etching a Φ100 nm spot on the surface of the passive film with Ar$^+$ ions, and the thermal oxidation of standard SiO$_2$/Si was adopted to determine the sputtering rate of 1 nm/min.

3. Results and Discussion

3.1. Microstructure Analysis

It can be seen from the scanning electron microscopy (SEM) in Figure 1a,a$_1$ that the microstructure of traditional cast-TC4 is mainly equiaxed. The metastable β phase and the equiaxial α phase are uniformly distributed on the matrix, the duplex α-phase microstructure volume fraction is about 73.65%, the microstructure dispersion is high, and the grain size is about 9.75 μm. Figure 1b,b$_1$ depicts the SEM morphology of PBF-LB/M-TC4.

The β-phase self-diffusion coefficient is higher, and the grain growth activation energy is lower, which leads to the epitaxial growth of grains. The martensitic lath in the β-phase columnar crystal has a preferred orientation, so the structure diagram shows an alternating phenomenon of light and dark. The longitudinal macro structure is an epitaxial growth columnar crystal with a length of up to a millimeter and a width of 27.34 μm. Upon further enlarging the longitudinal section of the metallographic structure, it is found that a large number of acicular martensite is distributed in the columnar crystal, which is basically parallel along the length direction and is composed of martensite α' and martensite α''.

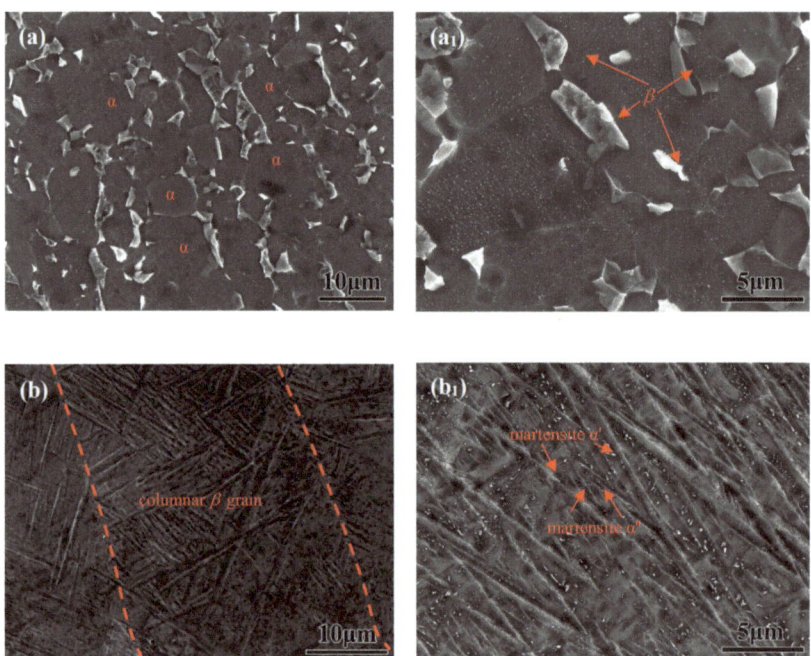

Figure 1. Scanning electron microscopy (SEM) of (**a,a₁**) cast- and (**b,b₁**) PBF-LB/M-TC4.

3.2. Electrochemical Analysis

3.2.1. Open Circuit Potential

Figure 2 depicts the OCPs of cast- and PBF-LB/M-TC4 after immersion in physiological solution for 1800 s.

The variation in OCP at the early stage of immersion can determine the passive film formation rate [24]. As can be seen from Figure 2a, the OCP of two TC4 increases in the positive direction rapidly. The passive film growth rate can be derived by Equation (1) [25]:

$$E = \text{const.} + 2.303 \, \delta^- / A \log t \tag{1}$$

wherein δ^- is the passivation film formation rate corresponding to $\log t$. A can be calculated by Equation (2):

$$A = \frac{nF}{RT}\alpha\delta\prime \tag{2}$$

where α and $\delta\prime$ represent the charge transfer coefficient ($\alpha = 0.5$) [26] and charge transfer process energy accumulation width ($\delta\prime = 1$), respectively. Numerous studies have demonstrated that TiO_2 is the main composition in the passive film of Ti-related alloys [2,19], and the results of XPS and AES will support that the passive film of two TC4is mainly TiO_2. Herein, the thickening of the passive film is assumed to be mainly through Ti^{4+} diffusion to the Ti and oxygen interface and $n = 4$ in Equation (2); the calculated A equals 78 nm/V. The early-stage formation rate of passive film $\delta\prime$ can be derived (see Figure 2b) and shows the following order: Cast-TC4 > PBF-LB/M-TC4 (see Figure 2c).

The long-term OCP of the two TC4 continuous monitoring for 168 h in physiological solution is depicted in Figure 2d. The OCP increases rapidly at the initial 0.5 h, then rises slowly at 0.5–48 h and stabilizes at 72 h. The 168 h OCP of the cast- and PBF-LB/M-TC4 is 101 mV_{SCE} and −7 mV_{SCE}, respectively. Here, the power function was applied to fitting the OCP vs. t [25]:

$$E = a \cdot \exp(-t/b) + c \cdot \exp(-t/d) + e^- \tag{3}$$

where $a - e$ are all constants. The E vs. t of two TC4 in 0.9 wt.% NaCl solution is fitted by Equation (3). It can be seen from the fitting results (see Figure 2e and Table 1) that the OCP of two TC4 alloy conforms well to the power function.

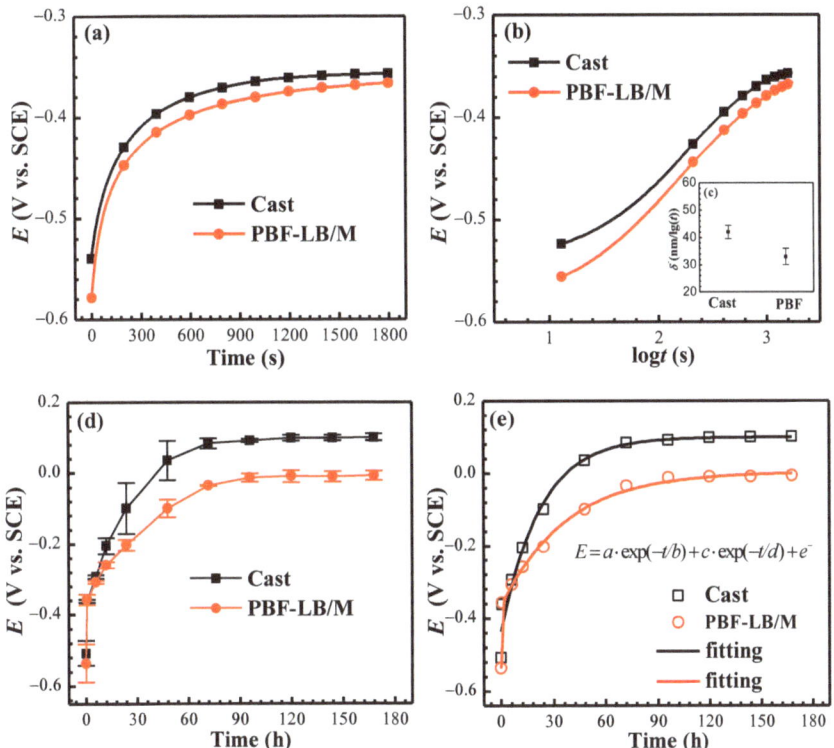

Figure 2. (**a**) The 1800 s OCPs of TC4 immersion in simulated physiological solution, (**b**) E vs. log$t$ (s), (**c**) passive film formation rate, (**d**) OCP of 168 h, (**e**) fitting results of OCP.

Table 1. OCP fitting results of two TC4 after continuous monitoring in physiological solution for 168 h.

Alloy	Fitting Result	R^2
Cast-TC4	$E = -0.268 \cdot \exp(-t/0.923) - 0.268 \cdot \exp(-t/0.923) + 0.099$	0.9666
PBF-LB/M-TC4	$E = -0.369 \cdot \exp(-t/1.528) - 0.170 \cdot \exp(-t/0.003) + 0.004$	0.9969

3.2.2. Potentiodynamic Polarization

In terms of corrosion thermodynamics, a completely stabilized OCP of two TC4 needs at least 72 h of immersion (see Figure 2). Herein, the test of potentiodynamic polarization curves chosen for 0.5 h and 120 h corresponds to the OCP (see Figure 3).

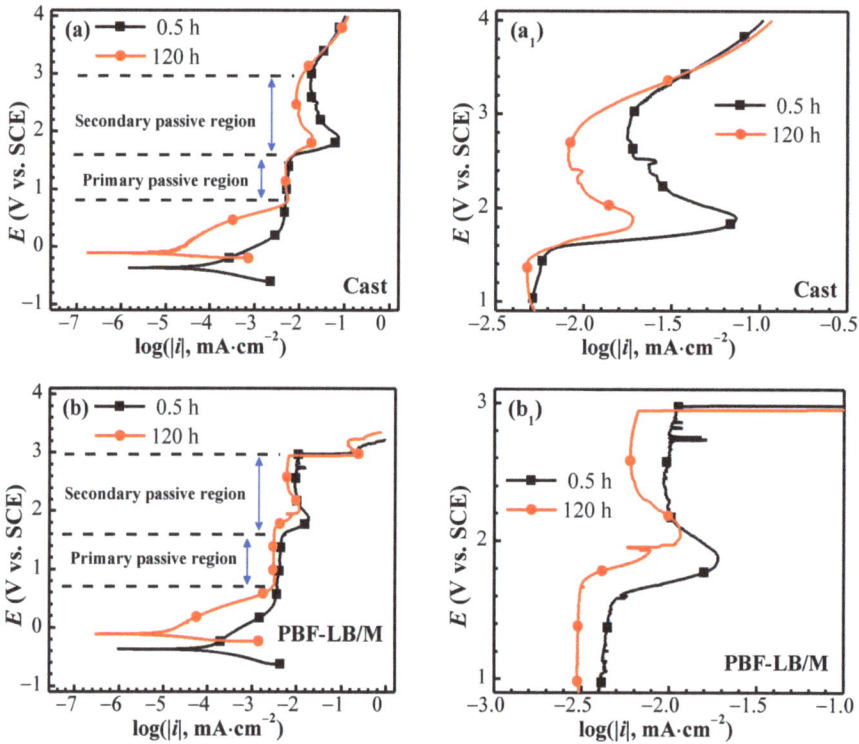

Figure 3. Potentiodynamic polarization curves of cast- and PBF-LB/M-TC4 immersed in physiological solution for 0.5 and 120 h. (**a**) cast-TC4, (**a₁**) magnification of anode curve of cast-TC4, (**b**) PBF-LB/M-TC4 and (**b₁**) magnification of anode curve of PBF-LB/M-TC4.

The anode polarization curves exhibit typical metal passivation characteristics, and two obvious passivation zones can be observed without obvious activation to passivation transition. The primary and the secondary passivation regions are 0.8–1.6 V_{SCE} and 1.6–3 V_{SCE}, respectively [27]. It can be seen from the fitting results derived from the potentiodynamic polarization curves shown in Table 2 that PBF-LB/M-TC4 has a relatively negative self-corrosion potential (E_{corr}). The self-corrosion current density (i_{corr}) from low to high at the first 0.5 h of immersion is as follows: cast-TC4 < PBF-LB/M-TC4 and inversely shows PBF-LB/M-TC4 < Cast-TC4 after 120 h of immersion. However, the maintaining passivity current density (i_{pass}) of PBF-LB/M-TC4 is all along lower than that of cast-TC4 at fixed 1 V_{SCE}.

Table 2. Fitting values of potentiodynamic polarization parameters of cast- and PBF-LB/M-TC4.

Sample	Time (h)	E_{corr} (mV$_{SCE}$)	b_c, mV·dec^{-1}	i_{corr}, μA·cm^{-2}	i_{pass}, μA·cm^{-2}
Cast-TC4	0.5	-383 ± 12	-144 ± 11	0.12 ± 0.03	6.5 ± 0.6
	120	-113 ± 8	-127 ± 9	0.04 ± 0.02	4.6 ± 0.4
PBF-LB/M-TC4	0.5	-385 ± 13	-191 ± 14	0.17 ± 0.03	4.4 ± 0.5
	120	-118 ± 7	-143 ± 6	0.02 ± 0.01	3.0 ± 0.3

The relatively limited integrity of the 0.5 h formed passive film may result in lower corrosion resistance. The metastable β phase and the equiaxial α phase of cast-TC4 promotes the passive film formation rate (Figure 3a). Nevertheless, the corrosion resistance of martensitic-based PBF-LB/M-TC4 (0.02 μA·cm^{-2}) is higher than cast-TC4 (0.04 μA·cm^{-2}) after immersion for 120 h and is positively correlated with the microstructure (Figure 1).

Furthermore, the i_{corr} of two TC4 reduces greatly after 120 h of immersion, which is much lower than 0.1 μA·cm^{-2} [24], indicating that a corrosion-resistant passive film could be formed under a longer period of immersion.

3.2.3. Cyclic Voltammetry

The cyclic voltammetry (CV) curves of two TC4 tested in different potential ranges are exhibited in Figure 4. Four different scanning potential ranges—(-2.0–2.5 V$_{SCE}$), (2.0–2.0 V$_{SCE}$), (-1.5–1.5 V$_{SCE}$) and (-1.0–1.0 V$_{SCE}$)—were selected to distinguish redox reactions as much as possible. In the first cycle of the wide potential range (-2–2.5 V$_{SCE}$) in Figure 4a,b, five distinct anode current peaks could be seen—a_1 (-1.38 V$_{SCE}$), a_2 (-1.1 V$_{SCE}$) and a_3 (-0.59 V$_{SCE}$)—corresponding to the oxidizing of Ti to Ti^{2+}, Ti^{2+} to Ti^{3+} and Ti^{3+} to Ti^{4+}, respectively [21,27]:

$$a_1 : Ti + H_2O \leftrightarrow TiO + 2H^+ + 2e^- \tag{4}$$

$$a_2 : 2TiO + H_2O \leftrightarrow Ti_2O_3 + 2H^+ + 2e^- \tag{5}$$

$$a_3 : Ti_2O_3 + H_2O \leftrightarrow 2TiO_2 + 2H^+ + 2e^- \tag{6}$$

The other two peaks (a_4 and a_5) could be seen when forward scan potential exceeds 0.6 V$_{SCE}$, indicating that the electrode enters an oxygen-controlled zone.

Four cathodic current peaks can be detected in the reverse scanning, among which c_1 (1.9 V$_{SCE}$), c_2 (-0.3 V$_{SCE}$) and c_3 (-0.87 V$_{SCE}$) are the reduction peaks of a_4 and a_5, and a_3 and a_2, respectively. Due to the dissolved oxygen having a strong reduction reaction and reaching its limit at -0.6 V$_{SCE}$ [28], part of the reduction process signal of Ti^{4+}/Ti^{2+}/Ti could be covered up, resulting in an insignificant reduction peak. The a_3 peak is the highest and the c_2 peak can be observed, indicating that the film is mainly Ti^{4+} [27]. The a_2 peak is relatively small, and a larger c_3 peak can be seen, indicating Ti^{3+} may be present in the passive film. Except for the relatively obvious a_2 peak, the other anode peaks are sharply reduced in the second scanning cycle for two alloys, but a new reduction peak c_4 (-1.4 V$_{SCE}$) corresponding to the oxidation peak a_1 can be observed, indicating that Ti^{2+} is in an unstable state.

For the test potential range of -2~2 V$_{SCE}$ in Figure 4a$_1$,b$_1$, the CV curve of cast-TC4 is similar to that in Figure 4a.

Nevertheless, a stable film could be detected for PBF-LB/M-TC4, and only the inconspicuous peaks a_2 and c_1 could be seen on the overlapped CV curves. Only the a_2 and a_3 peaks and the corresponding c_2 and c_3 peaks can be seen in the scanning range of -1.5–1.5 V$_{SCE}$ (see Figure 4a$_2$,b$_2$), indicating that Ti^{4+} and Ti^{3+} exist in the passive film. The c_3 peak of cast-TC4 is smaller, suggesting that Ti^{3+} is unstable, while the c_2 peak of PBF-LB/M-TC4 is relatively larger, implying that Ti^{4+} remains in the passive film and shows better corrosion resistance [24]. The a_3, a_4, c_1 and c_2 peaks can be seen while further shortening the potential range to -1–1 V$_{SCE}$ (see Figure 4a$_3$,b$_3$), manifesting that TiO$_2$ is mainly in the passive film.

The peaks of Al and V cannot be observed in the CV curves within the test potential ranges. Here, a preliminary judgement can be made that the two TC4 passive film is mainly Ti oxides, of which Ti^{4+} is the key component in the film. The stability of film from high to low is PBF-LB/M-TC4 > cast-TC4; the specific composition and change with thickness will be discussed in the XPS and AES section.

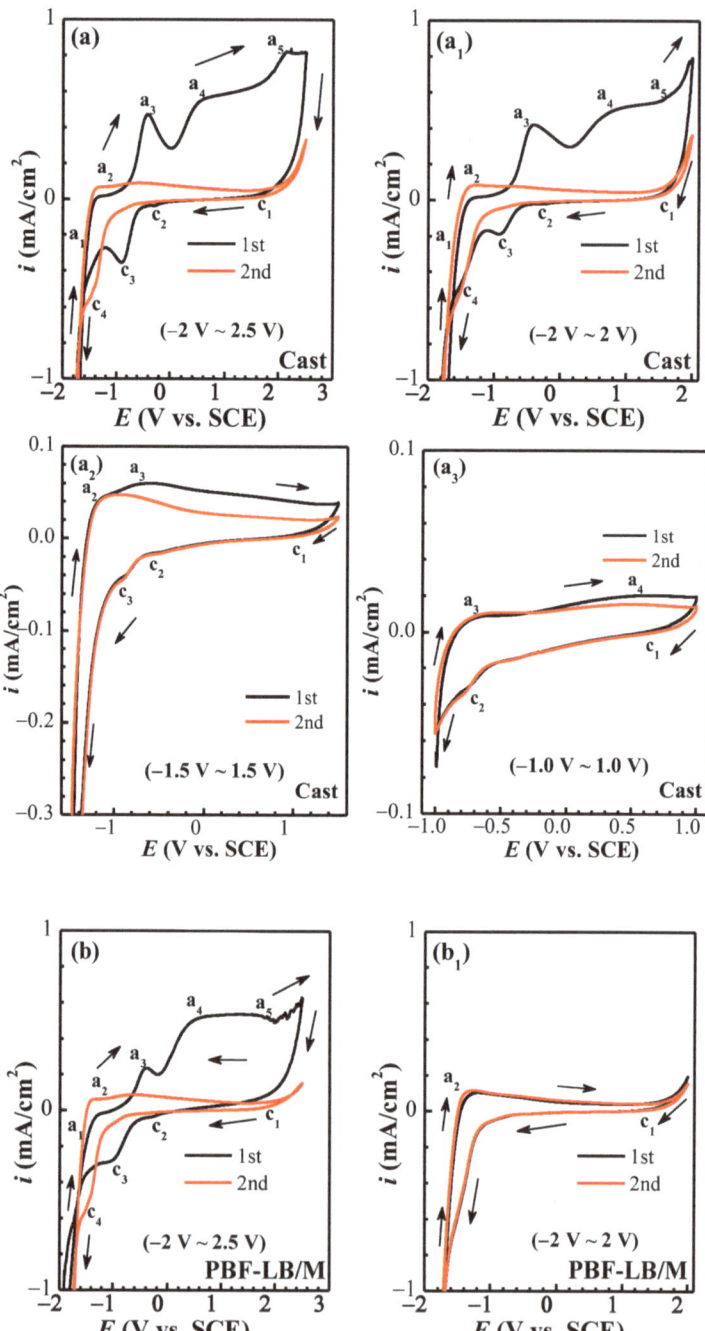

Figure 4. *Cont.*

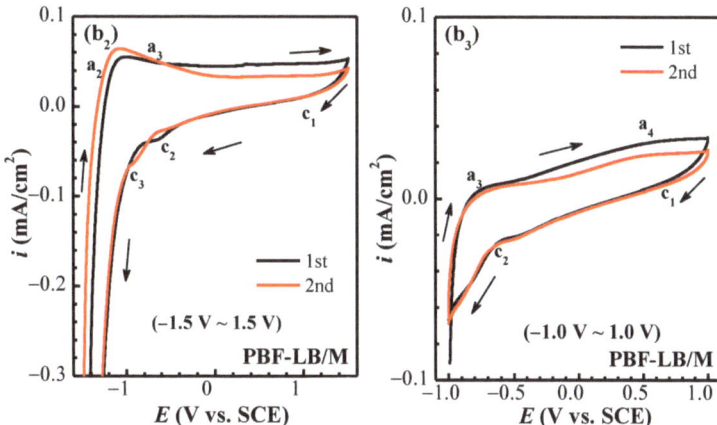

Figure 4. Cyclic voltammetry curves of TC4 with different scanning ranges of immersion in physiological solution: (**a–a3**) cast-TC4, (**b–b3**) PBF-LB/M-TC4.

3.2.4. Linear Polarization Curve

The linear polarization (LPR) curves and the fitting results after being immersed for different times are depicted in Figure 5.

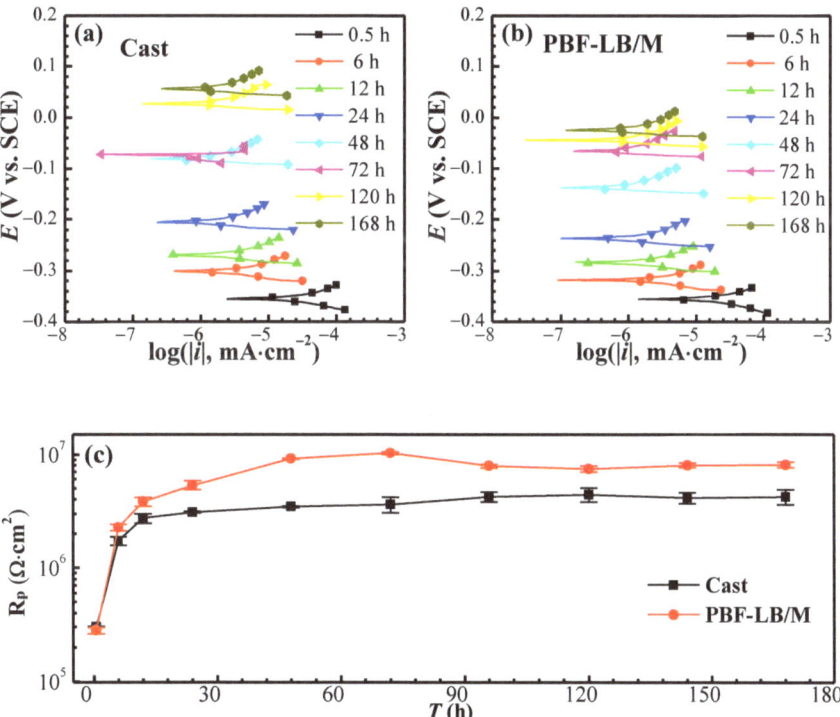

Figure 5. LPR curves of two TC4 immersed for different times: (**a**) cast-TC4, (**b**) PBF-LB/M-TC4 and (**c**) R_p fitting.

As can be seen from Figure 5a,b, the E_{corr} moves upward, and the overall potential of the cast-TC4 is more positive than that of the PBF-LB/M-TC4 under the same conditions, which is consistent with the OCP part (see Figure 2c). The polarization resistance (R_p) values after fitting the slope of the LPR curves is shown in Figure 5c. Despite the E_{corr} of the cast-TC4 being more positive than that of PBF-LB/M-TC4, the R_P shows the opposite trend except for immersion for 0.5 h. After 168 h of immersion, the R_P of PBF-LB/M-TC4 (8.2×10^6 $\Omega \cdot cm^2$) is twice that of cast-TC4 (4.1×10^6 $\Omega \cdot cm^2$), indicating that the passive film of PBF-LB/M-TC4 has excellent corrosion resistance.

3.2.5. Electrochemical Impedance Spectroscopy

The representative Nyquist (Figure 6a,b) and Bode (Figure 6a_1,b_1) plots of TC4 immersed in 0.9 wt.% NaCl solution are exhibited in Figure 6.

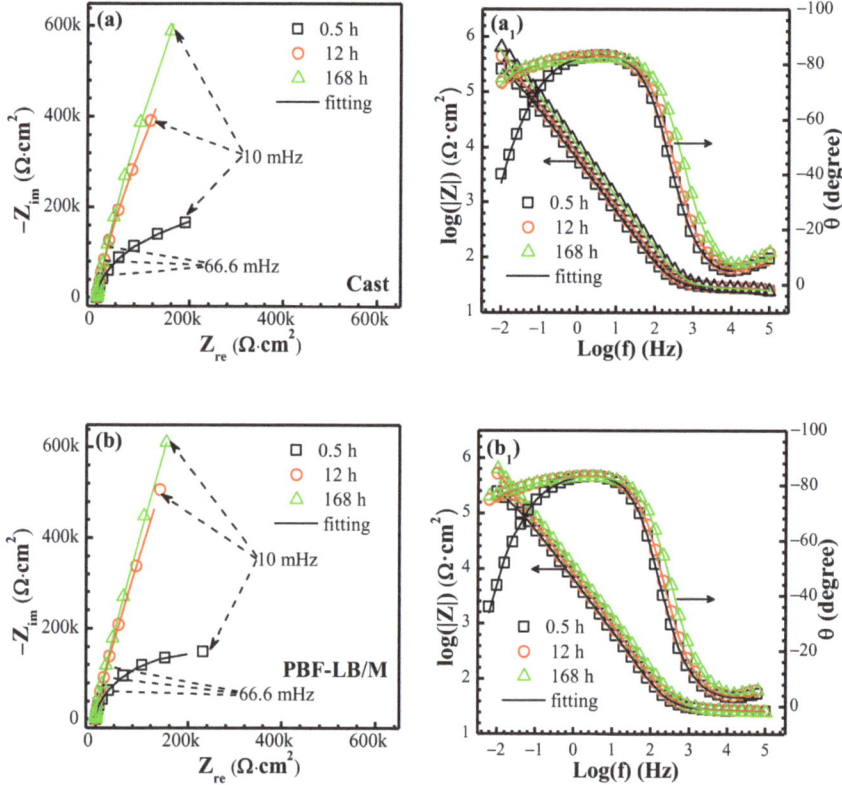

Figure 6. The EIS test results of two TC4 immersion in physiological solution: (**a,b**) Nyquist and (**a₁,b₁**) Bode.

The radius of the capacitance loop in the Nyquist plot of two TC4 increases remarkably from 0.5 h to 168 h. The Bode plots of the phase angle of two TC4 show a wide arc of capacitance in the frequency range (10^3–10^{-2} Hz), indicating that there are at least two superposed time constants and that after 12 h and 168 h of immersion, the phase angle gradually moves upward in the low-frequency region (10^0–10^{-2} Hz). The impedance modulus $|Z|_{0.01}$ is commonly applied to identify the corrosion resistance of the alloy [29]. The $|Z|_{0.01}$ from high to low of the two alloys is PBF-LB/M-TC4 > cast-TC4 (see Figure 6), indicating that PBF-LB/M-TC4 has better corrosion resistance.

The equivalent circuit (EEC) in Figure 7 is often chosen to fit the passive film corrosion process of Ti and its alloys [9,27]. The model considers the passive film layer (R_f, Q_f) and in series the charge transfer layer (R_{ct}, Q_{dl}), where R_s is the solution resistance, R_f and Q_f are the passive film resistance and related double-layer constant phase element and R_{ct} and Q_{dl} are the charge transfer resistance and related electrical double-layer constant phase element, respectively. In general, capacitance is non-ideal due to the electrode surface roughness [27–30], which can be written as [30–33]

$$Z_{Q_{dl}} = \frac{1}{Y_0(j\omega)^n} \tag{7}$$

where ω, j and n are the imaginary unit, angular frequency and exponent, respectively, and the constant phase element is equal to the capacitance while $n = 1$ [27–33].

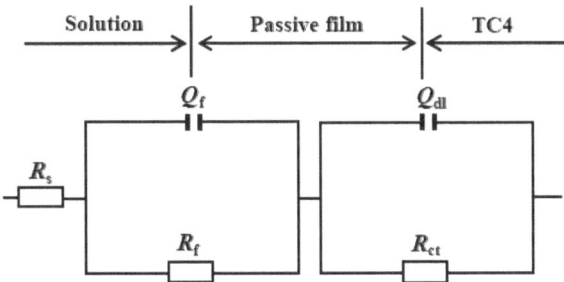

Figure 7. Equivalent circuit of TC4 immersion in physiological solution.

The EIS fitting results are shown in Figure 8. The R_s depicted in Figure 8b changes little in the range of 27–32 $\Omega \cdot cm^2$ (cast-TC4 < PBF-LB/M-TC4).

The R_f of two TC4 increases rapidly from 10^5 $\Omega \cdot cm^2$ (0.5 h) to $10^{6.3}$ $\Omega \cdot cm^2$ (12 h) and then fluctuates around $10^{6.3}$ $\Omega \cdot cm^2$ for cast-TC4 (see Figure 8c). However, the R_f of the PBF-LB/M-TC4 continues to increase, reaching the order of $10^{6.7}$ $\Omega \cdot cm^2$ after 24 h of immersion, and then fluctuates around this value in subsequent immersion. The R_{ct} depicted in Figure 8d of the two alloys is much smaller than that of the R_f and stabilizes at around $10^{4.6}$ $\Omega \cdot cm^2$ during the whole test period, indicating that the resistance of the reaction mainly comes from the passive film, which is consistent with CV and LPR tests and indicates that the EEC model (Figure 7) applied here is appropriate. The Q_f and R_f show an apparent opposite trend (see Figure 8c): the passive film thickness of the alloy is usually preserved inversely proportionally to Q_f [27,30], and the sharply decreased Q_f with the extension of immersion time indicates a thickened passive film. Nevertheless, the Q_{dl} increases rapidly after immersion for 0.5 h, then reaches a maximum value at 6 h and then decreases slowly (see Figure 8d). Due to the relatively small and minimal difference in the R_{ct} of two TC4, herein the 168 h R_f was adopted to differentiate the corrosion resistance. PBF-LB/M-TC4 ($10^{6.7}$ $\Omega \cdot cm^2$) is 2.5 times more than that of cast-TC4 ($10^{6.3}$ $\Omega \cdot cm^2$), indicating that PBF-LB/M-TC4 has excellent corrosion resistance. In addition, it can be seen from the results [5,21] that the passive film resistance of the Ti alloy in simulated body fluids is also on the order of 10^6 $\Omega \cdot cm^2$, indicating that the test results here have high reliability and that further in vivo research is also needed.

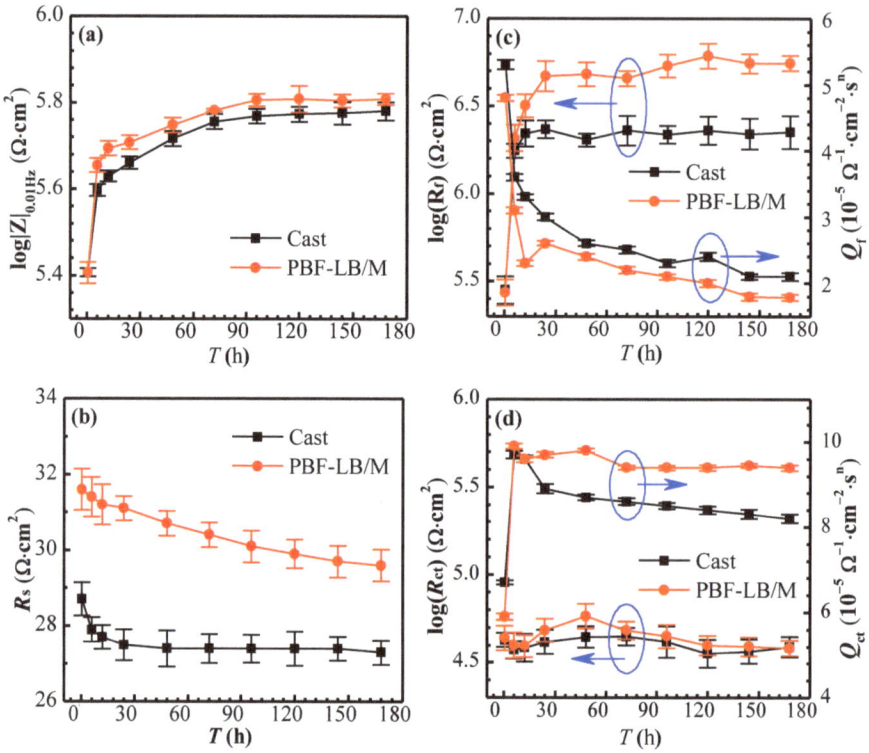

Figure 8. EIS fitting results of TC4 after immersion in physiological solution for different times. (**a**) $|Z|_{0.01}$, (**b**) R_s, (**c**) R_f and Q_f, (**d**) R_{ct} and Q_{ct}.

3.2.6. Mott–Schottky Analysis

The Mott–Schottky (M–S) curve after 168 h of immersion of two TC4 in physiological solution is exhibited in Figure 9.

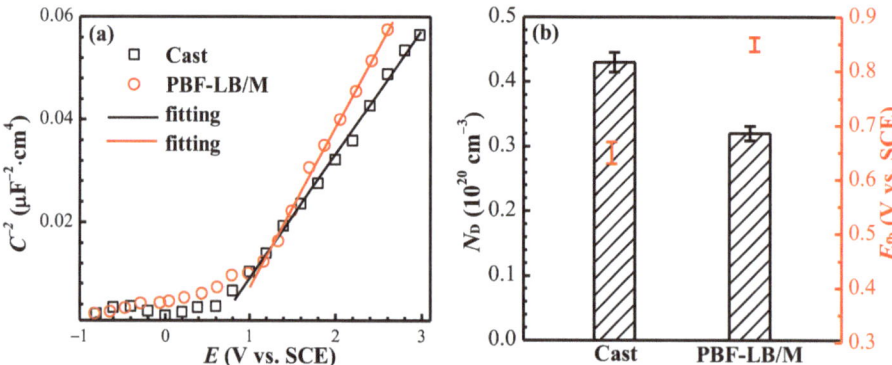

Figure 9. Mott–Schottky curves and fitting results of two TC4 immersed in physiological solution for 168 h: (**a**) M–S curve, (**b**) N_D and $\underline{E}_{fb}$.

The linear relation between $1/C^2$ and E (a) shown in Figure 9a can be described in Equation (8) [19,30]. The curves appear positive while the E is higher than the flat-band potential (E_{fb}), the curves are all positive, indicating that the film of two TC4 shows an n-type semiconductor. The donor carrier density N_{D} can be derived from Equation (8) [19]:

$$\frac{1}{C^2} = \frac{2}{\varepsilon\varepsilon_0 e N_{\text{D}}}(E - E_{\text{fb}} - \frac{kT}{e}) \tag{8}$$

where, C, ε, ε_0, e, E, E_{fb}, k and T are the space charge layer capacitance, dielectric constant of passive film (for TiO$_2$, $\varepsilon = 100$) [19,21], dielectric constant of vacuum ($\varepsilon_0 = 8.85 \times 10^{-12}$ F/m), number of electrons ($e = 1.602 \times 10^{-19}$ C), applied potential, flat-band potential, Boltzmann constant ($k = 1.38 \times 10^{-23}$ J/K) and the thermodynamic temperature.

The E_{fb} and N_{D} fitting results in Figure 9b of PBF-LB/M- and cast-TC4 are 0.852, 0.653 V_{SCE} and 0.31, 0.43 (10^{20} cm^{-3}), respectively. The defects of the passive film in cast-TC4 are 139% times those of PBF-LB/M-TC4, indicating that a more compacted passive film is formed on the PBF-LB/M-TC4 surface [19].

3.3. XPS Analysis

Figure 10 exhibits the XPS peaks analysis of the two TC4 passive film immersed in physiological solution for 168 h.

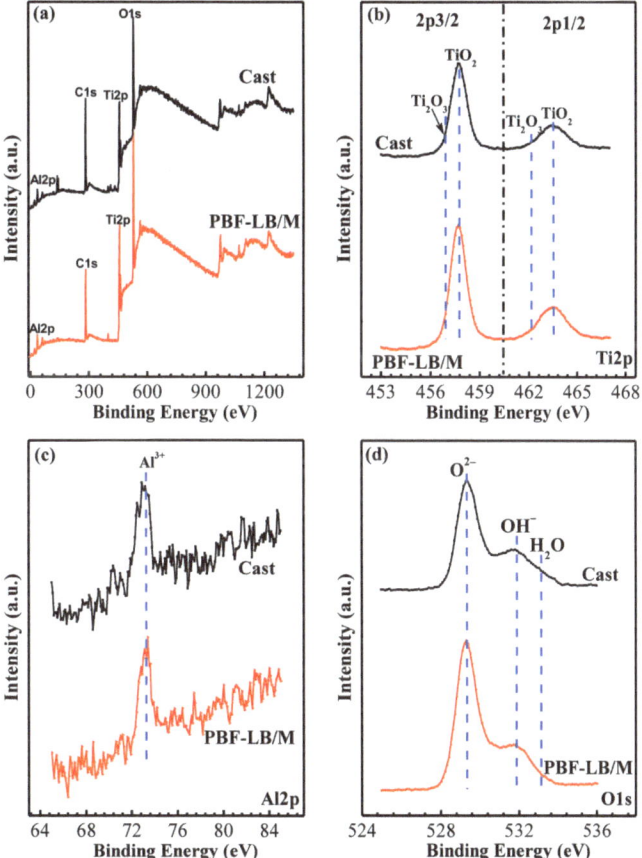

Figure 10. XPS peaks comparison of cast- and PBF-LB/M-TC4 immersed in physiological solution for 168 h. (**a**) full spectrum, (**b**) Ti 2p, (**c**) Al2p, (**d**) O1s.

For the full spectrum in Figure 10a, the film is mainly composed of Ti and O, weak Al, and hardly any V peak could be detected. Ti is mainly in TiO_2 2p3/2 (458.8 eV) and 2p1/2 (464.3 eV) in the film; small amounts of Ti_2O_3 $2p_{3/2}$ (456.8 eV) and $2p_{1/2}$ (462.0 eV) could also be detected (see Figure 10b). Our previous studies confirmed that Ti^{4+} mainly exists in the outer layer, and a low-priced Ti element mainly exists in the inner layer [7]. Only a weak Al peak can be detected (Figure 10c); the Al oxides are mainly $Al(OH)_3$ (75.1 eV) and Al_2O_3 (74.3 eV). In Figure 10d, O is mainly composed of O^{2-} (530.2 eV); a small amount of OH^- (531.8 eV) and H_2O (533 eV) can also be detected. The O^{2-} may be involved in the formation of Ti oxides (Ti_2O_3, TiO_2) and Al oxides (Al_2O_3), and the OH^- mainly participates in the formation of $Al(OH)_3$ and other compounds.

The valence states and contents of each major element of two TC4 alloys were summarized and analyzed, as shown in Figure 11.

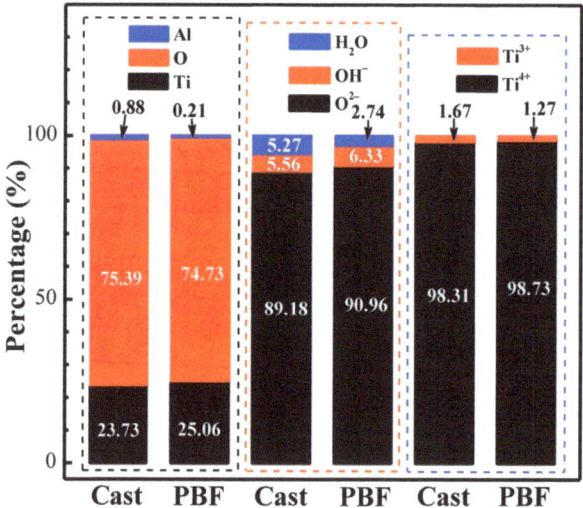

Figure 11. XPS element content comparison of passive film of cast- and PBF-LB/M-TC4 immersed in physiological solution for 168 h.

In general, the O element accounts for the largest proportion (cast- and PBF-LB/M-TC4 are 75.39% and 74.73%, respectively), followed by Ti (cast- and PBF-LB/M-TC4 are 23.73% and 25.06%, respectively), and Al accounts for a small proportion, less than 1% (cast- and PBF-LB/M-TC4 are 0.88% and 0.21%, respectively). O and Ti mainly in the form of O^{2-} and Ti^{4+} conform to TiO_2; the content of TiO_2 in PBF-LB/M-TC4 is higher than that of cast-TC4, indicating that PBF-LB/M-TC4 has better corrosion resistance. These findings match perfectly with CV (Figure 4), LPR (Figure 5) and EIS (Figure 6) results.

3.4. AES Analysis

Figure 12 depicts the AES passive film depth profile of two TC4 immersed in physiological solution for 168 h.

In Figure 12a, Ti content increases slowly when the sputtering depth is in the region of 0–7.1 nm, increases sharply between 7.1 and 20 nm and changes little and becomes stable after the sputtering depth exceeds 20 nm. The overall Al content does not change much and stays at a relatively low level (Figure 12b). The V content decreases slowly in the region of 0–7.1 nm, increases between 7.1 and 20 nm and remains relatively stable after 20 nm (Figure 12c). The change trend of O content is completely opposite to that of Ti, which decreases in the range of 0–7.1 nm, increases between 7.1 and 20 nm and tends to be stable after sputtering exceeds 20 nm (Figure 12d). This again verifies that the outer passive film is mainly TiO_2, while other low-priced Ti-oxide content increases in the inner layer, and the oxygen content decreases correspondingly. Despite the fact that the thickness of the PBF-LB/M-TC4 passive film (13.2 nm) is smaller than that of cast-TC4 (15.1 nm) (the location of the passive film thickness is defined as the oxygen content halved [34]), the passive film thickness is not a good criterion to evaluate corrosion resistance; on the contrary, the content of TiO_2 in the passive film shows a positive correlation to corrosion resistance, with a higher concentration of TiO_2 and fewer defects in the passive film enhancing better corrosion protection of martensitic based PBF-LB/M-TC4.

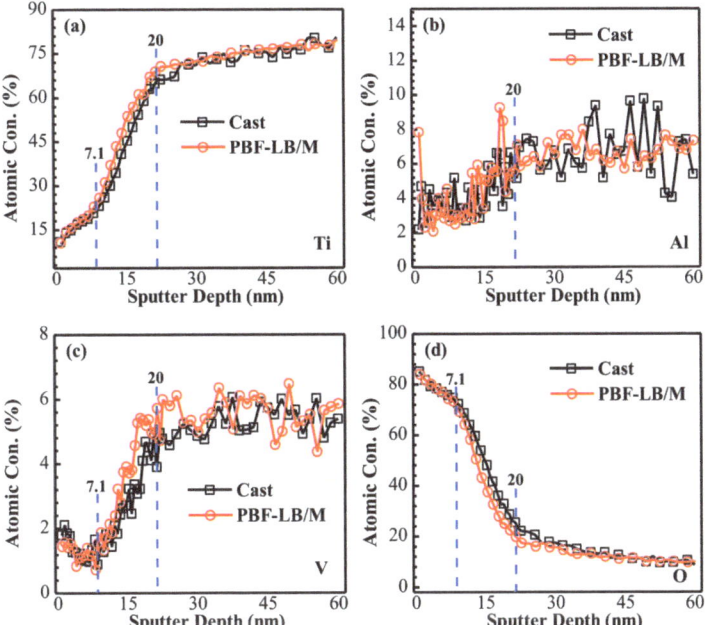

Figure 12. AES depth profile of cast- and PBF-LB/M-TC4 immersed in physiological solution for 168 h ((**a**)—**Ti,** (**b**)—Al, (**c**)—V, and (**d**)—O).

4. Conclusions

The formation and corrosion resistance of cast- and PBF-LB/M-TC4 passive films in physiological solution were studied by electrochemical techniques combined with surface analysis. The following main conclusions can be drawn:

1. The OCP of cast- and PBF-LB/M-TC4 conforms to the power function and increases rapidly with an extension of immersion time. Due to the large grain size of cast-TC4, the passive film formation rate shows the following order: cast-TC4 > PBF-LB/M-TC4.
2. The early-stage formed passive film shows a lower corrosion resistance. The metastable β phase and the equiaxial α phase of cast-TC4 promotes the passive film formation rate. A corrosion-resistant passive film could be formed during a longer period of immersion. The martensitic-based PBF-LB/M-TC4 shows better corrosion resistance than that of cast-TC4 after 120 h of immersion.
3. The R_P of LPR immersed for 168 h of PBF-LB/M-TC4 (8.2×10^6 $\Omega \cdot cm^2$) is twice of that cast-TC4 (4.1×10^6 $\Omega \cdot cm^2$), indicating the formed passive film of PBF-LB/M-TC4 has excellent corrosion resistance. The two TC4 passive film shows a typical n-type semiconductor, and the defect density in cast-TC4 is 139% times that of PBF-LB/M-TC4.
4. The passive film of two TC4 alloy is mainly Ti oxide. Ti^{4+} plays a dominant role, and the passive film's stability from high to low is PBF-LB/M-TC4 > cast-TC4. Compared to the passive film thickness, the content of TiO_2 in the passive film is a good criterion

to evaluate corrosion resistance. A higher TiO_2 concentration and fewer defects promote better corrosion protection of martensite-based PBF-LB/M-TC4.

Author Contributions: Conceptualization, M.L.; Methodology, Y.Z.; Formal analysis, M.L.; Resources, J.W. and X.G.; Data curation, M.L. and Z.L.; Writing—original draft, M.L.; Writing—review & editing, M.L. and Z.L.; Visualization, J.W. and Y.Z.; Supervision, M.L. and X.G. All authors have read and agreed to the published version of the manuscript.

Funding: This research was funded by Shaanxi Natural Science Foundation project (2024JC-YBMS-339), the doctoral initial funding for teachers of Xi'an University of Technology (101-451124004).

Institutional Review Board Statement: Not applicable.

Informed Consent Statement: Not applicable.

Data Availability Statement: The original contributions presented in the study are included in the article, further inquiries can be directed to the corresponding author.

Acknowledgments: The authors acknowledge the financial support by Xuechao Feng the manager of Shaanxi Gryffindor Software Technology Co., Ltd.

Conflicts of Interest: Jie Wang was employed by the company Shaanxi Zhou Doctor Dental Medical Co., Ltd. Yongqiang Zhang was employed by the company Xi'an QinTi Intelligent Manufacturing Technologies Co. Ltd. Xin Gao was employed by Beijing Med-Zenith Medical Scientific Corporation Limited. The remaining authors declare that the research was conducted in the absence of any commercial or financial relationships that could be construed as a potential conflict of interest.

Nomenclatures

PBF-LB/M	laser powder bed fusion of metals
TC4	Ti-6Al-4V
SCE	saturated calomel reference electrode
OCP	open circuit potential
CV	cyclic voltammetry
EIS	electrochemical impedance spectroscopy
M-S	Mott–Schottky
XPS	X-ray photoelectron spectroscopy
AES	Auger electron spectroscopy
SEM	scanning electron microscopy
δ^-	passivation film formation rate
α	charge transfer coefficient
$\delta\prime$	energy accumulation width
E_{corr}	self-corrosion potential
i_{corr}	self-corrosion current density
i_{pass}	maintaining passivity current density
LPR	linear polarization
EEC	equivalent circuit
R_s	solution resistance
R_f	passive film resistance
Q_f	passive film double-layer constant phase element
R_{ct}	charge transfer resistance
Q_{dl}	charge transfer double-layer constant phase element
C	space charge layer capacitance
ε	dielectric constant of passive film
ε_0	dielectric constant of vacuum ($\varepsilon_0 = 8.85 \times 10^{-12}$ F/m)
e	number of electrons ($e = 1.602 \times 10^{-19}$ C)
E	applied potential
E_{fb}	flat-band potential
k	Boltzmann constant ($k = 1.38 \times 10^{-23}$ J/K)
T	thermodynamic temperature

References

1. Du, H.; To, S.; Yin, T.; Zhu, Z. Microstructured surface generation and cutting force prediction of pure titanium TA2. *Precis. Eng.* **2022**, *75*, 101–110. [CrossRef]
2. Huang, G.; Fan, Z.; Li, L.; Lu, Y.; Lin, J. Corrosion Resistance of Selective Laser Melted Ti6Al4V3Cu Alloy Produced Using Pre-Alloyed and Mixed Powder. *Materials* **2022**, *15*, 2487. [CrossRef] [PubMed]
3. Wang, D.; Chen, G.; Wang, A.; Wang, Y.; Qiao, Y.; Liu, Z.; Qi, Z.; Liu, C.T. Corrosion behavior of single-and poly-crystalline dual-phase TiAl-Ti3Al alloy in NaCl solution. *Int. J. Miner. Metall. Mater.* **2023**, *30*, 689–696.
4. Hattingh, D.G.; Botha, S.; Bernard, D.; James, M.N.; du Plessis, A. Corrosion fatigue of Ti-6Al-4V coupons manufactured by directed energy deposition. *Fatigue Fract. Eng. Mater. Struct.* **2022**, *45*, 1969–1980. [CrossRef]
5. Chen, L.-Y.; Zhang, H.-Y.; Zheng, C.; Yang, H.-Y.; Qin, P.; Zhao, C.; Lu, S.; Liang, S.-X.; Chai, L.; Zhang, L.-C. Corrosion behavior and characteristics of passive films of laser powder bed fusion produced Ti–6Al–4V in dynamic Hank's solution. *Mater. Des.* **2021**, *208*, 109907. [CrossRef]
6. Feng, J.; Wang, J.; Yang, K.; Rong, J. Microstructure and performance of YTaO4 coating deposited by atmospheric plasma spraying on TC4 titanium alloy surface. *Surf. Coatings Technol.* **2022**, *431*, 128004. [CrossRef]
7. Liu, M.; Zhu, J.-N.; Popovich, V.; Borisov, E.; Mol, J.; Gonzalez-Garcia, Y. Passive film formation and corrosion resistance of laser-powder bed fusion fabricated NiTi shape memory alloys. *J. Mater. Res. Technol.* **2023**, *23*, 2991–3006. [CrossRef]
8. Cao, J.; Zhao, W.; Wang, X.; Zhao, Y.; Ma, P.; Jiang, W.; Song, G. The Microstructure and tribological behavior of ultrasonic electroless Ni-P plating on TC4 titanium alloy with heat-Treatment. *Ferroelectrics* **2022**, *589*, 1–11. [CrossRef]
9. Huo, W.; Zhao, L.; Zhang, W.; Lu, J.; Zhao, Y.; Zhang, Y. In vitro corrosion behavior and biocompatibility of nanostructured Ti6Al4V. *Mater. Sci. Eng. C* **2018**, *92*, 268–279. [CrossRef]
10. Kazemi, M.; Ahangarani, S.; Esmailian, M.; Shanaghi, A. Investigation on the corrosion behavior and biocompatibility of Ti-6Al-4V implant coated with HA/TiN dual layer for medical applications. *Surf. Coatings Technol.* **2020**, *397*, 126044. [CrossRef]
11. Ilani, M.A.; Khoshnevisan, M. An evaluation of the surface integrity and corrosion behavior of Ti-6Al-4 V processed thermody-namically by PM-EDM criteria. *Int. J. Adv. Manuf. Technol.* **2022**, *120*, 5117–5129. [CrossRef]
12. Liao, Y.; Bai, J.; Chen, F.; Xu, G.; Cui, Y. Microstructural strengthening and toughening mechanisms in Fe-containing Ti-6Al-4V: A comparison between homogenization and aging treated states. *J. Mater. Sci. Technol.* **2022**, *99*, 114–126. [CrossRef]
13. Cheng, R.; Luo, X.; Huang, G.; Li, C.J. Corrosion and wear resistant WC17Co-TC4 composite coatings with fully dense microstructure enabled by in-situ forging of the large-sized WC17Co particles in cold spray. *J. Am. Acad. Dermatol.* **2021**, *296*, 117231. [CrossRef]
14. Yao, J.; Wang, Y.; Wu, G.; Sun, M.; Wang, M.; Zhang, Q. Growth characteristics and properties of micro-arc oxidation coating on SLM-produced TC4 alloy for biomedical applications. *Appl. Surf. Sci.* **2019**, *479*, 727–737. [CrossRef]
15. Dai, N.; Zhang, L.-C.; Zhang, J.; Chen, Q.; Wu, M. Corrosion behavior of selective laser melted Ti-6Al-4 V alloy in NaCl solution. *Corros. Sci.* **2016**, *102*, 484–489. [CrossRef]
16. Zhang, H.; Man, C.; Dong, C.; Wang, L.; Li, W.; Kong, D.; Wang, L.; Wang, X. The corrosion behavior of Ti-6Al-4V fabricated by selective laser melting in theartificial saliva with different fluoride concentrations and pH values. *Corros. Sci.* **2021**, *179*, 109097. [CrossRef]
17. Zhao, B.; Wang, H.; Qiao, N.; Wang, C.; Hu, M. Corrosion resistance characteristics of a Ti-6Al-4V alloy scaffold that is fabricated by electron beam melting and selective laser melting for implantation in vivo. *Mater. Sci. Eng. C* **2017**, *70*, 832–841. [CrossRef] [PubMed]
18. Hamza, H.M.; Deen, K.M.; Haider, W. Microstructural examination and corrosion behavior of selective laser melted and conventionally manufactured Ti6Al4V for dental applications. *Mater. Sci. Eng. C* **2020**, *113*, 110980. [CrossRef] [PubMed]
19. Kong, D.; Dong, C.; Ni, X.; Li, X. Corrosion of metallic materials fabricated by selective laser melting. *npj Mater. Degrad.* **2019**, *3*, 24. [CrossRef]
20. Zhou, L.; Yuan, T.; Tang, J.; He, J.; Li, R. Mechanical and corrosion behavior of titanium alloys additively manufactured by selective laser melting—A comparison between nearly b titanium, a titanium and a + b titanium. *Opt. Laser Tech.* **2019**, *119*, 105625. [CrossRef]
21. Liu, M.; Zhu, J.-N.; Popovich, V.A.; Borisov, E.; Mol, J.M.C.; Gonzalez-Garcia, Y. Corrosion and passive film characteristics of 3D-printed NiTi shape memory alloys in artificial saliva. *Rare Met.* **2023**, *42*, 3114–3129. [CrossRef]
22. Qin, P.; Chen, Y.; Liu, Y.J.; Zhang, J.; Chen, L.Y.; Li, Y.; Zhang, X.; Cao, C.; Sun, H.; Zhang, L.C. Resemblance in corrosion behavior of selective laser melted and traditional monolithic b Ti-24Nb-4Zr-8Sn alloy. *ACS Biomater. Sci. Eng.* **2019**, *5*, 1141–1149. [CrossRef] [PubMed]
23. Chen, Y.; Zhang, J.; Dai, N.; Qin, P.; Attar, H.; Zhang, L.C. Corrosion behaviour of selective laser melted Ti-TiB biocomposite in simulated body fluid. *Electrochim. Acta* **2017**, *232*, 89–97. [CrossRef]
24. Liu, M.; Cheng, X.; Li, X.; Pan, Y.; Li, J. Effect of Cr on the passive film formation mechanism of steel rebar in saturated calcium hydroxide solution. *Appl. Surf. Sci.* **2016**, *389*, 1182–1191. [CrossRef]
25. El Haleem, S.A.; El Aal, E.A.; El Wanees, S.A.; Diab, A. Environmental factors affecting the corrosion behaviour of reinforcing steel: I. The early stage of passive film formation in Ca(OH)$_2$ solutions. *Corros. Sci.* **2010**, *52*, 3875–3882. [CrossRef]
26. Fischer, H.; Hauffe, K. *Passivierende Filme und Deckschichten Anlaufschichten*; Springer: Berlin/Heidelberg, Germany, 1956.

27. Yang, X.; Du, C.; Wan, H.; Liu, Z.; Li, X. Influence of sulfides on the passivation behavior of titanium alloy TA2 in simulated seawater environments. *Appl. Surf. Sci.* **2018**, *458*, 198–209. [CrossRef]
28. Wang, J.; Wang, Z.; Sui, Q.; Xu, S.; Yuan, Q.; Zhang, D.; Liu, J. A Comparison of the Microstructure, Mechanical Properties, and Corrosion Resistance of the K213 Superalloy after Conventional Casting and Selective Laser Melting. *Materials* **2023**, *16*, 1331. [CrossRef] [PubMed]
29. Chen, G.; Wen, S.; Ma, J.; Sun, Z.; Lin, C.; Yue, Z.; Mol, J.; Liu, M. Optimization of intrinsic self-healing silicone coatings by benzotriazole loaded mesoporous silica. *Surf. Coatings Technol.* **2021**, *421*, 127388. [CrossRef]
30. Jia, C.; Wang XHu, M.; Su, Y.; Li, S.; Gai, X.; Sheng, L. Corrosion Behavior of TiNi Alloy Fabricated by Selective Laser Melting in Simulated Saliva. *Coatings* **2022**, *12*, 840. [CrossRef]
31. Wang, Y.; Cheng, X.; Li, X. Electrochemical behavior and compositions of passive films formed on the constituent phases of duplex stainless steel without coupling. *Electrochem. Commun.* **2015**, *57*, 56–60. [CrossRef]
32. Chen, X.; Liao, Q.; Gong, M.; Fu, Q. Corrosion Performances of Selective Laser Melting Ti6Al4V Alloy in Different Solutions. *Metals* **2023**, *13*, 192. [CrossRef]
33. Liu, M.; Cheng, X.; Li, X.; Lu, T.J. Corrosion behavior of low-Cr steel rebars in alkaline solutions with different pH in the presence of chlorides. *J. Electroanal. Chem.* **2017**, *803*, 40–50. [CrossRef]
34. Luo, H.; Su, H.; Dong, C.; Xiao, K.; Li, X. Electrochemical and passivation behavior investigation of ferritic stainless steel in alkaline environment. *Constr. Build. Mater.* **2015**, *96*, 502–507. [CrossRef]

Article

Dynamic Marine Atmospheric Corrosion Behavior of AZ91 Mg Alloy Sailing from Yellow Sea to Western Pacific Ocean

Lihui Yang [1,2,4,*], Cong Liu [3], Ying Wang [1], Xiutong Wang [1] and Haiping Gao [2,*]

1. Key Laboratory of Advanced Marine Materials, Key Laboratory of Marine Environmental Corrosion and Bio-Fouling, Institute of Oceanology, Chinese Academy of Sciences, Qingdao 266071, China
2. National Key Laboratory of Marine Corrosion and Protection, Luoyang Ship Material Research Institute, Qingdao 266237, China
3. Southwest Technology and Engineering Research Institute, Chongqing 400039, China
4. Guangxi Key Laboratory of Marine Environmental Science, Institute of Marine Corrosion Protection, Guangxi Academy of Sciences, Nanning 530007, China
* Correspondence: lhyang@qdio.ac.cn (L.Y.); gaohp@sunrui.net (H.G.)

Abstract: In this work, the dynamic marine atmospheric corrosion behavior of AZ91 Mg alloy sailing from Yellow Sea to Western Pacific Ocean was studied. The corrosion rates were measured using the weight loss method. The microstructure, phase, and chemical composition of corroded samples were investigated by SEM, EDS, XRD, and XPS. The results show that the evolution of corrosion rates of AZ91 Mg alloy was divided into three stages: rapidly increasing during the first 3 months, then remaining stable for the next three months, and finally decreasing after 6 months. The annual corrosion rate of Mg alloy reached 32.50 μm/y after exposure for 12 months in a dynamic marine atmospheric environment, which was several times higher than that of the static field exposure tests. AZ91 magnesium alloy was mainly subjected to localized corrosion with more destructiveness to Mg parts, which is mainly due to the synergistic effect of high relative humidity, the high deposition rate of chloride ion, sulfur dioxide acidic gas produced by fuel combustion, and rapid temperature changes caused by the alternating changes in longitude and latitude during navigation. As the exposure time increased, the corrosion pits gradually increased and deepened. The maximum depth of the corrosion pit was 197 μm after 12 months of exposure, which is almost 6 times the average corrosion depth. This study provides scientific data support for the application of magnesium alloys in shipborne aircraft and electronic equipment. The results could provide guidance for the design of new magnesium alloys and development of anti-corrosion technologies.

Keywords: Mg alloy; dynamic atmospheric corrosion; sea voyage; Western Pacific Ocean

Citation: Yang, L.; Liu, C.; Wang, Y.; Wang, X.; Gao, H. Dynamic Marine Atmospheric Corrosion Behavior of AZ91 Mg Alloy Sailing from Yellow Sea to Western Pacific Ocean. *Materials* **2024**, *17*, 2294. https://doi.org/10.3390/ma17102294

Academic Editor: Tomasz Czujko

Received: 3 March 2024
Revised: 3 May 2024
Accepted: 7 May 2024
Published: 13 May 2024

1. Introduction

Magnesium alloys, which display excellent properties such as low density and high specific strength, have been widely used in aerospace, automobile, and 3C industries [1–3]. However, the application and development of magnesium alloys are limited owing to their poor corrosion resistance, especially in harsh corrosive environments [4,5].

In order to accelerate the evaluation of the corrosion resistance of magnesium alloys, there have been a large number of studies simulating the corrosion behavior of magnesium alloys, including immersion tests and salt spray experiments [6–10]. However, the corrosion behavior of Mg alloys in actual service atmospheric environments is inconsistent with that in the accelerated simulated corrosion tests mentioned above [10–12]. It is well known that Mg alloys are mainly exposed to atmospheric environments during practical application. In recent years, many studies have been conducted on the corrosion behavior of magnesium alloys in marine and industrial environments [13–20]. Liao et al. [12] found that the corrosion rate of AZ31B in the marine atmospheric environment (Shimizu, Japan) was much higher than that in urban areas (Osaka, Japan). Jönsson et al. [14] researched the corrosion behavior of AZ91D exposed to three typical types of atmospheric environment, namely marine, rural, and urban atmospheric environments. The result showed that the corrosion rate of AZ91D exposed in the marine atmospheric environment was about 2 times higher than that in the rural and urban atmospheric environments. Yu et al. [15] studied the average corrosion rates of AM60, ZE41, AZ91D, and pure Mg exposed on an island, and found all were higher than those of samples exposed on a site located 1000 m inland from the coast (Xiamen, China). Jiang et al. [16] found that AZ91D was more susceptible to corrosion in marine environments (Qingdao, China) than in an inland environment (Beijing, China) owing to the sea salt and the high relative humidity. These studies consistently indicated that magnesium alloys suffered from more severe corrosion in marine atmospheric environments compared to inland areas.

As mentioned above, many studies of the marine atmospheric corrosion behavior of magnesium alloys have been performed in static exposure stations.

However, the service environments of magnesium alloy are dynamic when applied in a marine equipment, especially naval aviation equipment, which may lead to different corrosion behaviors compared to those in static environments. Until now, only two studies from our institute have reported the corrosion behavior of AZ31 and rare earth magnesium alloys in a dynamic marine atmosphere [17,18]. Magnesium alloys exhibited varying corrosion behaviors due to their different composition. Therefore, it is necessary to increase the amount of research on different types of magnesium alloys in dynamic marine environments. It is significant to clarify the dynamic corrosion mechanism under the interaction of multiple factors to improve the safe surface performance of Mg alloy components.

The aim of this work was to study the corrosion behavior of AZ91 Mg alloy in a dynamic marine atmospheric environment from Yellow Sea to Western Pacific Ocean. The evolution of corrosion rates, corrosion product composition, corrosion morphology, and the effects of the dynamic marine atmospheric environment on corrosion behavior of AZ91 magnesium alloy were examined. Relevant research can clarify the corrosion mechanism of AZ91 Mg alloys in a dynamic marine atmosphere, which is of great significance for promoting the application of Mg alloys in marine engineering.

2. Materials and Methods

2.1. Material Preparation

The chemical composition of as-extruded AZ91 Mg alloy is listed in Table 1. The size of the experimental samples was 100 mm × 50 mm × 3 mm. All specimens were ground with successive grades of SiC grit paper from #400 to #1000. The specimens were cleaned using distilled water and degreased with acetone, dried with cool air, and stored in a desiccator. Five parallel samples were exposed for each period. Three samples were used to determine the weight loss of specimens and the remaining were used for corrosion performance evaluation.

Table 1. Chemical composition of AZ91 magnesium alloy (wt.%).

Material	Al	Zn	Mn	Si	Fe	Cu	Ni	Mg
AZ91	8.93	0.68	0.25	0.02	0.003	0.003	0.0006	Bal.

2.2. Atmospheric Exposure Experiment

The dynamic atmospheric exposure experiment was carried out on the deck of the Research Vessel KEXUE (Science) of the Institute of Oceanology, Chinese Academy of Sciences (Qingdao, China). AZ91 magnesium alloy was placed on the exposure rackwith an angle of 45° horizontal to the deck for one year (Figure 1a). The sailing route was from Yellow Sea (Qingdao, China) to Western Pacific Ocean (Figure 1b), which spanned from a northern temperate climate to a tropical climate. The exposure period was from September 2020 to September 2021 (5 ocean voyages were conducted), with four intervals at 1, 3, 6, and 12 months.

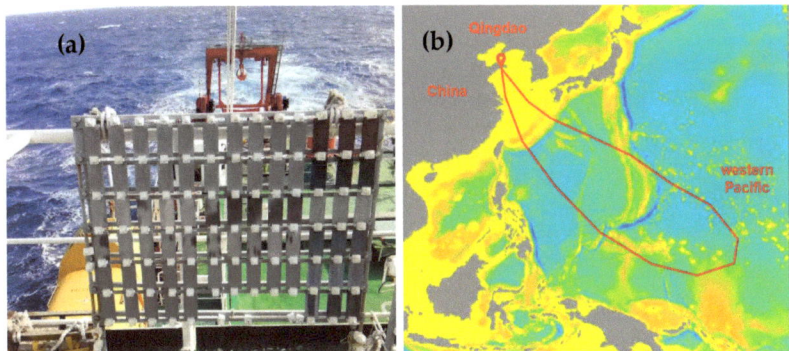

Figure 1. Atmospheric environmental corrosion test on Research Vessel KEXUE (**a**) exposure rack, (**b**) navigation route.

2.3. Environmental Factors Measurements

The meteorological factors (such as temperature and relative humidity (RH)) were measured by the automatic weather observation system of Chinese Research Vessel KEXUE (Science).

The deposition rate of chloride was measured based on GJB 8894.1-2017 [21]. Three parallel specimens of double-layer gauze with size of 100 cm^2 were exposed under the rain shelter on the deck for 7 days. The collected gauze specimens were cleaned, and the chloride ion concentration was measured using an Ion-Chromatography instrument (ICS-5000, ThermoFisher, Waltham, MA, USA).

The data for temperature, relative humidity, and chloride ion deposition rate during a one-year sea voyage are shown in Figure 2.

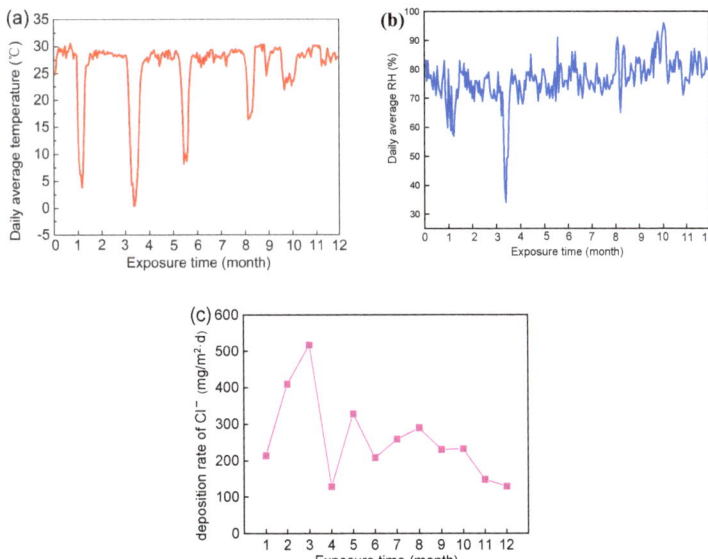

Figure 2. Environment factor of exposure site during the exposure time: (**a**) temperature, (**b**) relative humidity, and (**c**) chloride ion deposition rate [17].

2.4. Characterization and Analysis of the Exposed Samples

The surface and cross-section morphology of the corroded samples was observed by scanning electron microscope (SEM, Regulus 8100, HITACHI, Tokyo, Japan), energy dispersive spectrometer (EDS), metallographic microscope (Axio Vert.A1, Hanover, Germany), and laser confocal scanning microscopy (LCSM, OLS5000, Olympus, Tokyo, Japan). Phase composition was analyzed by X-ray diffraction (XRD, Ultime IV, Rigaku, Tokyo, Japan) with a Cu target and a monochromator, at 40 kV and 150 mA with a scanning rate of $10°/\text{min}$ and a step size of $0.02°$. The element types and valence states of the corrosion products were analyzed by X-ray photoelectron spectroscopy (XPS, ESCALAB 250Xi, Thermo, Waltham, MA, USA).

The corrosion products were removed using 200 g/L CrO_3 + 10 g/L $AgNO_3$ immersed for 5–10 min at 25 °C, and then the samples were rinsed with distilled water and alcohol, dried for 24 h, and weighted. The samples before and after exposure were weighed using an analytical balance with an accuracy of 0.1 mg.

The corrosion rate of AZ91 magnesium after exposure for different duration was calculated using the equation as follows:

$$v = (w_0 - w_1)/(S \cdot T \cdot \rho) \tag{1}$$

In the above formula, v is the corrosion rate of AZ91 Mg alloys. w_0 and w_1 are the initial and final mass (after removing the corrosion products), respectively. S represents the surface area. T is the exposure time, and ρ is the density of AZ91 Mg alloy.

3. Results and Discussion

3.1. Initial Microstructure of AZ91 Mg Alloy

Figure 3 shows the optical micrograph of AZ91 Mg alloy. The image clearly exhibits the α-Mg matrix and β-phase interdendritic network. As is well known, the Al content of AZ91 Mg alloy exceeds its solubility limit in magnesium, and β-phase $Mg_{17}Al_{12}$ intermetallic compounds are formed near grain boundaries during solidification [22,23].

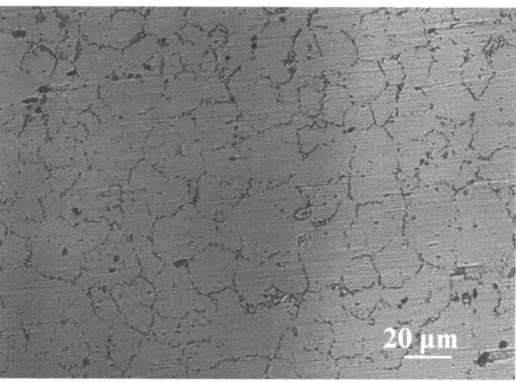

Figure 3. Optical micrograph of AZ91 Mg alloy.

3.2. Corrosion Morphologies of AZ91 Mg Alloys Exposed to Dynamic Marine Atmosphere

The macroscopic images of AZ91 magnesium alloy specimens with corrosion products exposed for different periods during an ocean-going voyage are shown in Figure 4. It can be clearly seen that the color of the surface gradually becomes darker with the increase in exposure time. Due to the increasing number of corrosion pits and formation of corrosion products, the surface of magnesium alloy gradually became rough and lost its metallic luster after exposure for 3 months. There are obvious traces of rain erosion on the surface of the samples after exposure for 6 months and 12 months, which may lead to some loose corrosion products and soluble compounds being washed away.

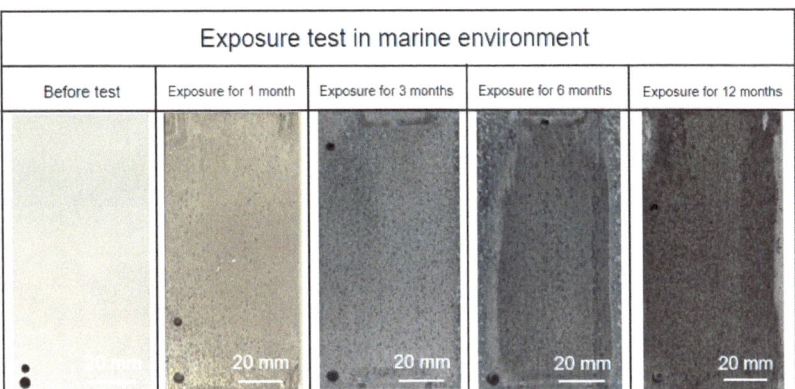

Figure 4. Macroscopic images of AZ91 magnesium alloy specimens exposed for different periods during an ocean-going voyage.

Figure 5 exhibits SEM images of AZ91 magnesium alloy specimens exposed for different periods during an ocean-going voyage. As shown in Figure 5a, some randomly distributed corrosion products formed on the surface of the sample after one month of exposure. The corrosion film was not compact, and many cracks and scratches on the surface of the substrate can be clearly seen. When the Mg alloy samples had been exposed for 3 to 12 months, it can be seen that the morphologies of the corrosion products were similar. The number of corrosion products gradually increased, and there were many cracks that may cause the infiltration of corrosive media such as chloride ions. In some areas, the corrosion products appeared as flower clusters, which were loose and easily washed away by rainwater or strong winds.

Figure 6 displays the EDS analysis of the cross-section of AZ91 Mg alloy exposed for 1 year on the deck of the Research Vessel KEXUE. After exposure for 12 months, the corrosion product layer with a thickness of more than 90 μm formed on the surface of the samples. The O element was mainly concentrated on the corrosion product layer, which indicated that the corrosion product layer formed on AZ91 magnesium alloy was composed of oxide. It can also be seen that there were some small cracks in the corrosion product layer, and the C element was concentrated in these cracks. This is mainly due to the thin electrolyte layer formed on the surface of specimens. CO_2 in the air and dissolved in the thin electrolyte layer could permeate into these cracks and generate carbonate-containing compounds [24,25]. In high relative humidity environments, the electrolyte solution formed on the surface could penetrate into the substrate surface through cracks, promoting further corrosion.

Figure 7 displays laser confocal scanning microscopy (LCSM) analysis of AZ91 magnesium alloy specimens exposed for different periods in a marine environment. It can be obviously seen that the number of pitting corrosions and the maximum pitting depth gradually increase with the increased exposure time. After 12 months of exposure, the maximum pitting depth was approximately 197 μm, which is about 6 times that of the average corrosion depth. Combined with the analysis of morphology of the corrosion products, the corrosion pits formed on the specimens may be related to the thin electrolyte layer, which contained a high concentration of chloride ions. The thin electrolyte layer having a high concentration of chloride ions might permeate into the matrix through cracks and react with the matrix, causing serious localized corrosion in these areas.

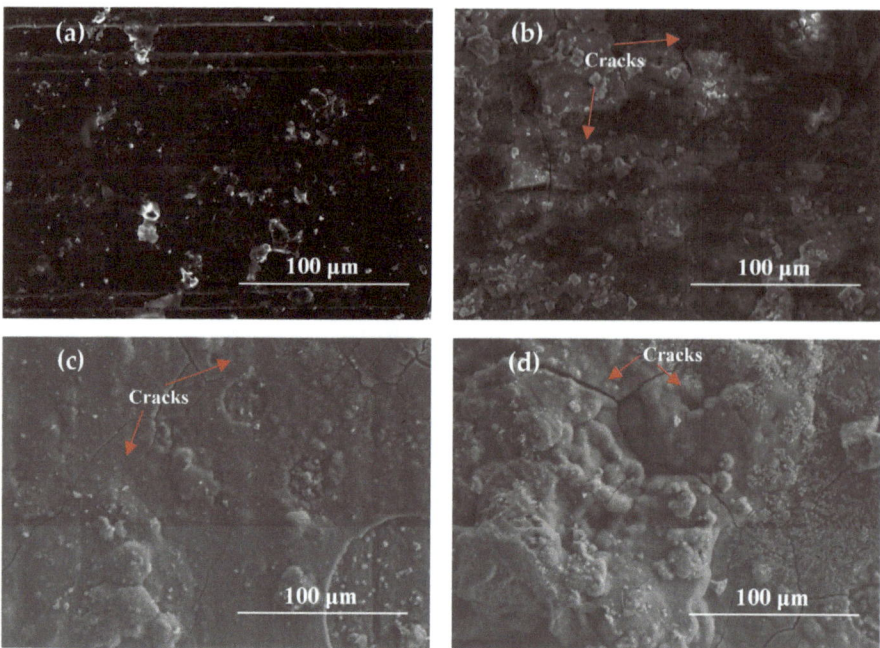

Figure 5. SEM images of AZ91 Mg alloy specimens exposed for different periods: (**a**) 1 month, (**b**) 3 months, (**c**) 6 months, (**d**) 12 months.

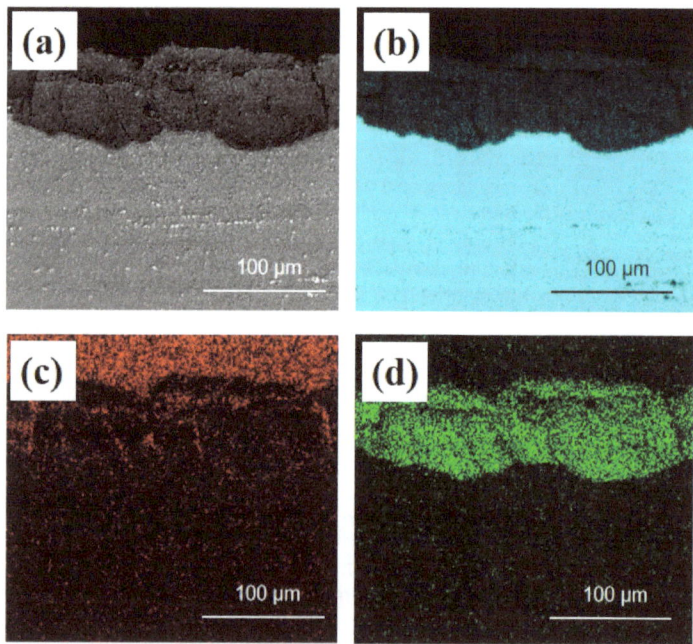

Figure 6. EDS analysis of the cross-section of AZ91 magnesium alloy exposed for 1 year on the deck of Research Vessel KEXUE: (**a**) electron image, (**b**) Mg map, (**c**) C map, (**d**) O map.

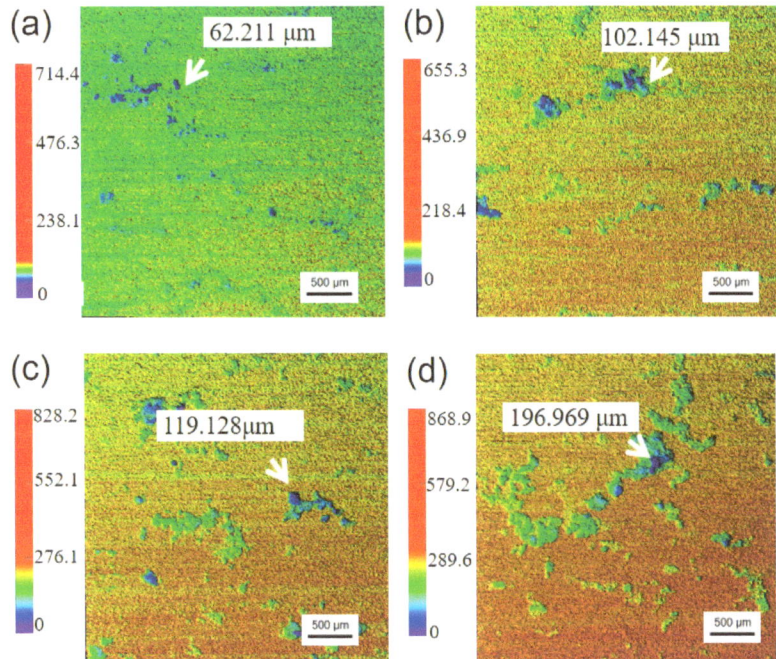

Figure 7. Laser confocal scanning microscopy (LCSM) of AZ91 magnesium alloy specimens without corrosion products exposed for different periods: (**a**) 1 month, (**b**) 3 months, (**c**) 6 months, (**d**) 12 months.

3.3. Corrosion Products Analysis

Figure 8 exhibits the XRD patterns of AZ91 magnesium alloy with corrosion products and matrix. The results show that the corrosion products on the surface of AZ91 magnesium alloys exposed in marine environments for 12 months were mainly composed of hydromagnesite ($Mg_5(CO_3)_4(OH)_2 \cdot 4H_2O$ [26] and the chloride-containing compound $Mg_2(OH)_3Cl \cdot 4H_2O$ [27]. In previous reports on the corrosion products of specimens exposed to different atmospheric environments, $Mg_{5(}CO_3)_4(OH)_2 \cdot 4H_2O$ was found together with $MgCO_3$ and/or $MgCO_3 \cdot xH_2O$ (x = 3, 5) [12,14]. However, there are no obvious peaks of $MgCO_3$ detected in the XRD analysis of this work, indicating that it may be low content or amorphous structures.

Figure 9 shows the XPS spectrum of corrosion products formed on AZ91 magnesium alloy exposed for 12 months in the dynamic marine atmospheric environment ((a) the whole spectrum, (b) narrow scan spectrum of O 1s). The results show that the main elements of the surface contained a large amount of Mg, O, and C, and a small amount of Al and Cl, as shown in Figure 9a. In Figure 9b, the O 1s spectrum consists of three main peaks. The peaks at 529.5 eV and 531.8 eV are respectively associated with MgO and $Mg(OH)_2$ [28,29]. Furthermore, the peak present on the O 1s high-resolution spectrum at 533 eV is ascribed to $MgCO_3$, a compound that is also commonly found in the Mg corrosion layer [30].

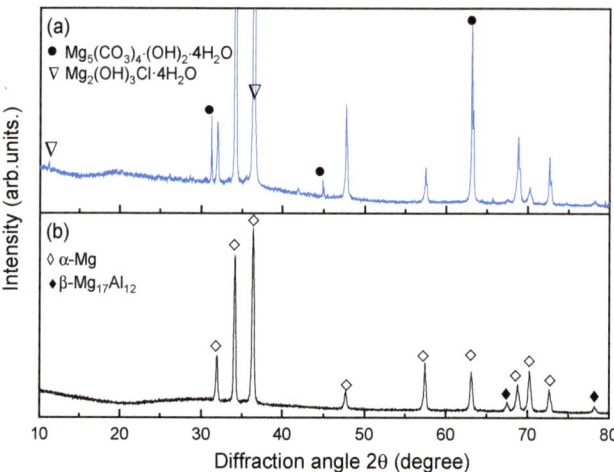

Figure 8. XRD patterns of AZ91 magnesium alloy with: (**a**) corrosion products, (**b**) matrix.

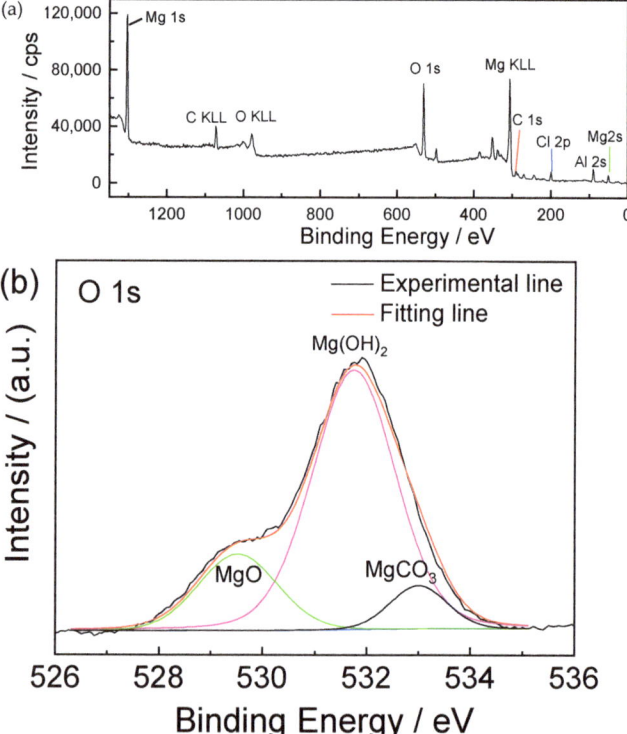

Figure 9. The XPS binding energy spectrum of corrosion products formed on AZ91 magnesium alloy exposed for 12 months in the marine environment of an ocean voyage: (**a**) whole spectrum, (**b**) narrow scan spectrum of O 1s. Different colors were used to distinguish different binding states of oxygen element.

3.4. Corrosion Rate of AZ91 Mg Alloy Exposed to Dynamic Marine Atmosphere for Different Durations

The variation curve of the corrosion rate of AZ91 Mg alloy on the deck of Research Vessel KEXUE after 1 year of exposure is shown in Figure 10. In the first three months of exposure, the corrosion rate rapidly increased with an approximately consistent slope. The corrosion rate in three months reached as high as 48.08 μm/y. The corrosion rate remained stable for the next three months, and after six months of exposure, the corrosion rate of the sample gradually decreased with the passage of exposure time. After 12 months of exposure, the corrosion rate of the sample was 32.50 μm/y. The reason for this is speculated to be the more positive potential of the second phase (β-$Mg_{17}Al_{12}$) compared to the matrix phase [31], resulting in micro-galvanic corrosion. After 3 months, the micro-galvanic corrosion and the hindrance of the second phase reached equilibrium, and the corrosion rate remained unchanged. After 6 months, with the dissolution of the matrix phase, the relative proportion of the second phase increased, and its hindrance became more obvious. The thick corrosion product film also hindered the infiltration of chloride ions, resulting in decreasing the corrosion rate of AZ91 Mg alloy.

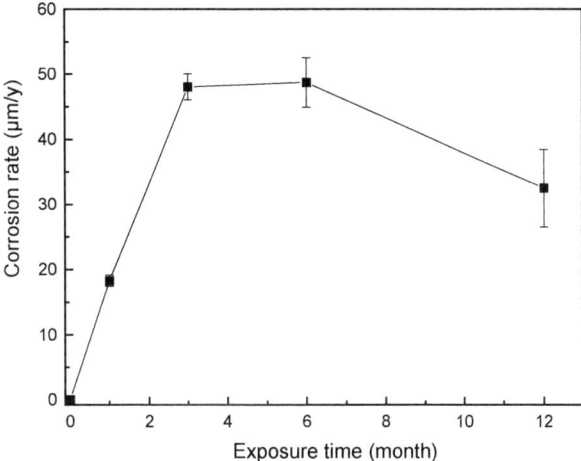

Figure 10. Variation in corrosion rate of AZ91 Mg alloy exposed to a dynamic marine atmosphere for different durations.

The corrosion rate of AZ91 magnesium alloy exposed on the deck of Research Vessel KEXUE was five to eight times that of the investigations conducted in static marine environments, such as the study conducted in a marine environment located 3–5 m away from the Atlantic shore, Brest, France [14], and the research conducted in a marine environment 600 m from the coast in Xiamen, China [15]. The more severe marine atmospheric corrosion behavior of AZ91 magnesium alloy may be due to the special exposure environment from the Yellow Sea to the Western Pacific Ocean, including rapid temperature changes, high humidity, and chloride ion deposition rate.

Compared with exposure tests conducted in static marine atmospheric environments, the temperature shows seasonal changes with changes in longitude and latitude in a real dynamic marine atmospheric environment. Sudden alternating changes in temperature may increase the cracks in corrosion products, leading to the infiltration of corrosive media such as chloride ions and accelerating the occurrence of corrosion [13]. During the exposure period, the marine scientific research vessel carried out five voyages from the Yellow Sea to the Western Pacific Ocean. After exposure for 3 months, the Research Vessel KEXUE returned to Qingdao from the Western Pacific Ocean. The average temperature in the Western Pacific Ocean was maintained at about 29 °C all year round. There were five sudden drops in temperature during the exposure period (Figure 2a), which were closely related to the voyages returning to the Yellow Sea.

The deposition rate of chloride is extremely high in a real dynamic marine atmospheric environment. During the ocean voyage, the monthly deposition rate of chloride was above 100 mg/m^2·d, and specifically above 1100 mg/m^2·d in the third month (Figure 2c). As is well known, NaCl has a strong corrosive effect on magnesium alloys, which is closely related to the high corrosion rate of magnesium alloys in dynamic marine atmospheric environments [17].

It was worth noting that relative humidity (RH) was high and, during the exposure period, the proportion of time when RH > 75% was 56% during the ocean voyage. Previous reports have revealed that NaCl can form a salt solution by absorbing water at relative humidity > 75% [32], which indicates a thin electrolyte layer having a high concentration of chloride could form on the surface of specimens for more than half of the sailing time. A thin electrolyte layer provides a near-solution environment for electrochemical corrosion, leading to large-scale electrochemical connections on the surface and increasing the corrosion process of Mg alloy. Therefore, AZ91 magnesium alloy suffers severe corrosion in real dynamic marine environments due to the special environmental conditions noted above.

3.5. Corrosion Mechanism of AZ91 Mg Alloy in Dynamic Marine Atmospheric

Figure 11 shows the corrosion process schematic of AZ91 magnesium alloy during exposure in the dynamic marine atmosphere.

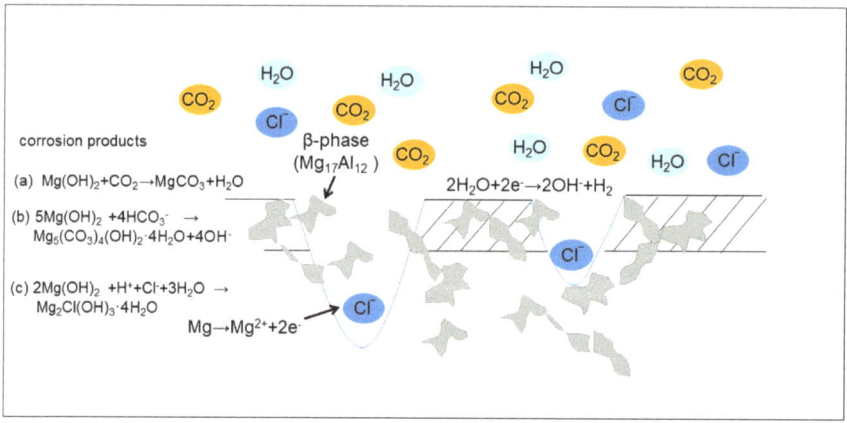

Figure 11. The corrosion process schematic of AZ91 magnesium alloy.

Due to the high deposition rate of chloride and high RH, AZ91 magnesium alloy was covered by a thin electrolyte layer having a high concentration of Cl$^-$ when exposed in a real dynamic marine atmosphere. The corrosion of AZ91 magnesium alloy was dominated by the chemical reaction process including oxidation and hydration reactions at the initial stage.

Anodic reaction:

$$Mg \rightarrow Mg^{2+} + 2e^- \tag{2}$$

Cathodic reaction:

$$2H_2O + 2e^- \rightarrow 2OH^- + H_2 \tag{3}$$

Because of the existence of the thin electrolyte layer, Mg(OH)$_2$ easily formed on the surface of the specimens [33].

Brucite reacted with CO_2 (which came from the marine atmosphere and gas generated by the combustion of fuel) to form $MgCO_3$ as follows [24]:

$$Mg(OH)_2 + CO_2 \rightarrow MgCO_3 + H_2O \tag{4}$$

CO_2 reacted with H_2O to form HCO_3^-, and then reacted with brucite [19]:

$$5Mg(OH)_2 + 4HCO_3^- \rightarrow Mg_5(CO_3)_4(OH)_2 \cdot 4H_2O + 4OH^- \tag{5}$$

Brucite reacted with H^+ (derived from the dissolution of acidic gases such as SO_2 from fuel combustion in thin liquid films), Cl^-, and H_2O to form $Mg_2Cl(OH)_3$ as follows [25]:

$$2Mg(OH)_2 + H^+ + Cl^- + 3H_2O \rightarrow Mg_2Cl(OH)_3 \cdot 4H_2O \tag{6}$$

A corrosion product layer formed on the surface of specimens experiencing a rapid change in temperature during the ocean-going voyage. The volume changes in the matrix and the corrosion product layer were different when temperature changed rapidly. Therefore, there was obvious stress at the interface between the matrix and the corrosion product layer, which accelerated the detachment of corrosion products and generation of cracks. The thin electrolyte layer having a high concentration of chloride ions could permeate into the matrix through these cracks and react with matrix, causing severe localized corrosion. The corrosion behavior of AZ91 Mg alloy was still dominated by localized corrosion after exposure for 12 months, which was mainly due to micro-galvanic corrosion between the α-matrix and β-phase.

Earlier research reported that the β-phase has more positive potential than the α-Mg matrix. When α-Mg is dissolved, the β-phase still remains on the surface. $Mg_{17}Al_{12}$ precipitates with high connectivity along the grain boundary and inside the alloy grains, functioning as a barrier to inhibit the corrosion.

4. Conclusions

The corrosion rate of AZ91 Mg alloy during navigation from the Yellow Sea to the Western Pacific Ocean can be divided into three stages: rapidly increasing in the first three months, then remaining almost the same for the next three months, and ultimately significantly decreasing after 6 months of exposure. The corrosion rate was 32.50 $\mu m/y$ after 12 months of exposure in a dynamic marine atmospheric environment during an ocean voyage, which is 5–8 times higher than that of static coastal field exposure.

In the studied dynamic marine atmospheric environment, AZ91 magnesium alloy was mainly subjected to localized corrosion. After 12 months of exposure, the maximum depth of the corrosion pit was 197 μm, which was almost 6 times that of the average corrosion depth. Localized corrosion poses greater risks to the service life of magnesium alloys. The pollutants of NaCl particles accelerate the localized corrosion with the coupling effect of meteorological factors and the microstructure of the AZ91 Mg substrate.

This study provides scientific data for the application of magnesium alloy in shipboard aircraft and other equipment. It is expected that this research will be of benefit to engineers and researchers developing new corrosion-resistant technology for marine applications.

Author Contributions: Writing—original draft, L.Y.; Resources, Environmental data analysis, C.L.; Data curation, Y.W.; Investigation, X.W.; Writing—review & editing, Corrosion performance evaluation, H.G. All authors have read and agreed to the published version of the manuscript.

Funding: This research was funded by the Program (No. HDH59010101), Overseas Science and education cooperation center deployment project (No. 121311KYSB20210005), National Science and Technology Resources Investigation Program of China (Grant No. 2019FY101400), and Wenhai Program of the S&T Fund of Shandong Province for Pilot National Laboratory for Marine Science and Technology (Qingdao) (NO. 2021WHZZB2304).

Institutional Review Board Statement: Not applicable.

Materials **2024**, *17*, 2294

Informed Consent Statement: Not applicable.

Data Availability Statement: Data are contained within the article.

Acknowledgments: Data Support from Oceanographic Data Center, Chinese Academy of Sciences (CASODC, https://www.casodc.com/) (9 September 2020–8 September 2021). The data and samples were collected by RV KEXUE.

Conflicts of Interest: Haiping Gao was employed by the Luoyang Ship Material Research Institute. Cong Liu was employed by the Southwest Technology and Engineering Research Institute. The remaining authors declare that the research was conducted in the absence of any commercial or financial relationships that could be construed as a potential conflict of interest.

References

1. Guo, J.; Cui, X.; Zhao, W.; Chi, C.; Cao, X.; Lin, P. The tensile deformation behavior of AZ31B magnesium alloy sheet under intermittent pulse current. *Proc. Inst. Mech. Eng. Part C J. Mech. Eng. Sci.* **2022**, *236*, 471–480. [CrossRef]
2. Sheng, Y.; Hou, R.; Liu, C.; Xue, Z.; Zhang, K.; Li, J.; Guan, S. Tailoring of Biodegradable Magnesium Alloy Surface with Schiff Base Coating via Electrostatic Spraying for Better Corrosion Resistance. *Metals* **2022**, *12*, 471. [CrossRef]
3. Xiong, Y.; Yu, Y.; Yang, J. Fatigue behavior after pre-corroded in a simulated body fluid for ZK60 magnesium alloy prepared by micro-arcoxidation. *Fatigue Fract. Eng. Mater. Struct.* **2022**, *45*, 239–258. [CrossRef]
4. Wang, D.; Pei, S.; Wang, Y.; Ma, K.; Dai, C.; Wang, J.; Wang, J.; Pan, F. Effect of magnesium-to-phosphate ratio on the corrosion resistance of magnesium alloy embedded in magnesium potassium phosphate cement. *Cem. Concr. Compos.* **2023**, *135*, 104826. [CrossRef]
5. Lv, X.; Deng, K.; Wang, C.; Nie, K.; Shi, Q.; Liang, W. The corrosion properties of AZ91 alloy improved by the addition of trace submicron SiCp. *Mater. Chem. Phys.* **2022**, *286*, 126143. [CrossRef]
6. Yang, L.; Lin, C.; Gao, H.; Xu, W.; Li, Y.; Hou, B.; Huang, Y. Corrosion Behaviour of AZ63 Magnesium Alloy in Natural Seawater and 3.5 wt.% NaCl Aqueous Solution. *Int. J. Electrochem. Sci.* **2018**, *13*, 8084–8093. [CrossRef]
7. Li, J.; Liu, X.; Zhang, J.; Zhang, R.; Wang, M.; Sand, W.; Duan, J.; Zhu, Q.; Zhai, S.; Hou, B. Effects of Inorganic Metabolites of Sulphate-Reducing Bacteria on the Corrosion of AZ31B and AZ63B Magnesium Alloy in 3.5 wt.% NaCl Solution. *Materials* **2022**, *15*, 2212. [CrossRef] [PubMed]
8. Li, Y.; Zhang, T.; Wang, F. Unveiling the effect of Al–Mn intermetallic on the micro-galvanic corrosion of AM50 Mg alloy in NaCl solution. *J. Mater. Res. Technol.* **2023**, *26*, 753–763. [CrossRef]
9. Zhao, T.; Wang, Z.; Feng, Y.; Li, Q. Synergistic corrosion inhibition of sodium phosphate and sodium dodecyl sulphate on magnesium alloy AZ91 in 3.5 wt.% NaCl solution. *Mater. Today Commun.* **2022**, *31*, 103568. [CrossRef]
10. Walton, C.; Martin, H.; Horstemeyer, M.; Wang, P. Quantification of corrosion mechanisms under immersion and salt-spray environments on an extruded AZ31 magnesium alloy. *Corros. Sci.* **2012**, *56*, 194–208. [CrossRef]
11. Song, W.; Martin, H.; Hicks, A.; Seely, D.; Walton, C.; Lawrimore, W.; Wang, P.; Horstemeyer, M. Corrosion behaviour of extruded AM30 magnesium alloy under salt-spray and immersion environments. *Corros. Sci.* **2014**, *78*, 353–368. [CrossRef]
12. Liao, J.; Hotta, M.; Motoda, S.; Shinohara, T. Atmospheric corrosion of two field-exposed AZ31B magnesium alloys with different grain size. *Corros. Sci.* **2013**, *71*, 53–61. [CrossRef]
13. Man, C.; Dong, C.; Wang, L.; Kong, D.; Li, X. Long-term corrosion kinetics and mechanism of magnesium alloy AZ31 exposed to a dry tropical desert environment. *Corros. Sci.* **2020**, *163*, 108274. [CrossRef]
14. Jönsson, M.; Persson, D.; Leygraf, C. Atmospheric corrosion of field-exposed magnesium alloy AZ91D. *Corros. Sci.* **2008**, *50*, 1406–1413. [CrossRef]
15. Yu, R.; Cao, F.; Zhao, C.; Yao, J.; Wang, J.; Wang, Z.; Zou, Z.; Zheng, D.; Cai, J.; Song, G. The marine atmospheric corrosion of pure Mg and Mg alloys in field exposure and lab simulation. *Corros. Eng. Sci. Technol.* **2020**, *55*, 609–621. [CrossRef]
16. Jiang, Q.; Zhang, K.; Li, X.; Li, Y.; Ma, M.; Shi, G.; Yuan, J. Atmospheric corrosion of Mg-rare earth alloy in typical inland and marine environments. *Corros. Eng. Sci. Technol.* **2014**, *49*, 651–655. [CrossRef]
17. Wang, Y.; Xu, W.; Wang, X.; Jiang, Q.; Li, Y.; Huang, Y.; Yang, L. Research on Dynamic Marine Atmospheric Corrosion Behavior of AZ31 Magnesium Alloy. *Metals* **2022**, *12*, 1886. [CrossRef]
18. Jiang, Q.; Lu, D.; Cheng, L.; Liu, N.; Hou, B. The corrosion characteristic and mechanism of Mg-5Y-1.5Nd-xZn-0.5Zr (x = 0, 2, 4, 6 wt.%) alloys in marine atmospheric environment. *J. Magnes. Alloy.* **2024**, *12*, 139–158. [CrossRef]
19. Cui, Z.; Li, X.; Xiao, K.; Dong, C. Atmospheric corrosion of field-exposed AZ31 magnesium in a tropical marine environment. *Corros. Sci.* **2013**, *76*, 243–256. [CrossRef]
20. Song, Y.; Dai, J.; Sun, S. A comparative study on the corrosion behavior of AZ80 and EW75 Mg alloys in industrial atmospheric environment. *Mater. Today Commun.* **2024**, *38*, 108263. [CrossRef]
21. *GJB 8894.1-2017*; Determination Method for for Natural Factors—Part 1: Atmospheric Environmental Factors. Equipment Development Department of Central Military Commission: Beijing, China, 2017.
22. Grimm, M.; Lohmüller, A.; Singer, R.; Virtanen, S. Influence of the microstructure on the corrosion behaviour of cast Mg-Al alloys. *Corros. Sci.* **2019**, *155*, 195–208. [CrossRef]

23. Kim, J.; Byeon, J. Quantitative relation of discontinuous and continuous $Mg_{17}Al_{12}$ precipitates with corrosion rate of AZ91D magnesium alloy. *Mater. Charact.* **2021**, *174*, 111015. [CrossRef]
24. Lin, C.; Li, X.G. Role of CO_2 in the initial stage of atmospheric corrosion of AZ91 magnesium alloy in the presence of NaCl. *Rare Met.* **2006**, *25*, 190–196. [CrossRef]
25. Yu, L.; Jia, P.; Zhao, B.; Song, Y.; Wang, J.; Cui, H.; Feng, R.; Li, H.; Cui, X.; Gao, Z.; et al. Effect of CO_2 on the microstructure and corrosion mechanism of Mg-Nd-Zn-Ca plasma electrolytic oxidation coatings. *Mater. Today Commun.* **2023**, *34*, 105081. [CrossRef]
26. Liao, J.; Hotta, M. Atmospheric corrosion behavior of field-exposed magnesium alloys: Influences of chemical composition and microstructure. *Corros. Sci.* **2015**, *100*, 353–364. [CrossRef]
27. Lojka, M.; Jiříčková, A.; Lauermannová, A.; Pavlíková, M.; Pavlík, Z.; Jankovský, O. Kinetics of formation and thermal stability of $Mg_2(OH)_3Cl\cdot4H_2O$. *AIP Conf. Proc.* **2019**, *2170*, 020009.
28. Felten, M.; Nowak, J.; Beyss, O.; Grünewald, P.; Motz, C.; Zander, D. The effect of time dependent native oxide surface conditions on the electrochemical corrosion resistance of Mg and Mg-Al-Ca alloys. *Corros. Sci.* **2023**, *212*, 110925. [CrossRef]
29. Zhang, Y.; Chen, X.; Zhang, S.; Ding, X.; Liu, B. In-situ formation $Mg(OH)_2$-ZrO_2 coating on AZ61 magnesium alloy for corrosion protection. *Mater. Lett.* **2024**, *358*, 135854. [CrossRef]
30. Pacheco, M.; Aroso, I.; Silva, J.; Lamaka, S.; Bohlen, J.; Nienaber, M.; Letzig, D.; Lima, S.; Barros, A.; Reis, R. Understanding the corrosion of Mg alloys in in vitro urinary tract conditions: A step forward towards a biodegradable metallic ureteral stent. *J. Magnes. Alloys* **2023**, *11*, 4301–4324. [CrossRef]
31. Li, Y.; Wu, Z.; Wei, J.; Wu, P.; Zhang, Y.; Wu, H.; Liang, S.; Yong, H.; Song, G.; Fang, D.; et al. Enhanced corrosion resistance of $Mg_{17}Al_{12}$ compounds by Ce modification. *Vacuum* **2023**, *218*, 112663. [CrossRef]
32. Esmaily, M.; Shahabi-Navid, M.; Svensson, J.; Halvarsson, M.; Nyborg, L.; Cao, Y.; Johansson, L. Influence of temperature on the atmospheric corrosion of the Mg-Al alloy AM50. *Corros. Sci.* **2015**, *90*, 420–433. [CrossRef]
33. Song, G.; Atrens, A. Recently deepened insights regarding Mg corrosion and advanced engineering applications of Mg alloys. *J. Magnes. Alloys* **2023**, *11*, 3948–3991. [CrossRef]

Article

Microstructure and Chlorine Ion Corrosion Performance in Bronze Earring Relics

Zhiqiang Song [1,2] and Ojiyed Tegus [1,2,*]

1   Institute for the History of Science and Technology, Inner Mongolia Normal University, 81 Zhaowuda Road, Hohhot 010022, China; songzhiqiang@imnu.edu.cn
2   College of Physics and Electronic Information, Inner Mongolia Normal University, 81 Zhaowuda Road, Hohhot 010022, China
*   Correspondence: tegusph@imnu.edu.cn; Tel.: +86-158-4714-9908

Abstract: Chlorine ions play an important role in the corrosion of bronzeware. This study employs techniques such as XRD, OM, SEM, EBSD, and electrochemical testing to analyze the microstructure, crystal structure, chemical composition, and corrosion performance of bronze earrings unearthed at the Xindianzi site in Inner Mongolia. The results indicate the presence of work-hardened structures, including twinning and equiaxed crystals, on the earrings' surface. With an increase in chloride ion concentration in NaCl solutions from 10^{-3} mol/L to 1 mol/L, the corrosion current density of the bronze earrings increased from 2.372×10^{-7} A/cm^2 to 9.051×10^{-7} A/cm^2, demonstrating that the alloy's corrosion rate escalates with chloride ion concentration. A 3-day immersion test in 0.5% NaCl solution showed the formation of a passivation layer of metal oxides on the earrings' surface. These findings underscore the significance of the impact chloride ions have on the corrosion of copper alloys, suggesting that activating the alloy's reactive responses can accelerate the corrosion process and provide essential insights into the corrosion mechanisms of bronze artifacts in chloride-containing environments.

Keywords: bronze earrings; alloy structure; NaCl solution; polarization curve; electrochemical corrosion

Citation: Song, Z.; Tegus, O. Microstructure and Chlorine Ion Corrosion Performance in Bronze Earring Relics. *Materials* **2024**, *17*, 1734. https://doi.org/10.3390/ma17081734

Academic Editor: Frank Czerwinski

Received: 11 March 2024
Revised: 31 March 2024
Accepted: 8 April 2024
Published: 10 April 2024

1. Introduction

Bronze, as one of the earliest alloys used in human history, played an irreplaceable role in the development of ancient civilizations due to its unique and exceptional chemical and physical properties [1]. The addition of tin and lead not only significantly improved the characteristic fluidity and formability of bronze, it established bronze as an indispensable material in modern industrial production [2,3]. China's bronze civilization, with its unique artistic style and highly developed craftsmanship, demonstrated the prosperity and progress of this ancient society. In the central and southern regions of Inner Mongolia, archaeologists have discovered a significant number of bronze artifacts from the Eastern Zhou period, reflecting characteristics of the northern nomadic tribes and their cultural exchanges and integration during this period [4]. Notably, ancient bronze artifacts were primarily cast, however some were further processed through hot and cold working to enhance the alloy's mechanical properties. Bronze earrings, in particular, often underwent intricate hot and cold processing, showcasing exquisite skills and craftsmanship of the ancient artisans' and their pursuit of beauty. However, these precious bronze relics inevitably suffered corrosion after being buried underground for considerable lengths of time [5]. There are multiple metallic elements present in bronzeware, many of which can cause severe electrochemical corrosion and breed significant bronze diseases, leading to mineralization and even disintegration of the bronzeware. The rust on the surface of bronzeware can be divided into harmless rust and harmful rust. Harmful rust, also known as "powdery rust", is predominantly composed of cuprous chloride and basic copper chloride. The chloride ions contained in it can cause corrosion, penetrating the surface and the inside of objects [6].

Bronze earrings, compared to other small copper items, have shown superior performance and integrity in resisting corrosion, sparking great interest among scholars.

The corrosion of ancient bronze artifacts has always been a focal point in archaeological materials science. Previous studies primarily focused on qualitative analysis of corrosion products [7–12]. Whereas these studies have provided valuable insights into the corrosion mechanisms of bronzes, the relationship between the corrosion behavior of bronze alloys and environmental factors, especially in specific geographic locations, climate types, and the combined effects of alloy composition, structure, phase distribution, and exposure time, requires further investigation [13]. The corrosion mechanisms of ancient bronze alloys are complex and diverse. These included preferential corrosion at impurities or non-pure grains, grain boundary corrosion, and pitting, amongst others. The impact of environmental conditions such as humidity, temperature, and anions on corrosion rate is also crucial [14–17]. Corrosion phenomena are closely related to the microenvironment of burial sites, particularly chloride ions, which are key factors in promoting the corrosion of bronze alloys [18]. Chloride ions in natural environments can promote the formation of copper oxides and chlorides on the surface of bronze alloys [19,20], leading to a corrosion layer known as patina, which not only is visually attractive but also acts as a stable protective layer [21]. Environmental changes can accelerate or decelerate the corrosion rate [22]. Studies in simulated buried and seawater environments have shown that the passivation layer of bronze alloys is related to the oxygen content in the environment [23]. Although the corrosion laws of bronze alloys have been widely studied [24]. However, the reason for the continued corrosion of the alloy after the excavation of bronze ware has always been a challenge. Some studies suggest that the study of the corrosion mechanism of bronze should be carried out on the surface covered with copper green, especially the CuCl layer [25]. Research has shown that CuCl, typically formed between the alloy and Cu_2O layer, plays an important role in the chloride ion corrosion of copper alloys [26]. When corrosive chloride ions corrode alloys, Cu as the anode will be corroded into CuCl. It is an unstable intermediate because chloride ions are released during the hydrolysis and oxidation process of CuCl. However, research on the impact of chloride ion concentration and the effect of surface processing on corrosion performance remains relatively limited.

Based on the context provided, this study conducts a comprehensive analysis of bronze earrings unearthed from the Xindianzi cemetery in Inner Mongolia. Utilizing modern scientific methods such as XRD, OM, SEM, and EBSD, the research delves into the structure and chemical composition of the bronze earrings. Through electrochemical polarization curves and electrochemical impedance spectroscopy, the study investigates the corrosion behavior of surface-treated bronze earrings in various concentrations of NaCl solutions. The goal is to uncover a pattern and thus show how chloride ion concentration affects the corrosion of bronze alloys, providing significant reference data and insights into the study of corrosion performance of ancient bronzes.

2. Materials and Methods

2.1. Materials

The M47 at Xindianzi Cemetery is a vertical pit tomb, with a length of 220 cm, a width of 80 cm, and a depth of 40 cm. The tomb owner is a North Asian male of around 45 years old. More than 1000 copper artifacts and 46 earrings have been unearthed from the Xindianzi Cemetery, and the alloy surface displays relatively stable corrosion. After excavation, the surface soil of the bronzeware was simply cleaned and then stored in the cultural relic storage room. The test samples originated from the Xindianzi cemetery in Inner Mongolia, from where the remains of a bronze earring from the Eastern Zhou Dynasty (Figure 1) M47 was unearthed. According to the cemetery briefing, it can be inferred that the Xindianzi cemetery dates back to the late Spring and Autumn period to the early Warring States period [27].

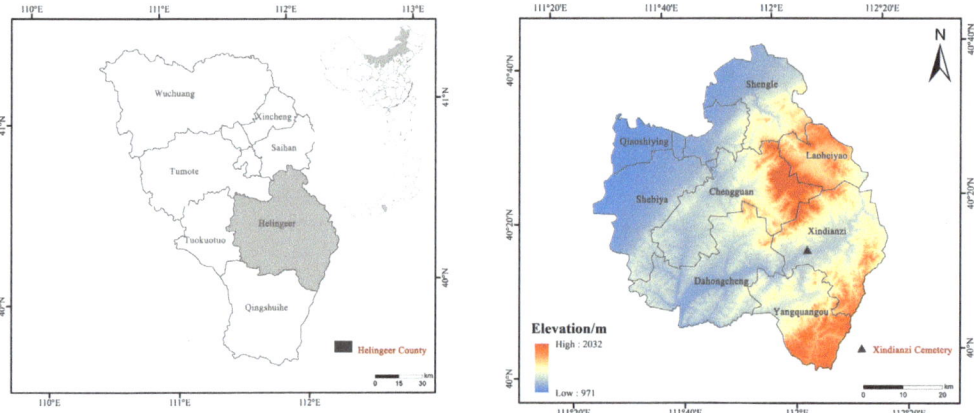

Figure 1. Schematic depiction of the location of the Xindianzi Cemetery during the Eastern Zhou Dynasty.

This earring, crafted from copper wire wound into a circular plane with an overlapped joint, has an external diameter of 3.18 cm. The cross-section of the copper wire is almost circular, with a diameter of 0.18 cm. The bronze earrings are shown in Figure 2a and the sampling and coordinate system are shown in Figure 2b.

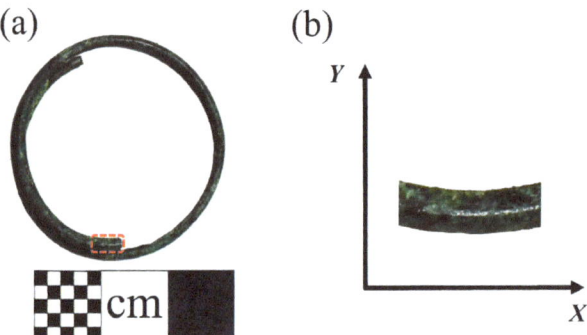

Figure 2. (**a**) Bronze earring sampling and (**b**) coordinate system.

2.2. Methods

Samples were sectioned along the longitudinal cross-section (Figure 2) and sequentially polished with metallographic sandpapers of 800#, 1200#, 2000#, and 4000# for rough and fine polishing. Subsequently, the samples were polished with Al_2O_3 particles of 2.5, 1.5, and 0.5 μm until no scratches were visible on the surface. The samples were then corroded using a solution prepared from 1 g $FeCl_3$, 3 mL HCl, and 12 mL C_2H_6O. Microstructures were observed using a metallographic microscope (ZEIZZ Observer A1m) and a thermal field emission scanning electron microscope (ZEISS Gemini SEM 300) equipped with an EDS detector from Oxford Instruments X-Max (Oxfordshire, UK). The analysis was conducted using Aztec 5.1 software from Oxford EDS, and a built-in virtual standard database was used. The analysis results were quantitative analysis without standard samples. The EBSD detector used was from Oxford Instruments Symmetry. The EBSD sample preparation process was based on the preparation of metallographic samples, followed by ion beam polishing. The data analysis software used was AZtecCrystal 2.1 from Oxford Instruments. Phase analysis of the samples was conducted with a PANalytical

Empyrean X-ray diffractometer under conditions of Cu-Kα radiation, 40 kV voltage, 40 mA current, a scanning speed of 5°/min, and a step size of 0.02°. Electrochemical measurements were performed with a CHI660E workstation from Shanghai Chenhua, Shanghai, China; NaCl (Tianjin Fuchen Chemical Reagents, Ltd., Tianjin, China); AgCl electrode (Shanghai Xianren Chemical Reagents, Ltd., Shanghai, China); platinum electrode (PE, homemade); and deionized water (homemade). Electrochemical impedance spectroscopy and polarization curves were utilized to analyze the effects of the different soaking times and chloride ion concentrations on the corrosion of bronze alloys. Electrochemical impedance spectroscopy tests were conducted at a stable open circuit potential, with frequencies ranging from 100 kHz to 0.1 Hz and a disturbance signal amplitude of 10 mV. Polarization curves were scanned at a rate of 20 mV/min, testing within a range of ±400 mV around the open circuit potential.

3. Results and Discussion

3.1. Structure and Composition

The microstructure, structure, and properties of the samples are closely related to their composition. Figure 3 shows the X-ray diffraction spectrum of the bronze earrings. Comprehensive analysis of the X-ray diffraction spectra and energy spectrum results of the alloy samples indicates that the samples are predominantly composed of face-centered cubic structured α-Cu(Sn) and Pb phases. This result not only reveals the microstructural characteristics of the samples but also provides an important material basis for further understanding the corrosion behavior of bronze earrings.

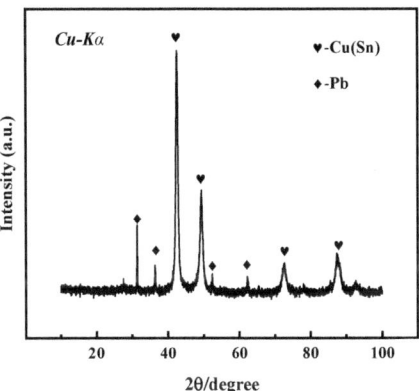

Figure 3. XRD pattern of bronze earrings.

Figure 4 shows the metallographic morphology of the bronze earring samples. It can be observed that the samples predominantly display structures formed during the casting process, with larger grains present in the central area, whereas areas closer to the surface are rich in equiaxed and twinned structures. The samples contain α-Cu(Sn) solid solution and second-phase Pb particles. Additionally, the images reveal defects and shrinkage porosity that may occur in the bronze alloys during casting, which is significant for understanding the microstructure and properties of the material [28].

To further investigate the mechanism of grain formation within the alloy, this study utilized Electron Backscatter Diffraction (EBSD) technology to reconstruct the grain structure of the bronze earrings' X-Y section in detail.

EBSD is an efficient technique for studying the crystallographic information of alloy materials in the micro region [29]. By using EBSD technology, the orientation imaging information of the sample can be obtained, thereby obtaining richer internal information of the sample. Combined with EDS, the morphology, composition, and structure of the sample can be obtained simultaneously [30]. This article uses EBSD technology to detect and analyze bronze earrings unearthed from the Xindianzi Cemetery, studying their phase, twinning, and grain size and shape. The composition of different phases and inclusions in the alloy was obtained by combining SEM-EDS technology [31].

Figure 4. Optical micrograph image of the bronze earring.

Through EBSD image (Figure 5), we could clearly observe the grain orientation distribution along the Z direction on the X-Y section, for which the attached orientation legend and 500 μm scale provides detailed visual references. The application of EBSD data processing software (AZtecCrystal 2.1) enabled us to deeply analyze the microstructural characteristics of the bronze earrings from various aspects such as pole figures, grain size and shape, orientation difference distribution, and phase distribution. Through the analysis of the poles in particular, shown in Figure 6, we found the earrings have a specific (214)<12-1> texture, indicating the crystals exhibit a certain preferential orientation in specific directions.

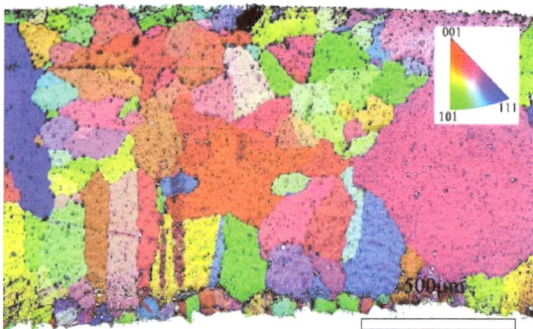

Figure 5. EBSD orientation map of the X-Y section of the bronze earring.

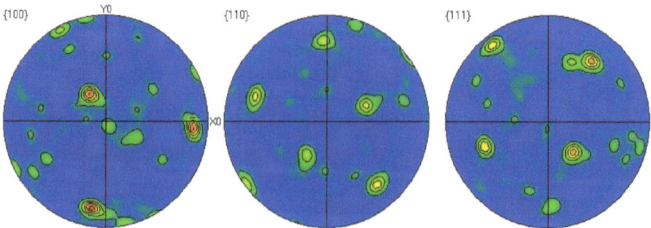

Figure 6. Polar diagram of the cross-section of the bronze earring.

Grain size, as a key parameter for assessing material performance, is meticulously analyzed using EBSD technology, which provides extensive information on grain boundaries, enabling precise analysis of grain size and shape within the tested area. The grain size distribution, illustrated in Figure 7a, reveals an average grain size of 35.3 μm for the alloy, ranging from a minimum of 7.1 μm to a maximum of 577.9 μm. Notably, larger grains predominantly occupy the central region of the alloy, displaying significantly larger grain sizes compared to typical bronze casting grains, without any observed segregation phenomena. Conversely, the alloy's lower surface vicinity contains a higher presence of smaller grains. According to the Hall-Petch mechanism, the material's yield strength is inversely proportional to the square root of the grain size, suggesting that the mechanical properties of the bronze earrings significantly improves in regions closer to the lower surface.

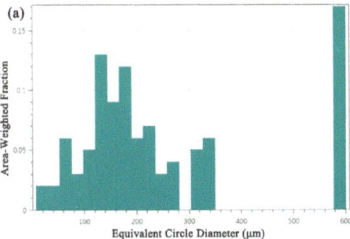 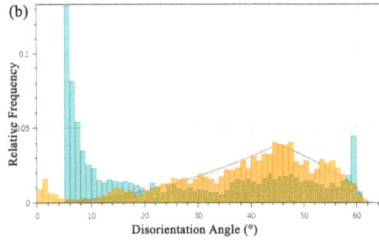

Figure 7. Microstructural characteristics of bronze earrings: (**a**) map of grain size analysis and (**b**) map of orientation difference distribution.

Figure 7b shows the distribution of the grain boundary orientation differences in the tested area of the bronze earring. A noticeable increase in the number of grain boundaries is observed when the orientation difference approaches 60°, which corresponds to the twin boundary orientation difference of 60° in face-centered cubic Cu(Sn) alloys, indicating the presence of twinning structures in the alloy. Analysis of the distribution of different orientation differences using EBSD data processing software shows that low-angle boundaries (orientation difference less than 15°) account for approximately 76.6% of the total grain boundaries, whereas high-angle boundaries account for about 23.4%. Compared to the theoretical random orientation difference distribution (generally considered to be around 45°), the proportion of low-angle grain boundaries in this alloy is higher.

Figure 8a shows the crystal orientation in a local area of the bronze earring, revealing the presence of equiaxed and twinned structures near the alloy surface. By measuring the orientation differences at the positions of twins, it was found that the orientation differences at the third and fifth positions of the alloy matrix are 59.28 degrees and 58.77 degrees, respectively, matching the twinning relationship of <111> crystal direction in α-Cu phase, whereas the orientation differences at other positions are less than 60 degrees. Additionally, EBSD technology was used not only for qualitative analysis of twins in the alloy but also for quantitative studies, revealing a higher twin boundary density in the lower surface area of the alloy. The introduction of KAM (kernel average misorientation) distribution maps further explored the degree of deformation in different areas inside the bronze earrings [32]. KAM distribution maps qualitatively describe the inhomogeneity of plastic deformation and the distribution of defect density, showing higher KAM values at the top and bottom surfaces of the alloy, indicating higher degrees of deformation in these areas. EBSD results (Figure 8a) indicate the presence of more twins and equiaxed crystals on the lower surface of the bronze earring, characteristic of annealing twins, suggesting that the alloy surface underwent heat treatment during its manufacturing process.

SEM-EDS was utilized to collect and test the distribution maps of different elements in the area, as shown in Figure 9. Quantitative analysis revealed the composition of the sample, with Cu accounting for 78.36%, Sn at 10.71%, and Pb at 10.82%. The EDS spectrum analysis results indicate that the alloy's matrix is a Cu-Sn solid solution constituted by Cu and Sn elements, whereas Pb elements in the alloy are distributed in the form of particles.

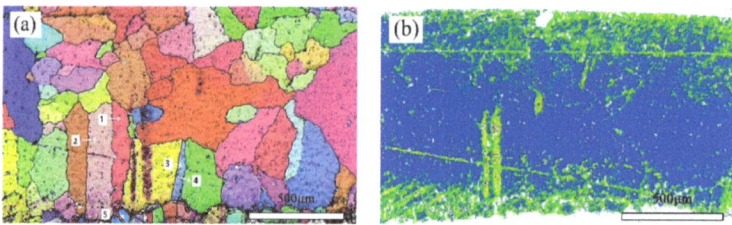

Figure 8. EBSD characterization of the bronze earring: (**a**) Local orientation and (**b**) KAM map.

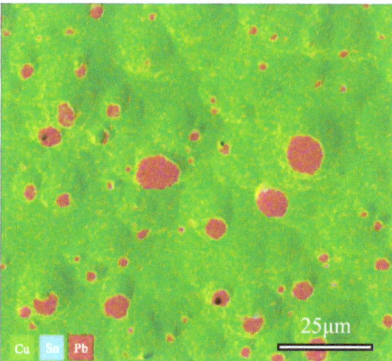

Figure 9. SEM-EDS and Cu, Sn, Pb elements distribution Map of bronze earrings (Cu78.36%, Sn10.71%, Pb10.82%).

Figure 10 shows the distribution information of local phases in the alloy, and further analysis confirmed the phase of the matrix and the second phase. The red region represents the face-centered cubic structure of α-Cu(Sn) solid solution, whereas the blue region refers to the face-centered cubic Pb element. These SEM-EDS analysis results, combined with EBSD and XRD analysis, collectively confirmed the structural information of the matrix and second phase of the bronze earrings. In the Cu-Sn-Pb ternary alloy, Pb does not form a solid solution with the Cu alloy but is distributed in a separated state, which positively affects the alloy's solidification properties, although adding Pb to Cu-Sn alloys may affect the mechanical properties such as the strength and hardness of the bronzeware [33].

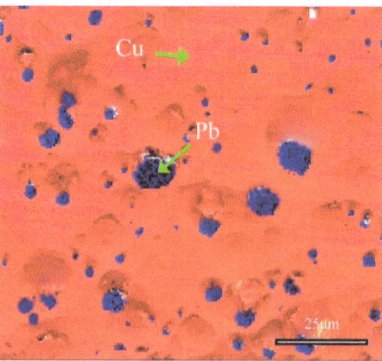

Figure 10. Phase distribution map of the local area of the bronze earring by EBSD.

3.2. Corrosion Products

To deeply understand the corrosion process and products of alloys, this study employed Scanning Electron Microscope-Energy Dispersive Spectroscopy (SEM-EDS) to analyze the pre- and post-corrosion states of bronze earrings. Regarding the issue of whether there are signs of active corrosion, analysis was conducted on the corrosion products using SEM-EDS. Figure 11 shows the SEM-EDS image of the corrosion layer near the surface of the earring before corrosion, as well as the distribution of Cu and O element lines. Through SEM-EDS analysis of rust products, the results showed that the corrosion near the sample surface was more severe, with a thickness of 10–40 microns. The oxygen content of the rust layer was significantly higher than that of the substrate, and the copper content of the rust layer was significantly lower than that of the substrate.

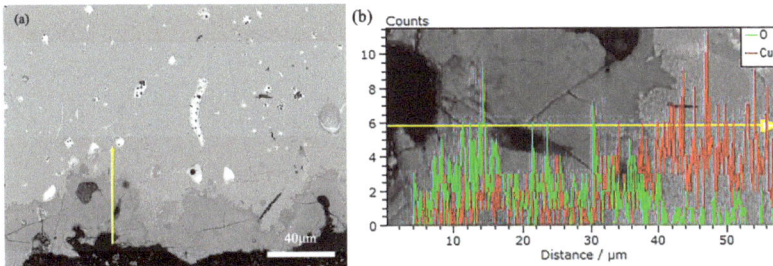

Figure 11. (**a**) SEM-EDS of Bronze Earrings and (**b**) the distribution of Cu and O Element Lines.

Figure 12a,b show the SEM images of the alloy before and after soaking in 0.5 mol/L NaCl solution for 3 days, with a significant increase in surface erosion confirmed by SEM-EDS compositional analysis, the results of which are shown in Table 1. Specifically, the copper (Cu) content decreased, whereas chloride (Cl), tin (Sn), and oxygen (O) contents increased significantly, indicating the progression of corrosion reactions. Electrochemical testing further revealed how an increase in NaCl solution concentration accelerates the dissolution of copper elements in Cu(Sn) alloys [34]. Chloride ions play a decisive role in the corrosion of copper alloys in sodium chloride solutions, with increased chloride ion concentration promoting the corrosion rate of the bronze alloy matrix. Copper elements are corroded as anodes to form CuCl [35,36], an unstable intermediate product that releases chloride ions during hydrolysis and oxidation, leading to continuous dissolution of the bronze alloy matrix [37], and forming a Cu_2O protective layer on the alloy surface [34,38]. According to Figure 12c, it can be seen that after soaking the alloy in a 0.5 mol/L NaCl solution, deep corrosion pits appear at the Pb particles, surrounded by granular and needle like products. The EDS results indicate that the position with high lead content has a higher Cl ion content than the CuSn solid solution, the CuSn phase did not undergo significant corrosion dissolution. The needle shaped corrosion products contain three elements: Pb, O, and Cl, with the highest Pb content. After corrosion of Pb, the Pb element transfers from the original position of Pb particles, forming corrosion pits, which can accelerate the corrosion of the sample [23].

Table 1. SEM-EDS composition analysis results of the bronze earrings before immersion (**a**) and after immersion for 3 days in 0.5 mol/L NaCl solution (**b**).

Position	Sn (wt.%)	Cu (wt.%)	Pb (wt.%)	O (wt.%)	Cl (wt.%)
Without corrosion	10.71	78.36	10.82	0.11	0
With corrosion	20.60	55.16	11.14	9.73	2.76

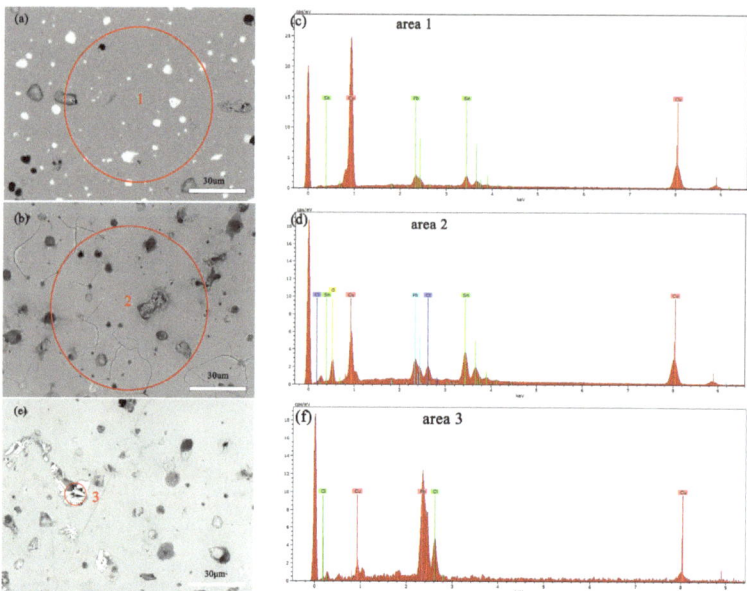

Figure 12. SEM-EDS images and elemental energy spectra of the bronze earrings before immersion (**a**,**c**) and after immersion for 3 days in 0.5 mol/L NaCl solution (**b**,**d**–**f**).

An alloy with a copper–tin–lead content similar to that of earrings was prepared by arc melting. The XRD diffraction patterns of the alloy before and after immersion in a 0.5 mol/L solution were compared, and it was found that the surface of the alloy after immersion contained CuCl and Cu_2O phases (Figure 13).

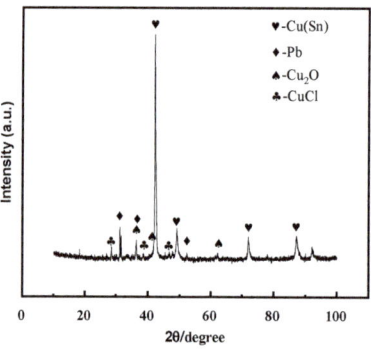

Figure 13. XRD pattern of copper–tin–lead alloy soaked in 0.5 mol/L NaCl solution for 3 days.

3.3. Electrochemical Corrosion Characteristics

Figure 14 shows the open circuit potential (OCP) curves of bronze earrings in NaCl solutions of different concentrations, revealing the trend that the self-corrosion potential changes with the concentration of NaCl. As the concentration of NaCl solution increases, the self-corrosion potential of the bronze alloy also increases accordingly, specifically from -0.09 V to -0.19 V. This change illustrates the direct impact of salt concentration on the corrosion rate of bronze alloys, with high concentrations of NaCl solution accelerating the corrosion process, making the protection of the alloy more difficult.

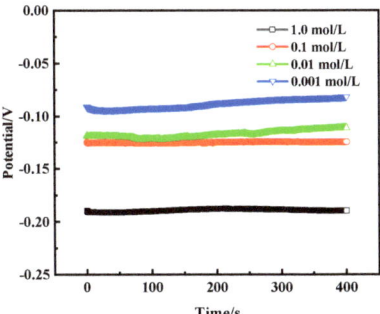

Figure 14. OCP curves of bronze earrings in NaCl solutions.

The Tafel curve, as a key tool for assessing alloy corrosion behavior, is crucial for analyzing the corrosion mechanism of bronze earrings in NaCl solutions of different concentrations. By plotting the Tafel curve, as shown in Figure 15, we observe that with the increase in NaCl concentration, the corrosion current density of the bronze earrings also increases correspondingly. This phenomenon intuitively demonstrates the positive correlation between NaCl concentration and the corrosion rate of the bronze earrings. The increase in corrosion current density signifies the enhanced corrosion activity of the alloy in environments with higher NaCl concentrations.

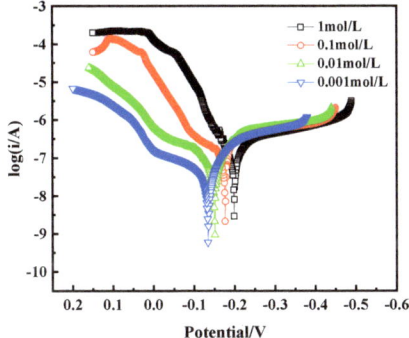

Figure 15. Tafel curves of bronze earrings in NaCl solutions.

The polarization curves were fitted using CHI660E software (CHI Version 14.05), and the results are shown in Table 2. The fitting indicates that the corrosion current density of bronze increases from 2.372×10^{-7} A/cm^2 to 9.051×10^{-7} A/cm^2 as the NaCl concentration increases from 10^{-3} mol/L to 1 mol/L. The polarization resistance (Rp) of the bronze alloy in different concentration solutions was calculated using the Stern equation, $Rp = Ba \times Bc / [2.3(Ba + Bc) \times I_{corr}]$ (in which Ba and Bc are the Tafel slopes of the anode and cathode, respectively, and Rp is the polarization resistance). The results showed that Rp decreases with increasing NaCl concentration, indicating that the bronze alloy exhibits better corrosion resistance in low-concentration NaCl solutions.

Table 2. Fitting parameters of polarization curves for bronze earrings in NaCl solutions.

Solution	E_{corr}/V	I_{corr}/A·cm^2	Ba/V^{-1}	Bc/V^{-1}	Rp/KΩ
1 mol/L	−0.196	9.051×10^{-7}	2.155	16.3	91.432
10^{-1} mol/L	−0.172	6.204×10^{-7}	3.353	5.816	149.044
10^{-2} mol/L	−0.145	5.795×10^{-7}	9.626	3.948	210.041
10^{-3} mol/L	−0.130	2.372×10^{-7}	10.519	3.776	509.386

Figure 18a illustrates the equivalent circuit diagram for bronze alloys in NaCl solution, modeling the electrochemical kinetics of the corrosion process on the bronze surface. In this circuit, Rs represents the solution resistance of the NaCl solution, indicating its conductivity; RCT represents the charge transfer resistance, inversely related to the corrosion rate of the bronze alloy surface, meaning a higher RCT indicates better corrosion resistance; and the constant phase element (CPE), denoted as Qdl, simulates the double-layer capacitance and its non-ideal behavior between the electrolyte and bronze surface. Fitting the electrochemical parameters of bronze earrings in various NaCl concentrations (Table 3) revealed that as NaCl concentration increases, Rs decreases, reflecting increased solution conductivity; meanwhile, changes in RCT suggest the increase in chloride ion concentration promotes the dissolution of copper elements in the alloy, accelerating the corrosion process of the bronze matrix.

Table 3. EIS fitting results of bronze earrings in NaCl solutions.

Solution	Rs/($\Omega \cdot cm^2$)	Qdl/($\Omega \cdot cm^2$)	n	Rct ($\Omega \cdot cm^2$)
1 mol/L	47.200	5.0805×10^{-8}	0.742	115,590
10^{-1} mol/L	71.570	3.5577×10^{-6}	0.737	123,000
10^{-2} mol/L	3347	1.8811×10^{-6}	0.687	230,200
10^{-3} mol/L	3454	1.9862×10^{-6}	0.705	319,460

Electrochemical Impedance Spectroscopy (EIS) analysis further revealed the corrosion behavior of bronze earrings in various concentrations of NaCl solutions. Figure 16a presents the Nyquist plots of the electrochemical impedance spectra for the bronze earrings in different NaCl solution concentrations. The Nyquist plots demonstrate that with an increase in NaCl concentration, the capacitive loop of the alloy decreases, suggesting a reduction in charge transfer resistance and an increase in corrosion rate. Analysis of the Bode plot in Figure 16b reveals that after soaking for 1 h in NaCl concentrations of 10^{-3}, 10^{-2}, 10^{-1}, and 1 mol/L, no oxide film formed on the surface of the bronze alloy, as indicated by the presence of a single time constant. Moreover, the Bode magnitude plot in Figure 16c indicates that the impedance values Z decrease progressively with increasing NaCl concentration, further confirming an increase in the corrosion rate.

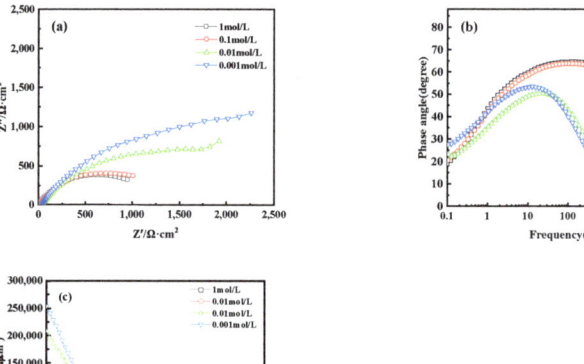

Figure 16. (a) Nyquist plot and (b,c) Bode of bronze earrings in NaCl solutions.

Figure 17 shows the electrochemical Nyquist and Bode plots of bronze alloy in 0.5 mol/L NaCl solution over different immersion times, including 1 h, 48 h, 36 h, and

30 days. Within 36 h, the capacitive arc of the alloy increases with immersion time, indicating the formation of a protective oxide film during the corrosion process. However, when the immersion time extends to 30 days, the capacitive arc decreases, indicating a reduction in corrosion rate. The Bode plot shows that at the initial stage of immersion (1 h), the alloy primarily exhibits a time constant in the high-frequency region, suggesting the activation reaction of the metal matrix without oxide film formation. Increasing the immersion time to 36 h, the Bode plot reveals two time constants: the high-frequency region corresponds to the reaction at the passive film interface, whereas the mid-to-low frequency region corresponds to the corrosion process of the metal matrix, indicating that the formation of the oxide film to some extent inhibits corrosion. As immersion time increases, the impedance magnitude of the alloy first increases and then decreases, reflecting the dynamic changes in the corrosion process. These results reveal the time-dependent corrosion behavior of bronze alloys in sodium chloride solutions and the formation and stability of protective oxide films. In addition, chlorides are a significant reason for the corrosion of iron cultural relics and the instability of iron cultural relics. Determining the types and properties of chlorides, and then selecting effective methods for control, dechlorination, or conversion treatment, is crucial in the protection process of iron cultural relics [39]. These studies contribute to understanding the long-term corrosion mechanisms of bronze alloys.

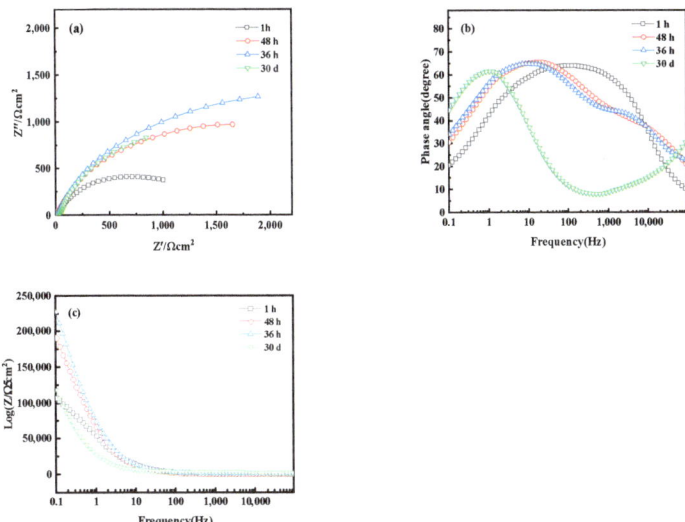

Figure 17. (**a**) Nyquist plot and (**b**,**c**) Bode of bronze earrings after immersion for different times in 0.5 mol/L NaCl solution.

From the Electrochemical Impedance Spectroscopy (EIS) tests and their fitting results conducted in NaCl solutions of different concentrations, it was observed that with an increase in chloride ion concentration in the NaCl solution, the charge transfer resistance (RCT) significantly decreases, indicating that the rise in chloride ion concentration notably accelerates the dissolution process of copper. The consistency among Open Circuit Potential (OCP), Tafel curves, and EIS test results further confirms the role of increased chloride ion concentration in promoting the corrosion rate of the bronze matrix. Particularly, in a 0.5 mol/L NaCl solution, the corrosion behavior of bronze was initially uniform, but after 3 days, a protective film formed on the alloy surface. By comprehensively analyzing these experimental results, we can detail the corrosion behavior of bronze in NaCl solution over time and thus draw the corresponding equivalent circuit diagram, as shown in Figure 18. The experimental outcomes indicate that at the initial stage of immersion in NaCl solution, a protective film has not yet formed on the alloy surface, but as the immersion time extends, a protective film gradually forms. Chloride ions play a decisive role in the corrosion mechanism of copper alloys, not only accelerating the corrosion process of bronze alloys but also significantly affecting the rate and morphology of corrosion through activating the reactivity of the bronze alloy surface. This process involves the facilitation of electrochemical reactions, in which chloride ions act as catalysts, promoting the formation of copper oxides and chlorides on the alloy surface [40,41]. Therefore, the prevention of bronze corrosion should focus on the stability of CuCl and the removal of chloride ions.

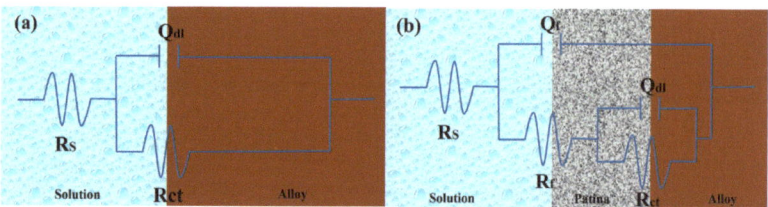

Figure 18. Equivalent circuits show the bronze earrings soaked in 0.5 mol/L NaCl initially (**a**) and for 3 days (**b**) [34].

4. Conclusions

This study thoroughly analyzes the structural characteristics of the bronze earrings unearthed from Tomb M47 at Xindianzi, as well as their electrochemical corrosion behavior in NaCl solutions of various concentrations. The comprehensive analysis of the bronze earrings revealed that their matrix is mainly composed of face-centered cubic Cu(Sn) solid solution with a certain amount of Pb particles. EBSD techniques showed that the earrings' grains exhibit specific orientations, namely, the (214)<12-1> texture, with more twins and equiaxed grains at the lower surface position, indicating annealing twins. Furthermore, this indicates that the bronze alloy undergoes thermal processing, resulting in specific (214)<12-1>textures on the earrings. Moreover, this suggests that the samples underwent thermal treatment during their manufacturing process, affecting their microstructure and corrosion behavior. Electrochemical testing results indicated that the corrosion rate of the bronze alloy significantly increases with the NaCl solution concentration from 10^{-3} to 1 mol/L, highlighting the key role of chloride ions in promoting the corrosion process of bronze alloys. Notably, immersion experiments in 0.5% NaCl solution showed that the loss of Cu in the alloy became more severe after 3 days of immersion, with the alloy surface forming an oxide film containing CuCl and Cu_2O. Research has shown that reducing the conversion of CuCl and removing chloride ions is an important factor for improving the stability of bronze. These results provide important scientific evidence for revealing the corrosion mechanism of bronze earrings in different environments.

Author Contributions: Conceptualization, O.T. and Z.S.; methodology, Z.S.; formal analysis, Z.S.; resources, Z.S.; data curation, Z.S.; writing—original draft preparation, Z.S.; writing—review and editing, O.T.; visualization, Z.S.; supervision, O.T.; funding acquisition, O.T. All authors have read and agreed to the published version of the manuscript.

Funding: This work is supported by the Natural Science Foundation of Inner Mongolia Autonomous Region (2022QN05002).

Institutional Review Board Statement: Not applicable.

Informed Consent Statement: Not applicable.

Data Availability Statement: The data that support the findings of this study are available from the corresponding author upon request (due to privacy).

Acknowledgments: The authors appreciate the help of Wei Wei, Chen Shuai and Tian Xiao for their advice on the data analysis. Special thanks to Cao Jian'en and Sun Jinsong from the Institute of

Archaeology of Inner Mongolia Autonomous Region for their strong support and guidance, which enabled the smooth development of this research paper.

Conflicts of Interest: The authors declare no conflicts of interest.

References

1. Wu, Z.; Zhu, Z.; Wu, R. Thermodynamic analysis of nanocrystalline solid solutions formation in copper-lead-tin ternary immiscible system during mechanical alloying. *Mater. Werkst.* **2021**, *52*, 1328–1337. [CrossRef]
2. Chang, T.; Herting, G.; Goidanich, S.; Amaya, J.S.; Arenas, M.; Le Bozec, N.; Jin, Y.; Leygraf, C.; Wallinder, I.O. The role of Sn on the long-term atmospheric corrosion of binary Cu-Sn bronze alloys in architecture. *Corros. Sci.* **2019**, *149*, 54–67. [CrossRef]
3. Dong, B.; Jie, J.; Peng, B.; Qu, J.; Wu, Z.; Wang, Q.; Zhou, J.; Liu, S.; Zou, Q.; Wang, T. Revealing the Relationship between Morphology of Pb-Rich Secondary Phases and Mechanical Properties of Laminated Cu–Pb–Sn/Steel Composite Through CALPHAD and FEA. *Metall. Mater. Trans. A* **2022**, *53*, 1462–1478. [CrossRef]
4. Cao, J.; Sun, J. 21st-Century Archaeological Discoveries of the Early Nomadic Cultural Remains–Centered in the Middle Section of the Great Wall Area in Inner Mongolia Autonomous Region. *J. Sib. Fed. Univ.* **2021**, *14*, 1121–1138.
5. Han, H.; Bai, W.; Liu, D.; Han, J. Comprehensive Research on the Layered Corrosion of Archaeological Copper Wares: A Case Study from Qinghai, China. *Stud. Conserv.* **2023**, 1–9.
6. Li, N.; Zhou, X.; Tian, X. Research progress on the application of corrosion inhibitors for bronze cultural relics. *Cult. Relics Apprais. Apprec.* **2023**, *24*, 22–27.
7. Liu, L.; Zhong, Q.; Jiang, L.; Li, P.; Xiao, L.; Gong, Y.; Zhu, Z.; Yang, J. Metallurgical and corrosion characterization of warring states period bronzes excavated from Pujiang, Chengdu, China. *Herit. Sci.* **2022**, *10*, 36. [CrossRef]
8. Privitera, A.; Corbascio, A.; Calcani, G.; Della Ventura, G.; Ricci, M.A.; Sodo, A. Raman approach to the forensic study of bronze patinas. *J. Archaeol. Sci. Rep.* **2021**, *39*, 103115. [CrossRef]
9. Li, H.; Zuo, Z.; Cui, J.; Tian, J.; Yang, Y.; Yi, L.; Zhou, Z.; Fan, J. Copper alloy production in the Warring States period (475-221 BCE) of the Shu state: A metallurgical study on copper alloy objects of the Baishoulu cemetery in Chengdu, China. *Herit. Sci.* **2020**, *8*, 67. [CrossRef]
10. Mu, D.; Luo, W.; Song, G.; Qiao, B.; Wang, F. The features as a county of Chu State: Chemical and metallurgical characteristics of the bronze artifacts from the Bayilu site. *Archaeol. Anthropol. Sci.* **2019**, *11*, 1123–1129. [CrossRef]
11. Li, B.; Jiang, X.; Wu, R.; Wei, B.; Hu, T.; Pan, C. Formation of black patina on an ancient Chinese bronze sword of the Warring States Period. *Appl. Surf. Sci.* **2018**, *455*, 724–728. [CrossRef]
12. Armetta, F.; Saladino, M.L.; Scherillo, A.; Caponetti, E. Microstructure and phase composition of bronze Montefortino helmets discovered Mediterranean seabed to explain an unusual corrosion. *Sci. Rep.* **2021**, *11*, 23022. [CrossRef] [PubMed]
13. Liu, Y.; Yu, G.; Cao, G.; Wang, C.; Wang, Z. Characterization of corrosion products formed on tin-bronze after 29 years of exposure to Shenyang, China. *J. Mater. Res. Technol. JmrT* **2023**, *23*, 5270–5279. [CrossRef]
14. Manti, P.; Watkinson, D. Corrosion phenomena and patina on archaeological low-tin wrought bronzes: New data. *J. Cult. Herit.* **2022**, *55*, 158–170. [CrossRef]
15. Wang, X.; Zhou, H.; Song, J.; Fan, Z.; Zhang, L.; Shi, J.; Chen, J.; Xiao, K. Mechanism of corrosion behavior between Pb-rich phase and Cu-rich structure of high Sn–Pb bronze alloy in neutral salt spray environment. *J. Mater. Res. Technol.* **2024**, *29*, 881–896. [CrossRef]
16. Petitmangin, A.; Guillot, I.; Chabas, A.; Nowak, S.; Saheb, M.; Alfaro, S.C.; Blanc, C.; Fourdrin, C.; Ausset, P. The complex atmospheric corrosion of α/δ bronze bells in a marine environment. *J. Cult. Herit.* **2021**, *52*, 153–163. [CrossRef]
17. Piccardo, P.; Mödlinger, M.; Ghiara, G.; Campodonico, S.; Bongiorno, V. Investigation on a "tentacle-like" corrosion feature on Bronze Age tin-bronze. *Appl. Phys. A* **2013**, *113*, 1039–1047. [CrossRef]
18. Hu, Y.; Wei, Y.; Li, L.; Zhang, J.; Chen, J. Same site, different corrosion phenomena caused by chloride: The effect of the archaeological context on bronzes from Sujialong Cemetery, China. *J. Cult. Herit.* **2021**, *52*, 23–30. [CrossRef]
19. Wallinder, I.O.; Zhang, X.; Goidanich, S.; Le Bozec, N.; Herting, G.; Leygraf, C. Corrosion and runoff rates of Cu and three Cu-alloys in marine environments with increasing chloride deposition rate. *Sci. Total Environ.* **2014**, *472*, 681–694. [CrossRef]
20. Liang, Z.; Jiang, K.; Zhang, T. Corrosion behaviour of lead bronze from the Western Zhou Dynasty in an archaeological-soil medium. *Corros. Sci.* **2021**, *191*, 109721. [CrossRef]
21. Wu, J.; Wang, J. The effects of UV and visible light on the corrosion of bronze covered with an oxide film in aqueous solution. *Corros. Sci.* **2019**, *154*, 144–158. [CrossRef]
22. Dermaj, A.; Hajjaji, N.; Joiret, S.; Rahmouni, K.; Srhiri, A.; Takenouti, H.; Vivier, V. Electrochemical and spectroscopic evidences of corrosion inhibition of bronze by a triazole derivative. *Electrochim. Acta* **2007**, *52*, 4654–4662. [CrossRef]
23. Walker, R. Aqueous Corrosion of Tin-Bronze and Inhibition by Benzotriazole. *Corrosion* **2000**, *56*, 1211–1219. [CrossRef]
24. Kwon, H. Corrosion Behaviors of Artificial Chloride Patina for Studying Bronze Sculpture Corrosion in Marine Environments. *Coatings* **2023**, *13*, 1630. [CrossRef]
25. Di Carlo, G.; Giuliani, C. Artificial patina formation onto copper-based alloys: Chloride and sulphate induced corrosion processes. *Appl. Surf. Sci.* **2017**, *421*, 120–127. [CrossRef]

26. Grayburn, R. Tracking the progression of bronze disease–a synchrotron X-ray diffraction study of nantokite hydrolysis. *Corros. Sci.* **2015**, *91*, 220–223. [CrossRef]
27. Cao, J.; Sun, J.; Hu, X. Excavation of Xindianzi Cemetery in Helingeer County, Inner Mongolia. *Archaeology* **2009**, *16*, 195–205.
28. Han, R.; Sun, S.; Li, X.; Qian, W. Microstructure of Ancient Chinese Copper Artefacts. *J. Univ. Scieice Technol. Beijing* **2002**, *24*, 219–230.
29. Randle, V. Applications of electron backscatter diffraction to materials science: Status in 2009. *J. Mater. Sci.* **2009**, *44*, 4211–4218. [CrossRef]
30. Yang, P. Electron backscatter diffraction technology, geometric crystallography, and materials science. *J. Chin. Electron Microsc. Soc.* **2008**, *27*, 7.
31. Garbacz-Klempka, A. Bronze Age Raw Material Hoard from Greater Poland: Archaeometallurgical Study Based on Material Research, Thermodynamic Analysis, and Experiments. *Materials* **2023**, *17*, 230. [CrossRef]
32. Saraf, L. Kernel average misorientation confidence index correlation from FIB sliced Ni-Fe-Cr alloy surface. *Microsc. Microanal.* **2011**, *17*, 424. [CrossRef]
33. Ahmed, N.; Medhat, E.H. Microstructure and Mechanical Behavior of Hot Pressed Cu-Sn Powder Alloys. *Adv. Mater. Sci. Eng.* **2016**, *2016*, 9796169.
34. Song, Z.; Tegus, O. The Corrosion Properties of Bronze Alloys in NaCl Solutions. *Materials* **2023**, *16*, 5144. [CrossRef] [PubMed]
35. Wang, T.; Wang, J.; Wu, Y. The inhibition effect and mechanism of L-cysteine on the corrosion of bronze covered with a CuCl patina. *Corros. Sci.* **2015**, *97*, 89–99. [CrossRef]
36. Wang, J.; Xu, C.; Lv, G. Formation processes of CuCl and regenerated Cu crystals on bronze surfaces in neutral and acidic media. *Appl. Surf. Sci.* **2006**, *252*, 6294–6303. [CrossRef]
37. Yang, X.; Wu, W.; Chen, K. Investigation on the electrochemical evolution of the Cu-sn-Pb ternary alloy covered with CuCl in a simulated atmospheric environment. *J. Electroanal. Chem.* **2022**, *921*, 116636. [CrossRef]
38. Chiavari, C.; Bernardi, E.; Martini, C.; Passarini, F.; Ospitali, F.; Robbiola, L. The atmospheric corrosion of quaternary bronzes: The action of stagnant rain water. *Corros. Sci.* **2010**, *52*, 3002–3010. [CrossRef]
39. Zhang, R. Formation and Hazards of Corrosion Product β-FeOOH on Iron Cultural Relics. *Corros. Prot.* **2021**, *42*, 1–11.
40. Chang, T.; Herting, G.; Jin, Y.; Leygraf, C.; Wallinder, I.O. The golden alloy Cu5Zn5Al1Sn: Patina evolution in chloride-containing atmospheres. *Corros. Sci.* **2018**, *133*, 190–203. [CrossRef]
41. Wu, L.; Ma, A.; Zhang, L.; Zheng, Y. Intergranular erosion corrosion of pure copper tube in flowing NaCl solution. *Corros. Sci.* **2022**, *201*, 110304. [CrossRef]

 materials

 MDPI

Review

Hydrogen Impact: A Review on Diffusibility, Embrittlement Mechanisms, and Characterization

Qidong Li [1], Hesamedin Ghadiani [1], Vahid Jalilvand [1], Tahrim Alam [2], Zoheir Farhat [1,*] and Md. Aminul Islam [3]

1   Department of Mechanical Engineering, Dalhousie University, Halifax, NS B3H 4R2, Canada; qidong98@student.ubc.ca (Q.L.); hs460468@dal.ca (H.G.)
2   Enbridge Gas Inc., Ottawa, ON K1K 2C7, Canada; tahrim.alam@enbridge.com
3   Mining Wear and Corrosion Laboratory, National Research Council Canada, Vancouver, BC V6T 1W5, Canada; mdaminul.islam@nrc-cnrc.gc.ca
*   Correspondence: zoheir.farhat@dal.ca

Abstract: Hydrogen embrittlement (HE) is a broadly recognized phenomenon in metallic materials. If not well understood and managed, HE may lead to catastrophic environmental failures in vessels containing hydrogen, such as pipelines and storage tanks. HE can affect the mechanical properties of materials such as ductility, toughness, and strength, mainly through the interaction between metal defects and hydrogen. Various phenomena such as hydrogen adsorption, hydrogen diffusion, and hydrogen interactions with intrinsic trapping sites like dislocations, voids, grain boundaries, and oxide/matrix interfaces are involved in this process. It is important to understand HE mechanisms to develop effective hydrogen resistant strategies. Tensile, double cantilever beam, bent beam, and fatigue tests are among the most common techniques employed to study HE. This article reviews hydrogen diffusion behavior, mechanisms, and characterization techniques.

Keywords: hydrogen embrittlement; hydrogen diffusion; damage mechanisms; mechanical properties

Citation: Li, Q.; Ghadiani, H.; Jalilvand, V.; Alam, T.; Farhat, Z.; Islam, M.A. Hydrogen Impact: A Review on Diffusibility, Embrittlement Mechanisms, and Characterization. *Materials* **2024**, *17*, 965. https://doi.org/10.3390/ma17040965

Academic Editor: Ming Liu

Received: 13 December 2023
Revised: 20 January 2024
Accepted: 16 February 2024
Published: 19 February 2024

1. Introduction

Hydrogen embrittlement (HE) corresponds to the abrupt degradation of mechanical properties of materials in the presence of hydrogen. HE failure in metals was first recognized by Johnson in 1875 [1] and has been observed in many metallic materials such as steels, aluminum alloys, titanium alloys, and superalloys [2–8]. This problem in metals has been of great concern in various industries including chemical, petrochemical, power, and marine industries. Hydrogen embrittlement can lead to catastrophic failure in oil and gas pipelines as a result of the presence of sour gas or as a result of the blending of natural gas with hydrogen. It has been generally established that hydrogen may reduce the macroscopic and microscopic tensile strength [9–14], fatigue strength [15–17], and fracture toughness [18–22], while its effect on the rate of fatigue crack growth is still debated, depending on the stress ratio level or frequency [23]. Although extensive studies on the hydrogen embrittlement of metals have been carried out, many issues are yet to be understood. The phenomenon of hydrogen damage is a challenging basic research problem. One main reason for the damage caused by hydrogen in metals and alloys is the extremely small size of the hydrogen atom, which makes it move very fast in the metallic lattice. It is therefore not surprising that over the years, a considerable research effort has been directed toward obtaining an understanding of this phenomenon.

Hydrogen-induced failures arise because cracks are able to grow to critical dimensions, with the initial stress intensity level increasing to the point under the requirement that $K = K_{IC}$, where K is stress intensity factor and K_{IC} is the critical stress intensity factor. Such crack extension can occur through a number of processes. Subcritical flaw growth mechanisms involving a cooperative interaction between a stress and the environment, leading to hydrogen embrittlement, and the final failure typically occurs after a period of time, rather than when exposure begins. This damage mechanism affects many important alloy systems, most notably high-strength steels. When atomic hydrogen is introduced into an alloy, the toughness and ductility can be reduced dramatically, and subcritical crack growth can occur. Body-centered cubic and hexagonal close-packed metals are most susceptible to hydrogen embrittlement. Face-centered cubic metals are not generally susceptible to hydrogen embrittlement. Hydrogen has a very high mobility in the BCC lattice of carbon and low-alloy steels [24].

Recently, there has been a renewed interest in the hydrogen embrittlement of metals as a result of the ever-increasing demand from world governments for cleaner energy. Global gas utility companies are exploring ways to blend natural gas with hydrogen as a cleaner energy source. However, the effect of hydrogen on existing infrastructure, including existing and new pipe networks, needs to be assessed prior to injecting hydrogen into the system, and the maximum hydrogen addition for safe operation needs to be determined. This is especially urgent and essential for older distribution pipeline networks, as pipe steels are known to be susceptible to hydrogen embrittlement, which may lead to catastrophic failures.

In this review, three aspects of the HE behavior of metals are discussed: hydrogen diffusion behavior, hydrogen embrittlement mechanisms, and HE characterization methods.

2. Entry of Gaseous and Aqueous Hydrogen into Metals

The exploration in this section is centered around the diffusion process of hydrogen from aqueous and gaseous media into metals. Metal surfaces exhibit a tendency of adsorption. This tendency stems from the fact that while the metal atoms inside the metal are in equilibrium with each other, the metal atoms located on the surface of the metal are not, leading to the manifestation of surface energy on the surface of the metal [25,26]. According to the second law of thermodynamics, the energy of all systems is inherently tilted toward lower values, so that on the surface of metals, hydrogen-containing substances tend to be adsorbed to reduce the overall energy of the system [27].

The primary gateway for hydrogen into metals is through surface adsorption, a process where certain solids selectively concentrate particular substances from a solution (gas or liquid) onto their surfaces [28]. Hydrogen diffusion into metals involves three principal mechanisms. These are physisorption (physical adsorption), chemisorption (chemical adsorption), and hydrogen uptake. Physisorption is typically created by van der Waals forces between hydrogen and the metal surface. This adsorption process is reversible and is primarily influenced by the conditions of the environment, such as pressure and temperature. The second mechanism is chemisorption, which involves the formation of a typically covalent chemical bond between molecules or atoms and is generally irreversible [29,30]. The final mechanism is hydrogen uptake, where hydrogen diffuses into the metal through desorption, leading to its incorporation into the metal lattice. Figure 1 provides an illustration of this concept.

The aqueous hydrogen diffusion process can be primarily depicted by the Volmer–Tafel–Heyrovsky reaction mechanism, which includes several significant stages [31–34]:

Electrochemical reduction: the initial stage of the reaction involves the reduction of hydronium ions (H_3O^+) by gaining electrons to produce water and atomic hydrogen. The reaction is represented as follows:

$$H_3O^+ + e^- \rightarrow H_2O + H \tag{R1}$$

Volmer reaction (chemisorption): Subsequently, the atomic hydrogen produced interacts with the metal surface, resulting in chemisorption. This process, also known as the Volmer reaction, generally occurs when the overpotential is relatively low due to a limited surface coverage of hydrogen [35].

$$H + M \rightarrow MH_{ad} \tag{R2}$$

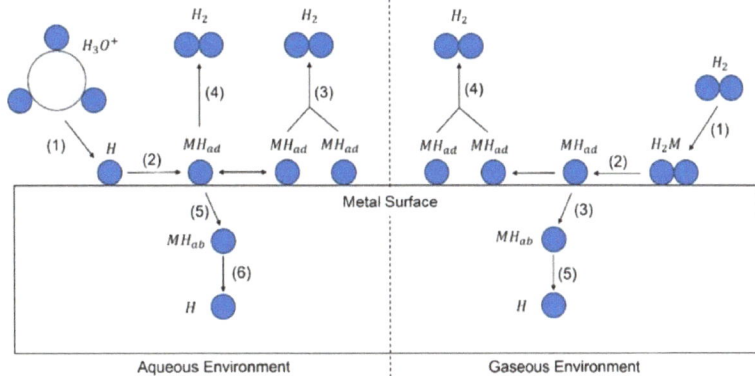

Figure 1. Possible ways of hydrogen transport and interaction in steel.

Tafel reaction: adsorbed hydrogen (MH_{ad}) can recombine and create molecular hydrogen.

$$2MH_{ad} \rightarrow 2M + H_2 \tag{R3}$$

Heyrovsky reaction: When overpotential increases due to substantial hydrogen presence on the metal surface, the Heyrovsky reaction prevails [35]. In this scenario, most atomic hydrogen generates gaseous hydrogen and leaves the metal surface. This reaction can be expressed as follows:

$$MH_{ad} + H_2O + e^- \rightarrow M + H_2 + OH^- \tag{R4}$$

Hydrogen absorption: while the reactions expressed in (3) and (4) are taking place, the absorption process is also in progress, where the atomic hydrogen is absorbed into the inner surface of the metal, represented by the following:

$$MH_{ad} \rightarrow MH_{ab} \tag{R5}$$

Desorption and dissolution: The final stage involves the desorption of the absorbed atomic hydrogen inside the metal. It is worth noting that this step is less about desorption and more about the dissolution of hydrogen into the metal lattice. This process can induce various microstructural changes and potentially lead to hydrogen embrittlement.

$$MH_{ab} \rightarrow M + H \tag{R6}$$

The process of hydrogen diffusion in gaseous media parallels that of aqueous hydrogen diffusion, and it unfolds as follows [30,36,37]:

Physisorption: The initial stage of the process begins when gaseous hydrogen comes into contact with the metal surface, resulting in physisorption. During this phase, no chemical bonds are formed between the hydrogen and metal, making it a weaker form of adsorption. Hence, the adsorption process is reversible.

$$H_2 + M \rightarrow H_2M \tag{R7}$$

Chemisorption: following physisorption, the H_2M and the metal surface form chemical bonds, marking an irreversible process.

$$H_2M + M \rightarrow 2MH_{ad} \tag{R8}$$

Absorption: subsequent to the chemisorption phase, the adsorbed hydrogen (MH_{ad}) is absorbed into the subsurface of the metal, resulting in the formation of absorbed hydrogen (MH_{ab}).

$$MH_{ad} \leftrightarrow MH_{ab} \tag{R9}$$

Recombination: simultaneously, the adsorbed hydrogen can also recombine to generate molecular hydrogen (H_2).

$$2MH_{ad} \rightarrow 2M + H_2 \tag{R10}$$

Desorption and dissolution: the absorbed hydrogen (MH_{ab}) undergoes a desorption process, transforming into atomic hydrogen dissolved within the metal, as given in R6.

The above only describes how hydrogen enters the metal in the aqueous and gaseous environments; knowledge about the reaction kinetics is excluded in this paper, and readers who are interested in this knowledge can refer to the work of Popov et al. [34].

3. Mechanisms of Hydrogen Diffusion

Due to their very small atomic size, hydrogen atoms have a greater tendency to diffuse or dissolve into steel compared to other atoms. Two primary mechanisms that govern hydrogen diffusion in metals are interstitial diffusion and quantum mechanical tunnel diffusion [38]. Given the high diffusivity of hydrogen in steel and the significant alteration it can cause in the metal's mechanical properties, it is critical to understand the mechanisms of hydrogen diffusion in metals in detail. Firstly, the process of interstitial diffusion of hydrogen is examined.

For the metals considered in this section, it is vital to emphasize that they are devoid of defects, meaning hydrogen atoms are not held in hydrogen traps. The effect of hydrogen traps on hydrogen diffusion is elaborated upon in Section 5.1 of this article. As shown in Figure 2, in the three main crystal structures—face-centered cubic (fcc), hexagonal close-packed (hcp), and body-centered cubic (bcc)—there are two types of interstitial sites, octahedral (O) and tetrahedral (T), that can house hydrogen. The absorption capacity of O sites and T sites varies and primarily hinges on the size of their interstitial gaps. An approximation based on the diameter of hydrogen atoms and the size of these gaps suggests that O sites are primarily populated in fcc and hcp structures, while T sites are more frequent in bcc structures. Elevated mobilities are particularly prevalent in body-centered cubic (bcc) metals due to the presence of adjacent interstitial sites that are in close proximity [38]. As hydrogen atoms dissolve in the metal, these sites progressively fill, displacing metal atoms and leading to elastic distortion and changes in crystal entropy. This effect translates to an alteration in enthalpy. The system's energy (ΔG) is at its minimum when hydrogen atoms reside in the interstitial sites and peaks when the hydrogen atoms are positioned between metal atoms. This peak value can be represented in terms of this enthalpy change. Consequently, the diffusion of hydrogen atoms occurs only when they are thermally activated and the energy that they possess exceeds ΔG.

Nonetheless, quantum mechanics suggests that diffusion can still occur when the energy of a hydrogen atom is lower than this energy barrier, a phenomenon known as quantum mechanical tunnel transport [39]. Due to its diminutive size, treating the hydrogen atom solely as a classical particle is inaccurate [40]. This topic has been the subject of extensive research, with scholars finding that this quantum effect becomes less significant at higher temperatures, with normal diffusion mechanisms taking precedence. At these temperatures, hydrogen atoms must surpass the energy barriers to diffuse. However, the impact of quantum mechanical tunnel transport becomes more pronounced at lower temperatures. Some researchers have postulated this temperature threshold to be around 250 K [41]. This suggests that quantum mechanical tunnel transport is not thermally activated, leading to the possibility that diffusion models based on the Arrhenius equation might be incorrect [42].

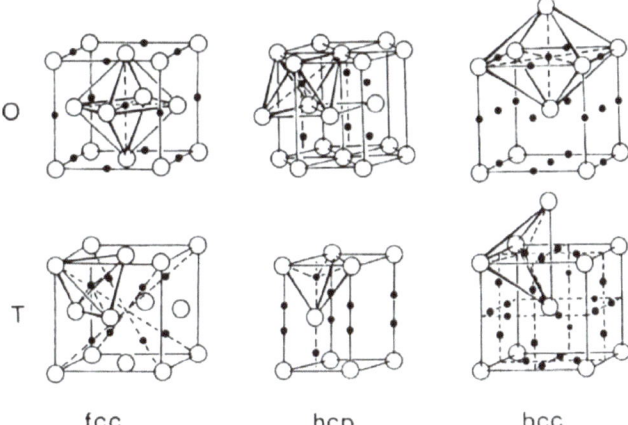

Figure 2. Interstitial sites (octahedral (O) sites and tetrahedral (T) sites) in fcc, hcp, and bcc lattices [43].

4. Characterization Techniques for the Measurement of Hydrogen Diffusion

4.1. Hydrogen Microprint Technique

The hydrogen microprint technique (HMT), a method commonly used for visualizing hydrogen diffusion in metals, was developed by Ovejero-García [44]. The foundational principle of this technique involves generating silver (Ag) microparticles via the reduction of silver bromide ($AgBr$) on the surface of a coated specimen. These silver micropartiscles can then be observed using scanning electron microscopy (SEM), providing insights into the behavior of hydrogen diffusion. The schematic and underlying principle of the experimental apparatus are illustrated in Figure 3.

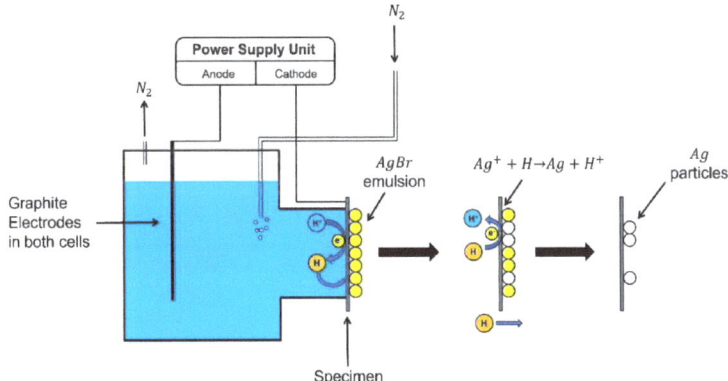

Figure 3. Schematic of the HMT test setup [45].

The implementation of the hydrogen microprint technique (HMT) initiates with specific preparatory steps: Initially, one surface of the specimen is ground to achieve a 600-grit finish, while the side exposed to air is polished to a 1 μm scale using diamond paste. If examination of the steel's microstructure is intended, the polished side of the specimen requires etching for a duration of 30 s using a 2% nital solution. Following this, a layer of silver bromide ($AgBr$) nuclear emulsion is applied to the polished surfaces, which is then left to dry for a period of 20 min or longer. This emulsion comprises 5 g of $AgBr$ powder and a 10 mL solution of 1.4 M sodium nitrite ($NaNO_2$). The employment of sodium nitrite serves a specific purpose, which is to mitigate corrosion of the etched side during the experiment.

The specimen is positioned in the apparatus as shown in Figure 3, with the side devoid of $AgBr$ exposed to the electrolyte for charging. This initiates a process analogous to the reaction occurring on the charging side during permeation tests, where hydrogen atoms diffuse to the surface of the specimen and permeate through it. Hydrogen atoms diffusing from the opposite side then interact with silver ions (Ag^+) according to the reaction $Ag^+ + H \rightarrow Ag + H^+$. This results in the reduction of Ag to visible silver particles, as depicted in Figure 3. Following a charging duration of 80 min or longer, the specimen is retrieved and immersed for 1 min in a photographic fixing solution composed of 0.6 M sodium thiosulfate $(Na_2S_2O_3)$ and 1.4 M sodium nitrite $(NaNO_2)$. This crucial step serves to eliminate any unreacted silver bromide $(AgBr)$ crystals. After a final rinse with deionized water and dehydration using a dryer, the resultant silver particles become visible. These particles, which appear as white spheres on the microstructure, are then ready for observation. Consequently, the diffusion of hydrogen to the steel surface can be effectively visualized using scanning electron microscopy (SEM) [44–47]. Jack et al.'s [45] SEM observations are depicted in Figure 4. Their experiments, conducted on two samples simultaneously, reveal distinct white spheres in the microstructure, corresponding to the reduced Ag particles. These particles signify hydrogen permeation through grain boundaries and phase interfaces. Figure 4c,f potentially exhibit matrix distortion and high local misorientation, commonly observed around inclusions.

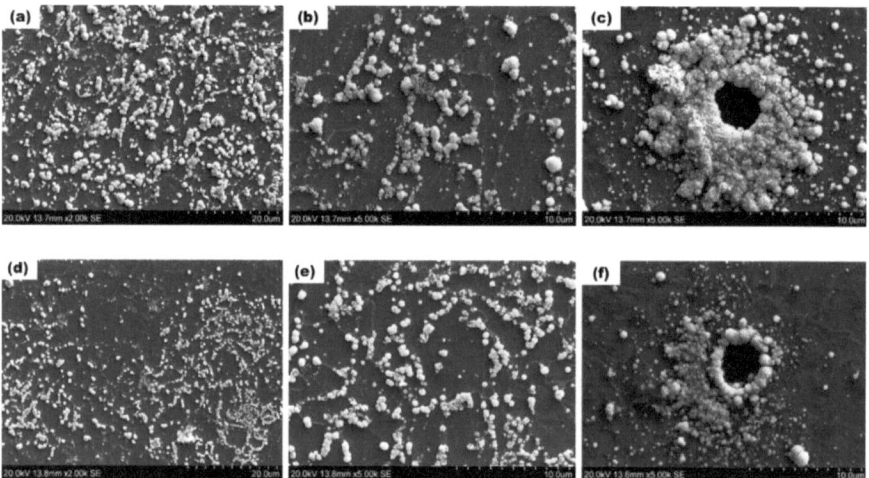

Figure 4. SEM images of the HMT test [45].

4.2. Hydrogen Permeation Tests

4.2.1. Electrochemical Permeation

While the hydrogen microprint technique (HMT) allows for the visualization of hydrogen distribution within a steel microstructure, it falls short in quantifying the hydrogen permeation process. To address this shortcoming, many researchers have adopted the methodology pioneered by Devanathan and Stachurski [48]. A representation of their experimental setup is illustrated in Figure 5. Below is a step-by-step breakdown of the procedure.

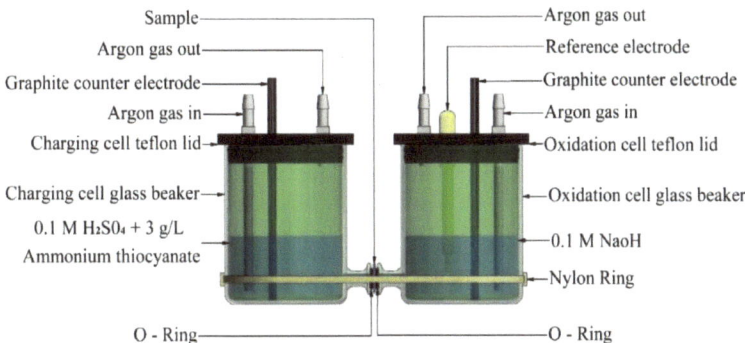

Figure 5. Schematic of electrochemical permeation test [46].

Initially, the test specimen is polished and cleaned. Subsequently, materials such as palladium or other suitable alternatives (like nickel) are used to coat the side of the specimen exposed to the oxidation cell, thereby preventing oxidation and reaching a steady permeation state in a shorter time, and improving the stability of current measurement [49,50]. Once the specimen is prepared, it is clamped between the charging and oxidation cells.

For the charging cell, the electrolyte can be either acidic or basic. Widely used electrolytes include 0.1 M H_2SO_4 or 0.1 M $NaOH$, but the pH of the electrolyte can influence the experimental results [51]. Additionally, a recombination poison is introduced to the charging side to prevent hydrogen atoms in the charging cell from recombining into H_2. Ammonium thiocyanate (NH_4SCN) is commonly used, with its concentration affecting the permeation current [52]. Traditionally, a concentration of 3 g/L is used [46,53]. Lu et al. have also incorporated 0.2 g/L thiourea (CH_4N_2S) [32]. Apart from the aforementioned conditions, there have been instances, such as the work of Fallahmohammadi et al., where a solution containing 0.2 M CH_3COOH and 0.4 M CH_3COONa (pH = 4.2) was employed [54]. The power supply unit (PSU) has its anode connected to a graphite electrode or any other inert electrode such as platinum (Pt) and its cathode connected to the specimen. As for charging conditions, both galvanostatic and potentiostatic power supply units (PSUs) can be utilized. The chosen currents and voltages will dictate the quantity of hydrogen atoms generated. Evidently, a higher current or potential result in an increased production of hydrogen, which leads to a higher steady-state current, as illustrated in Figure 6. Additionally, using potentiostatic charging allows for the analysis of the equivalence between gaseous hydrogen permeation and electrochemical hydrogen permeation. Further discussions on this topic can be found in Section 4.2.3 of this article.

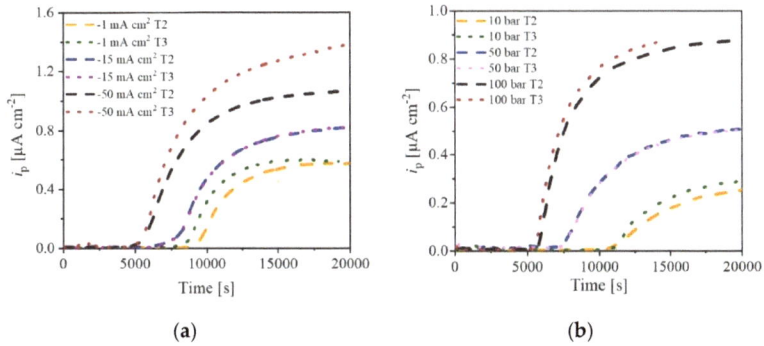

Figure 6. (**a**) Electrochemical hydrogen permeation curves of different charging currents; (**b**) gaseous hydrogen permeation curves of different charging hydrogen pressure [55].

In the oxidation cell, studies typically use 0.1 M $NaOH$ as the electrolyte. This cell comprises a three-electrode system with an electrochemical working station. The test specimen acts as the working electrode, a saturated calomel electrode (SCE) serves as the reference electrode, and an inert electrode functions as the counter electrode. The primary role of the electrochemical working station is to monitor the oxidation current.

After assembly, the electrolyte is first introduced into the oxidation cell, and an inert gas is passed through. The charging cell is kept idle until the measured current decreases below 0.1 $\mu A/cm^2$, which signifies the background current [50]. This current is later subtracted from the oxidation current during the analysis to obtain the actual permeation current. This step aims to remove any residual hydrogen present in the metal from prior processing. Once the process is finalized, the electrolyte mixed with recombination poison is added into the charging cell and the power supply is activated. Depending on the pH of the electrolyte, acidic or alkaline, H_2O or H^+ is reduced. This process involves these molecules gaining electrons to produce hydrogen atoms. When these atoms permeate through the specimen and reach the surface of the oxidation side, they are oxidized, generating an oxidation current measured by the electrochemical working station. The experiment is typically conducted at room temperature.

4.2.2. Gaseous Permeation

While electrochemical hydrogen charging predominates due to its relative simplicity and safety, gaseous hydrogen charging presents its own set of unique benefits and challenges [2,49,55]. The primary challenges associated with gas-phase charging include the following:

1. The need for a dedicated gas line system, with specialized valves and pressure gauges.
2. An intricate, tightly sealed arrangement to prevent any potential hydrogen leakage.
3. All components must be resilient against hydrogen embrittlement and able to withstand test pressures.
4. Presence of hydrogen in the charging cell prior to the completion of pressurization, can lead to inaccurate measurements.
5. For experiments across a spectrum of temperatures, the test apparatus must be robust enough to withstand such temperature fluctuations.

One of the main advantages of the gaseous hydrogen permeation test is its capability to mimic real-world environments. The experimental outcomes are often more representative of practical applications. Figure 7 showcases a typical setup for this method, which bears significant resemblance to the electrochemical permeation test.

In this experiment, an inert gas, such as nitrogen (N_2), is used to purge the test chamber prior to testing. This is followed by a background current-obtaining procedure, after which hydrogen gas is introduced to generate the permeation current. The pressure at which hydrogen is introduced directly impacts the quantity of hydrogen atoms formed on the charging side. Consequently, a higher pressure leads to a larger resultant permeation current. As depicted in Figure 6, this procedure is similar to electrochemical hydrogen charging. The hydrogen gas introduced can be varied in terms of its concentration or pressure. It can also be mixed with other gases to yield different blend concentrations.

For instance, Zhang et al. performed a permeation test using different concentrations of hydrogen and nitrogen. They also performed fracture toughness and fatigue tests using a similar blend ratio. It was found that 3% hydrogen in the blended gas can cause 67.7% reduction in fatigue life [56]. Exploring an alternative approach, Zhao et al. created a simulated coal gas by combining nitrogen, CO_2, and hydrogen. Using this simulated gas, they conducted tests to measure hydrogen permeation and explored slow strain rate tension tests (SSRT). Their investigation identified the coarse-grained heat-affected zone (CGHAZ) of X80 as the region with the highest hydrogen diffusion velocity. This increased velocity led to a rapid accumulation of hydrogen near the crack's front, enhancing the material's susceptibility to hydrogen embrittlement [49].

In conclusion, although both of these procedures are different, they also share core similarities, with both ultimately producing a permeation current. Depending on the test procedure used, it is possible to gain profound insights into the hydrogen permeation behavior.

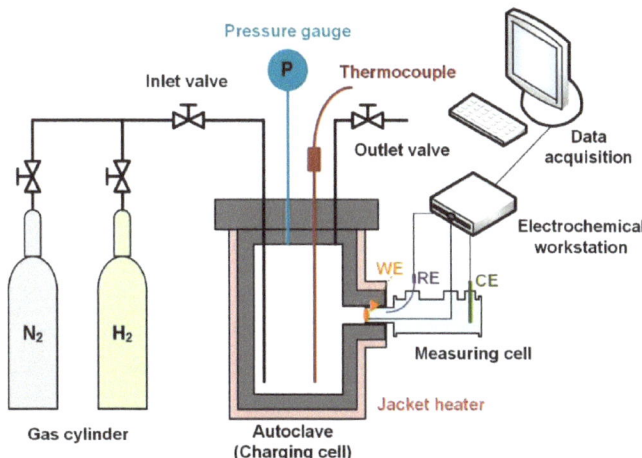

Figure 7. Schematic of gaseous hydrogen charging chamber [57].

4.2.3. Permeation Test Results Analysis

Whether in water or gaseous environments, hydrogen diffuses due to the concentration gradient. During this process, hydrogen atoms migrate from areas of high concentration to low concentration. Understanding the concentration gradient is essential for the study of hydrogen diffusion dynamics. The principles of this model can be formulated via Fick's First and Second Laws. Fick's First Law describes the diffusion flux, i.e., the relationship between the rate of substance flow and the concentration gradient, which is expressed by Equation (1):

$$J = -D\frac{\partial c}{\partial x} \tag{1}$$

where J represents the diffusion flux, D represents the diffusion coefficient, and $\partial c/\partial x$ is the concentration gradient. The negative sign in Equation (1) indicates that the diffusion flux always flows from regions of higher concentration to lower concentration, in opposition to the direction of the increasing concentration gradient.

Fick's Second Law, on the other hand, accounts for the changes in concentration over time and across the diffusion distance. It is expressed by Equation (2), where $\partial c/\partial t$ represents the change in concentration over time, D is the diffusion coefficient, and $\partial^2 c/\partial x^2$ is the second derivative of the concentration with respect to the distance, capturing the spatial variation in the concentration gradient.

$$\frac{\partial c}{\partial t} = D\frac{\partial^2 c}{\partial x^2} \tag{2}$$

Figure 8 shows a schematic of permeation test results after the first and second cycle. Hydrogen concentration on both sides of the specimen is zero prior to the experiment. After achieving a steady state (represented by the plateau), the hydrogen concentration on the entry side (previously referred to as the charging side) is denoted as c_{0R}, while on the exit side (previously referred to as the oxidation side), it is zero. Equations (3)–(8) can be derived by applying Fick's First and Second Laws [49,54,58]:

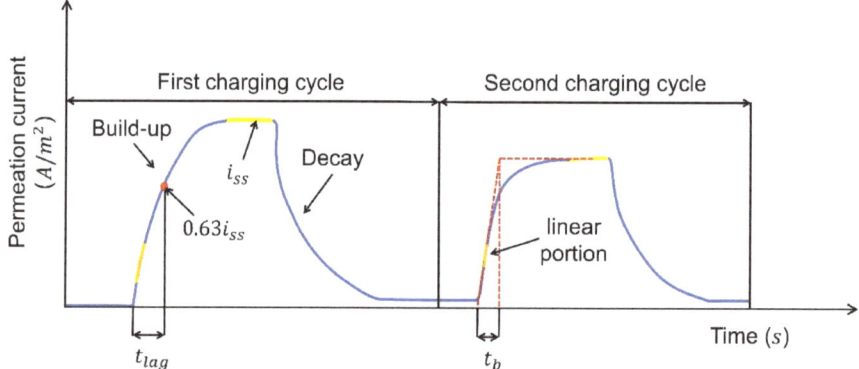

Figure 8. Typical hydrogen permeation curve during first and second charging cycles [46].

Time lag method for D_{eff} calculation:

$$D_{eff} = \frac{L^2}{6t_{lag}} \tag{3}$$

Breakthrough time method for D_{eff} calculation:

$$D_{eff} = \frac{L^2}{15.3t_b} \tag{4}$$

Fourier method for D_{eff} calculation:

$$\frac{i_t}{i_{ss}} = 1 - 2\exp\left(-\frac{\pi^2 D_{eff} t}{L^2}\right) \tag{5}$$

Laplace method for D_{eff} calculation:

$$\frac{i_t}{i_{ss}} = \frac{2}{\sqrt{\pi}} \frac{L}{\sqrt{D_{eff} t}} \exp\left(-\frac{L^2}{4D_{eff} t}\right) \tag{6}$$

Hydrogen permeation flux in the steady state:

$$J_{SS} = \frac{i_{ss}/A}{F} = \frac{D_{eff} C_{0R}}{L} \tag{7}$$

Subsurface hydrogen concentration:

$$c_{0R} = \frac{J_{SS} L}{D_{eff}} \tag{8}$$

In Equations (3)–(8), i_t (μA) is the transient current and i_{ss} (μA) denotes the steady-state current. t_{lag} (s) is the time elapsed from the beginning of the experiment until the ratio of transient-to-steady-state current (i_t/i_{ss}) reaches 0.63. t_b (s) is the time calculated by extrapolating the linear portion of the rising hydrogen permeation current transient. All of these parameters are marked in Figure 8. D_{eff} ($m^2 \cdot s^{-1}$) refers to the effective diffusion coefficient. J_{SS} ($mol \cdot m^{-2} s^{-1}$) corresponds to the permeation flux in the steady state, which reflects the rate of hydrogen permeation through the sample under steady conditions. C_{0R} ($mol \cdot m^{-3}$) is the summation of the subsurface concentration of hydrogen in interstitial lattice sites and reversible trap sites on the charging side of the sample [59].

Apart from calculating D_{eff} using Equations (3) and (4), this coefficient value can alternatively be estimated using Equation (5) or Equation (6) [58]. This involves constructing a linear graph and ascertaining its slope. For instance, in the case of Equation (5), the equation could be reformulated as Equation (9):

$$\ln\left(1 - \frac{i_L}{i_{ss}}\right) = -\frac{\pi^2 D_{eff} t}{L^2} + \ln 2 \tag{9}$$

In that case, a plot of $\ln\left(1 - \frac{i_L}{i_{ss}}\right)$ against t can be established. The effective diffusion coefficient D_{eff} can subsequently be determined by calculating the slope, $-\frac{\pi^2 D_{eff}}{L}$. In some studies, researchers also turn to fitting the build-up and decay curves described in Equations (10) and (11) to evaluate D_{eff} [54,55,60]. i_0 (μA) is the background current. The build-up and decay curves are noted in Figure 8.

Build-up curve:

$$\frac{i_t - i_0}{i_{ss} - i_0} = \frac{2L}{\sqrt{\pi D t}} \sum_{n=0}^{\infty} exp\left(-\frac{(2n+1)^2 L^2}{4 D_{eff} t}\right) \tag{10}$$

Decay curve:

$$\frac{i_t - i_{ss}}{i_0 - i_{ss}} = 1 - \frac{2L}{\sqrt{\pi D t}} \sum_{n=0}^{\infty} exp\left(-\frac{(2n+1)^2 L^2}{4 D_{eff} t}\right) \tag{11}$$

The calculated values of the D_{eff}, derived from these distinct methods, do not coincide, a discrepancy that could potentially be attributed to the phenomenon of short-circuit diffusion that occurs during the hydrogen transport process [61]. To better calculate the D_{eff}, the built-up curve in the first charging cycle is typically not preferred in these calculations. This preference stems from the understanding that during the build-up phase of the initial hydrogen charging, hydrogen is ensnared by both irreversible and reversible traps. For the subsequent curves, it is primarily the reversible traps that trap or release the hydrogen. Given that irreversible hydrogen traps are seldom considered in some applications, the D_{eff} ascertained from the first decay curve or second permeation curve offers greater relevance and accuracy for research [49,57,59,62]. Thus, employing the decay curve or second charging curve as the reference often leads to a more exact determination of the D_{eff}.

A second charging curve can not only determine a better D_{eff}, but also can give an estimation of reversible and irreversible hydrogen trap density. This can provide a differential impact of reversible and irreversible hydrogen traps on hydrogen permeation or HE susceptibility [45,53]. Detailed mechanisms of hydrogen trapping is covered in Sections 5.1 and 5.2. The estimation of these traps can be calculated from Equation (12), where N_T is the total density of hydrogen trap, N_A is Avogadro's number (6.022×10^{23} mol^{-1}), and D_l is the lattice diffusion coefficient of hydrogen [63].

$$N_T = \frac{N_A C_{0R}}{3}\left(\frac{D_l}{D_{eff}} - 1\right) \tag{12}$$

If the binding energy (E_b) is considered, N_T can also be calculated by Equation (13), where N_L is the density of the interstitial sites in the steel, R is the gas constant (8.314 J·K^{-1}·mol^{-1}), and T is the temperature [64–66].

$$N_T = N_L \times \left(\frac{D_l}{D_{eff}} - 1\right) \times e^{-\frac{E_b}{RT}} \tag{13}$$

From prior discussions, it is evident that the D_{eff} derived from the permeation curve of the first hydrogen charging cycle, when applied to either Equation (12) or Equation (13), offers a measure of the total hydrogen trap density (N_T), given all traps are actively involved hydrogen trapping. On the other hand, using the D_{eff} from the permeation curve of the second hydrogen charging in Equation (12) or Equation (13) provides the density of just the reversible hydrogen traps (N_r), as only reversible traps are involved in hydrogen trapping. The density of irreversible traps (N_{ir}) can be deduced by computing the difference, $N_T - N_r$ [63].

Apart from Equation (8), the concentration of hydrogen dissolved in a metal can also be quantified using Sieverts' law:

$$c_H = S \times \sqrt{p_{H_2}} \tag{14}$$

$$S = Ae^{-\frac{\Delta H}{RT}} \tag{15}$$

In Equations (14) and (15), p_{H_2} represents the partial pressure of gaseous hydrogen. The term S remains steady when the temperature, denoted by T, is constant. The symbol ΔH represents the dissolution enthalpy and A is the constant. These two parameters can be determined from the graph slope of $\ln(c_H)$ against $1/T$. However, when the hydrogen pressure exceeds a threshold of around 200 atm (approx. 20 MPa), the hydrogen concentration is determined by Equations (16) and (17). f_{H_2} is the fugacity, which is defined as the pressure of an ideal gas with the same chemical potential as the real gas. The calculation of fugacity aligns with Equation (17), recognized as the Able–Noble relationship. In this equation, b stands as a constant with a value set at 1.584×10^{-5} m^{3}mol^{-1} [67–69]:

$$c_H = S \cdot \sqrt{f_{H_2}} \tag{16}$$

$$f_{H_2} = p_{H_2} \cdot \exp\left(\frac{p_{H_2}b}{RT}\right) \tag{17}$$

A group of researchers established a relationship between gaseous charging and electrochemical charging, and they believe that the two charging conditions can be treated as analogous processes if two conditions can produce the same fugacity of hydrogen in metal [32,35,55,70–72]. For electrochemical charging, hydrogen fugacity can be calculated according to Equations (18)–(20). In these equations, η is the overpotential of the hydrogen evolution reaction; A and ζ are constants which are determined by the mechanism of hydrogen evolution reaction. E_H^0 stands for the equilibrium potential at the surface of the steel in the hydrogen evolution reaction's charging solution under a fugacity of 1 atm. Meanwhile, E_C represents the potential that has been applied to the specimen and counter electrode. It should be noted that f_{H_2} in Equation (20) is 1 atm, so $\log(f_{H_2})$ equals 0 [35,68].

$$f_{H_2} = A \cdot \exp\left(\frac{-\eta F}{\zeta RT}\right) \tag{18}$$

$$\eta = E_C - E_H^0 \tag{19}$$

$$E_H^0 = -0.0591 \times \text{pH} - 0.0295 \log(f_{H_2}) \tag{20}$$

5. Factors Affecting Hydrogen Diffusion into Metals

5.1. Hydrogen Trapping

5.1.1. Hydrogen Trapping Mechanism

Hydrogen traps fundamentally refer to a variety of crystal defects in metals capable of binding with hydrogen. This binding prolongs the interaction between hydrogen and the metal, affecting the metal's hydrogen permeability and consequently its vulnerability to hydrogen embrittlement. The range of crystal defects in metals is broad, starting with point defects which include vacancies and diverse solute atoms. Beyond point defects, there are line defects, embodied by entities such as dislocations. Metals also feature plane defects, predominantly grain boundaries. Finally, volumetric defects are observed, comprising inclusions, precipitation phases, voids, and different crystallographic phases. It is important to note that hydrogen atoms trapped in vacancies can lead to the creation of voids, and subsequently, molecular hydrogen can be formed [73]. This mechanism is illustrated in Figure 9.

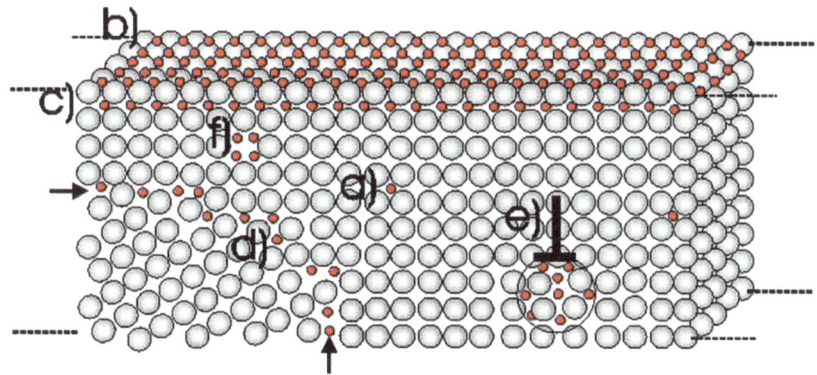

Figure 9. Schematic of hydrogen trapped in steel: (**a**) interstitial sites, (**b**) surface, (**c**) subsurface, (**d**) boundary sites, (**e**) edge dislocations (position indicated by ⊥), and (**f**) vacancies [38].

To better describe the behavior of hydrogen trapping, Lee et al. provided the reaction equation for the process of trapped hydrogen detaching from the trap and entering the lattice interstitial positions [74]:

$$\boxed{H}_{trap} \leftrightarrow \boxed{}_{trap} + H_{lattice} \tag{R11}$$

In this equation, $\boxed{}_{trap}$ represents the hydrogen trap, $\boxed{H}_{trap}$ denotes the hydrogen in the trap with its concentration represented by C_T, and $H_{lattice}$ indicates the hydrogen at lattice interstitial sites, with its concentration represented by C_L. Lu et al. [32] formulated equations to describe this process, as shown in Equations (21)–(26):

$$\theta_L = \frac{C_L}{N_L} \tag{21}$$

$$\theta_{trap} = \frac{C_T}{N_T} \tag{22}$$

$$N_L = \frac{N_A \beta \rho}{A} \tag{23}$$

$$\frac{\theta_{trap}}{1-\theta_{trap}} = \theta_L \cdot exp\left(-\frac{E_b}{RT}\right) \tag{24}$$

$$\frac{C_T}{C_L} = \frac{N_T}{N_L} \cdot exp\left(-\frac{E_b}{RT}\right) \tag{25}$$

$$\frac{C_L}{N_L} \approx exp\left(-\frac{E_L}{RT}\right) \tag{26}$$

In the provided equations, θ_L denotes the occupancy of lattice sites and N_L represents the number of lattice sites per unit volume. Other parameters include N_A, which is Avogadro's number defined as 6.02×10^{23} mol^{-1}; β, the number of interstitial sites per atom; ρ, the alloy's density; and A, the atomic weight of atoms in the alloy. Furthermore, E_b stands for the binding energy of the hydrogen trap, and E_L is the solubility energy of interstitial hydrogen. It is notable that if θ_{trap} is much less than 1, then Equation (24) simplifies to Equation (25). The concentration of lattice hydrogen is shown to follow a straightforward statistical distribution, as demonstrated in Equation (26). E_b and E_L can be determined by plotting $ln\left(\frac{C_T}{C_L}\right)$ and $ln\left(\frac{C_L}{N_L}\right)$ against the reciprocal of temperature, $\frac{1}{T}$ [8]. In

Equation (25), it is not difficult to find that the lower the E_b or the higher the T, the lower the C_T [75].

5.1.2. Classification of Hydrogen Traps

Hydrogen traps can be classified into reversible and irreversible categories, a classification based on the magnitude of the binding energy, E_b. Some researchers suggest that this categorization is driven more by practical implications, not solely by whether the binding energy surpasses or is less than a certain threshold [59]. For example, according to Equation (25), there is a decrease in the concentration of hydrogen contained within traps as the temperature escalates. This suggests that while certain traps might exhibit reversibility at elevated temperatures, they act irreversibly at typical operational temperatures. Consequently, such traps should be categorized under irreversible hydrogen traps. Indeed, this phenomenon underscores one of the pivotal factors highlighting the temperature's role in influencing the hydrogen permeation process [76]. On the other hand, some researchers advocate that traps with a binding energy (E_b) greater than 55 kJ/mol should be defined as irreversible traps [77].

At present, many researchers posit that reversible hydrogen traps play a more significant role in elevating the susceptibility to hydrogen-induced cracking (HIC) [45,78]. In contrast, irreversible traps, due to their reduced accumulation of free hydrogen at the sites of crack initiation, are perceived to possess a lower likelihood of instigating HIC [59,79].

5.1.3. Effects of Hydrogen Traps on Hydrogen Diffusion Behavior
Solute-Atom Hydrogen Traps

Different types of hydrogen traps manifest distinct diffusion behaviors, as extensively explored by previous researchers. Primarily, in the context of solute-atom hydrogen traps, alloys exhibit a lower rate of diffusion compared to pure metals due to the presence of a larger number of hydrogen traps. Fu and colleagues have demonstrated that the influence of different atomic dopants on hydrogen diffusion varies. Specifically, the addition of elements such as C, Si, and Mo can enhance the solubility of hydrogen in Fe. In contrast, the incorporation of Mn and Cr has the opposite effect; these elements decrease the solubility of hydrogen. Therefore, they propose that by adjusting the proportions of metal elements, the solubility of hydrogen in metals can be effectively reduced, which can further decrease the susceptibility of the metal to hydrogen embrittlement [80]. Similarly, Beck et al. discovered that a significant amount of hydrogen is trapped in nickel alloys compared to pure iron [81]. Fukuda and colleagues found that in martensite, the quantity of trapped hydrogen elevates with increasing carbon content [82].

Additionally, Pressouyre et al. established that the hydrogen diffusivity in ferrite decreases with an increase in nickel, chromium, and titanium content [83]. Conversely, Zaw et al. argued that the role of titanium in hydrogen trapping is not strictly linear. Their research into the impact of titanium content in vanadium–titanium alloys on hydrogen trapping revealed a complex relationship. With the addition of 0.5% titanium, the amount of trapped hydrogen reduced, whereas with the addition of 1% and 5% titanium, the trapped hydrogen amount increased. However, an addition of 10% titanium predominantly resulted in a decrease in the amount of captured hydrogen [84].

From these studies, it becomes clear that the influence of solute atoms on hydrogen diffusion is not as straightforward as merely reducing the diffusion rate. Rather, it is a multifaceted process significantly impacted by the type and concentration of the solute atoms. Therefore, these variables should be comprehensively considered when conducting research in this area.

Grain Size and Grain Boundaries

The influence of grain size and grain boundaries on hydrogen diffusion has also been thoroughly explored. As the grain size decreases, the density of grain boundaries increases. When hydrogen diffuses through a metal, these grain boundaries can often expedite the diffusion process. This is primarily because hydrogen tends to travel along these grain boundaries, a phenomenon referred to as short-circuit diffusion [85]. Even though grain boundaries are often considered hydrogen traps, they can sometimes facilitate faster diffusion, as reported by Oudriss et al. in their study of hydrogen short-circuit diffusion in polycrystalline nickel [86].

However, an increase in grain boundaries has also been observed to raise the density of nodal points, which can subsequently slow down hydrogen diffusion. For instance, Ichimura et al. developed a model of hydrogen diffusivity in aluminum calculated based on different grain sizes. Their findings indicated that for smaller grains, the diffusivity decreases because the grain boundary nodes participate more extensively in trapping at this point. When the grains are larger, hydrogen tends to diffuse along the grain boundaries, leading to an increase in diffusivity [87].

Apart from this, intracrystalline or lattice diffusion also plays a crucial role in the overall diffusion behavior. Therefore, all of these factors—grain size, grain boundary characteristics, and intracrystalline diffusion—should be considered when analyzing hydrogen diffusion in metals.

Dislocations

Due to its low binding energy, dislocations can be considered as weak hydrogen traps [88]. Dislocations in metals are primarily generated due to external stresses or cold working, as well as uneven rates during the crystal cooling process [89].

In terms of dislocations produced by external stresses or cold working, when dislocations serve as hydrogen traps, the hydrogen diffusion rate generally decreases. Yunjian and colleagues applied cold work to metals through the method of laser peening (LP) and studied the subsequent permeation rate. It was observed that metals treated with LP exhibited more tortuous grain boundaries and increased dislocations, and the amount of hydrogen permeating into the metal diminished [90]. However, as highlighted by Martin et al., hydrogen demonstrates a pronounced responsiveness to the elastic stress field surrounding dislocations [33]. When external force is exerted, it results in the creation of dislocations in the crystal, prompting hydrogen to migrate along these dislocations—a phenomenon termed as hydrogen–dislocation drag [91,92]. In such scenarios, the interaction between dislocations and hydrogen in metals is multifaceted. Hydrogen capture by dislocations might curtail the diffusion rate, while hydrogen migration with the movement of dislocations could potentially accelerate it. These two mechanisms coexist and compete internally within the metal.

Another significant source of dislocations is the segregation induced by metal cooling. There are primarily two reasons leading to this type of segregation during metal cooling: the differential cooling rates between inclusions and the metal matrix, and the disparate cooling rates at the metal surface compared to its centerline. For example, in casting processes, the exterior of the metal solidifies more rapidly, resulting in a finer grain structure and elevated dislocation density. As a result, the hydrogen diffusion efficacy at the metal's surface tends to be inferior to that at the centerline [93].

5.2. Effect of Microstructure

The intricate microstructures within steel play a pivotal role in dictating its hydrogen permeation characteristics. By subjecting steel to diverse heat treatments, a variety of microstructures emerge. Park et al. [94] treated API X65 grade pipeline steel thermally, obtaining samples spanning various microstructures. Through meticulous hydrogen permeation tests, they gauged the diffusivity of these uniquely structured samples, unveiling an escalating hydrogen permeation rate sequence: acicular ferrite < bainite < degenerated pearlite [94]. This observation aligns with the findings of Thomas et al., who concluded that hydrogen diffusion in X70 pipes primarily occurs through grains, grain boundaries, triple junctions, and cementite. Furthermore, they posited that traversing through cementite in degenerated pearlite represents the least complex diffusion pathway for hydrogen [46]. Thomas' result is also consistent with the findings of Haq et al. [65].

Another area warranting keen interest is the interplay of phase interfaces amid distinct microstructures. Turnbull et al. pinpointed the austenite/martensite phase juncture as a potent hydrogen-trapping locus [95]. Venturing further, Rudomilova et al. worked with samples enriched with ferrite, martensite, and a trace of residual austenite. Their observations underscored that specimens characterized by a relatively coarse ferrite microstructure exhibited peak diffusivity, potentially influenced by their pronounced grain size and sparser phase interfaces [62]. Offering a comprehensive assessment, Zhou et al. [66] examined the hydrogen permeation dynamics across an array of microstructures: single-phase, dual-phase, and intricate multi-phase configurations. Their study illuminated that phase boundaries such as martensite/ferrite, martensite/austenite, and ferrite/austenite acted as reversible hydrogen traps, impeding hydrogen's spread. Additionally, they proposed that while filmy retained austenite behaves as a reversible hydrogen trap, its blocky counterpart operates as an irreversible trap. Elevating the austenite phase fraction can potentiate hydrogen diffusion [44]. Taking the study further, Van den Eeckhout et al. dissected the permeability shifts in post-cold and heat-treated specimens. Their findings revealed a dip in permeability post-cold treatment, which saw an uptick following heat treatment. This flux, they hypothesized, likely stemmed from heat treatment's ability to purge certain point defects and instigate a reconfiguration of dislocation layouts [96].

6. Proposed Mechanisms of Hydrogen Embrittlement

6.1. Hydrogen-Enhanced Decohesion Mechanism (HEDE)

The H-enhanced decohesion mechanism was first introduced in 1926 by Pfeil et al. [97], and it is a simple mechanism. They proposed that hydrogen decreases the cohesive strength across lattice planes and grain boundaries. Troiano in 1959 proposed that [98] the increasing interatomic repulsive forces and thus the decreasing atomic bond strength were due to the fact that the 1 s electron from the hydrogen tends to enter the unfilled 3 d shell of the iron atoms. However, apart from a few elements like Pd, the hydrogen solubility in metals is too low to cause a significant decohesion effect; in that case, hydrogen atoms are homogenously distributed in the microstructure [99,100]. Hence, a sufficiently high concentration of hydrogen needs to be accumulated for decohesion to occur. It has been proposed that high hydrogen concentrations can occur due to high hydrostatic stresses including strain gradient hardening [101]. A variety of locations for decohesion have been suggested [37,102]: (1) adsorbed hydrogen atoms at crack tips, (2) dislocation shielding regions at crack tips, (3) grain boundaries and interphase boundaries at crack tips, (4) sites of maximum hydrostatic stresses, and (5) particle/matrix interfaces (Figure 10).

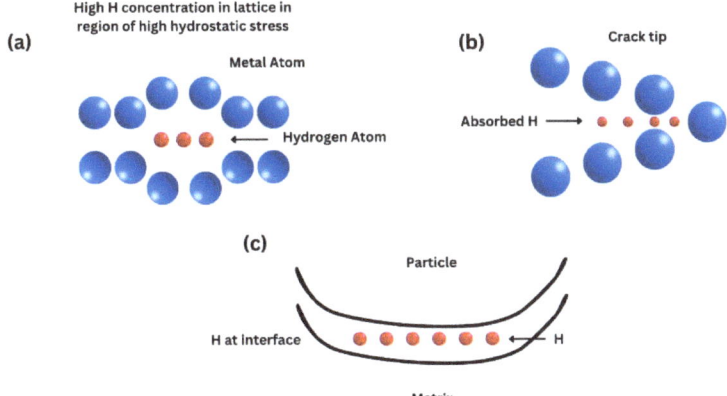

Figure 10. Schematic diagrams showing the HEDE mechanism, including tensile separation of atoms due to weakening of interatomic bonds by (**a**) hydrogen in the lattice, (**b**) hydrogen adsorbed at crack tips, and (**c**) hydrogen at particle–matrix interfaces.

Decohesion happens when the critical crack tip opening displacement (CTOD) is reached [103–105]. When hydrogen atoms are present in the microstructure and stresses are applied, then hydrogen atoms diffuse into the lattice structure and result in a reduction in cohesive strength at the crack tip and brittle cleavage-like fracture occurs. The surface energy of a material is decreased by reducing its cohesive strength so that fracture stress is also decreased and brittle fracture occurs below its allowable stress values. A major difficulty in proving this model is measuring the cohesive forces [104,106].

6.2. Hydrogen Pressure Theory

Zapffe et al. [107] presented a hydrogen pressure theory in 1941 suggesting that hydrogen atoms preferentially segregate at defect positions in the materials, such as micropores and inclusions. Then, locally accumulated hydrogen atoms gather to form hydrogen molecules. A high internal pressure is generated by the increase in hydrogen molecules. When the stress generated by the hydrogen gas pressure exceeds the yield strength of the material, hydrogen-induced cracking occurs. The concept of irreversible hydrogen embrittlement can be well explained by the hydrogen pressure theory.

6.3. Hydrogen-Enhanced Localized Plasticity (HELP)

This model was first suggested by Beachem [108] in 1972 and it is the most widely accepted mechanism. In this mechanism, hydrogen atoms accumulate near a crack tip. It also decreases the resistance to dislocation motion, increasing the mobility of dislocations. Therefore, dislocations act as carriers of plastic deformation in a metal lattice [106,109]. The presence of hydrogen around the dislocations results in a local drop in yield stress, and thus, a local movement of dislocations occurs at a low stress level (Figure 11). This implies that the fracture surfaces exhibit high localized plastic deformation near crack tips in embrittled materials and slip bands in those areas [110].

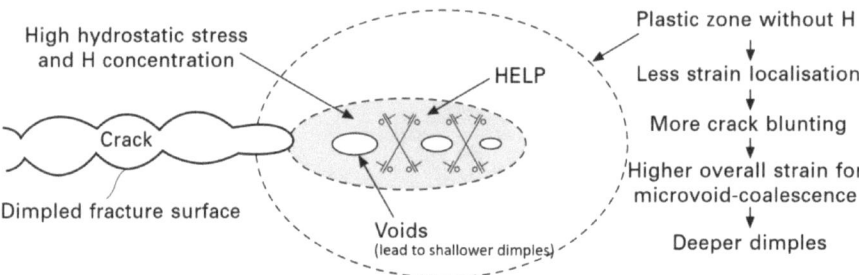

Figure 11. Schematic diagram illustrating the HELP mechanism [100].

Large increases in dislocation mobility in the presence of hydrogen have also been observed by in situ transmission electron microscopy (TEM) observations [8,111–114]. Two reasons have commonly been postulated to cause this increased dislocation mobility. (1) Hydrogen reduces the repulsive interactions between dislocations and obstacles (e.g., secondary phases, solute atoms, and other dislocations) by creating a shielding effect. This reduction in interaction energy increases the mobility and slip positioning of dislocations and decreases the stress value required for local plastic deformation. The hydrogen-induced shielding effect applies more to edge dislocations than screw dislocations. (2) Hydrogen can reduce the yield strength of the material. This phenomenon is called the "softening effect". The influence of hydrogen on the reduction in yield strength depends on the material, its purity, strain rate, temperature, and other factors [37,103,114,115]. For example, the degree of hydrogen-induced softening is sometimes large at low temperatures and low strain rates for pure iron single crystals, but is usually quite small for aluminum and nickel.

Nonetheless, this mechanism is also challenged by some experimental observations. For instance, tensile test results confirm that dislocations in IN718 alloys and pure aluminum are dragged by hydrogen [116]. In addition, it has been suggested that hydrogen impedes dislocation mobility according to simulation results [110,117]. Hence, it has commonly been assumed that the HELP system needs to combine with other systems to ultimately deteriorate material performance under a hydrogen atmosphere [118].

6.4. Adsorption-Induced Dislocation Emission (AIDE)

The adsorption-induced dislocation emission (AIDE) model was first proposed by Lynch [119] in 1976 and is a combination of both HEDE and HELP. In this model, the hydrogen atoms are adsorbed adjacent to a stress concentration area such as crack tips. The adsorption of hydrogen at crack tips weakens the interatomic bond energy and cohesive strength of materials through the HEDE mechanism and facilitates the subsequent emission of dislocations, then crack propagation by a slip step, and the generation of microvoids through the HELP mechanism [104,106,109,120]. The AIDE mechanism involves decohesion and dislocation injection from a crack tip facilitated by hydrogen adsorption, leading to nucleation and the growth of cracks (Figure 12). The formation of a slip step at the crack tip combined with microvoid coalescence results in crack propagation and fracture.

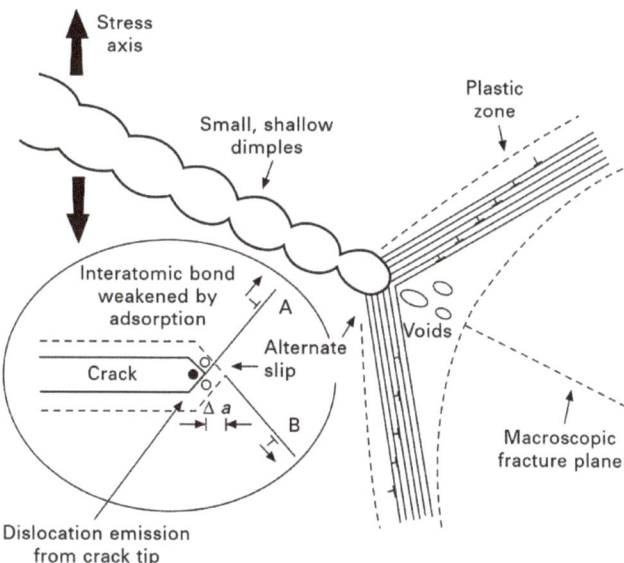

Figure 12. Schematic diagram illustrating the AIDE mechanism [100].

6.5. Hydride Formation

Westlake (in 1969) [121] was the first to suggest a mechanism based on the formation and fracture of brittle hydrides at crack tips. Hydrides are generally responsible for cleavage fractures in specific materials such as Zr, V, Nb, Ti, and Ta [122,123]. The combination of these materials with hydrogen enables the formation of brittle hydrides because of their large bond energies. This mechanism consists of four stages: (1) hydrogen diffusion to crack tips, (2) formation and growth of a hydride phase, (3) cracking the hydride along a specific cleavage plane when it reaches a critical size, and (4) crack arrest at the matrix/hydride interfaces (Figure 13). As a result, crack propagation occurs through the repetition of the above sequence.

The hydrides can be divided into thermodynamically stable hydrides and stress-induced hydrides, considering the hydrogen concentration of the alloys. At high hydrogen concentrations, specific metals and their alloys can combine with hydrogen to form thermodynamically stable hydrides in the absence of stress. For stress-induced hydrides, a sufficiently high applied stress can act to redistribute the initial low hydrogen concentration. In these systems, hydrides are formed when the local hydrogen concentration reaches the solubility limit of the materials.

Stress axis

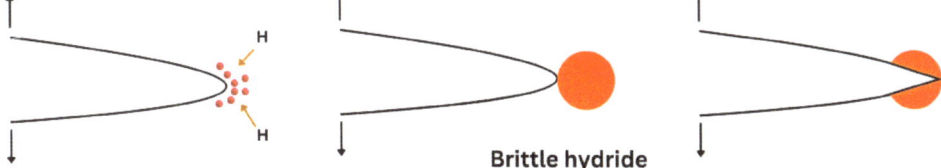

Figure 13. Schematic diagram showing subcritical crack growth including hydrogen diffusion to hydrostatically stressed regions, then formation and fracture of a brittle hydride at a crack tip.

6.6. Hydrogen-Induced Reduction in Surface Energy

This theory was proposed by Uhlig [124] in 1967 based on the Griffith criterion for fracture in ideally brittle solids. This theory assumes that the adsorption of hydrogen reduces the surface energy and thus decreases the force needed to form new crack surfaces, and that the existence of a crack occurs where the hydrogen is adsorbed. The crack can more easily grow under lower mechanical load because of this decrease. Nevertheless, it is noteworthy that the magnitude of the reduction in the surface energy by hydrogen is quite small (e.g., 7% in the case of ferrite and 9% for austenite [125]), and considering this phenomenon, along with the plastic work of separation, renders the overall effect negligible [126].

6.7. Hydrogen-Enhanced Macroscopic Plasticity (HEMP)

This mechanism is also called hydrogen-enhanced macroscopic ductility and is related to the decrease in the yield strength due to hydrogen, attributed to solid solution softening by hydrogen atoms. It is certain that the beginning of yielding is accompanied by the movement of a significant number of dislocations. Therefore, the reduction in yield strength due to hydrogen indicates the easier macroscopic motion of significant dislocation masses facilitated by the presence of hydrogen. HEMP is quite different from the subcritical cracking mechanism of HELP. This is because there is no subcritical crack propagation involved in the reduction in the yield strength, and also, the plastic deformation is not localized but rather uniform throughout the whole gauge section [127].

6.8. Hydrogen Assisted Microvoid Coalescence

Microvoid coalescence is primarily a ductile fracture system and is attributed to the preferential trapping influences of microstructural heterogeneities on hydrogen atoms in front of the crack tip. Crack generation and growth happens in different stages such as void nucleation, void growth, void coalescence, and extension of the crack and eventual breaking of remaining existing ligaments by shear stress [104,128]. Due to the hydrogen impact, dislocation and localized plastic deformation take place in the material. Due to the joining of voids present in the crack growth direction, crack propagation takes place in a zig-zag pattern.

A summary of possible corrosion–deformation interactions which could lead to hydrogen-induced cracking is presented in Figure 14.

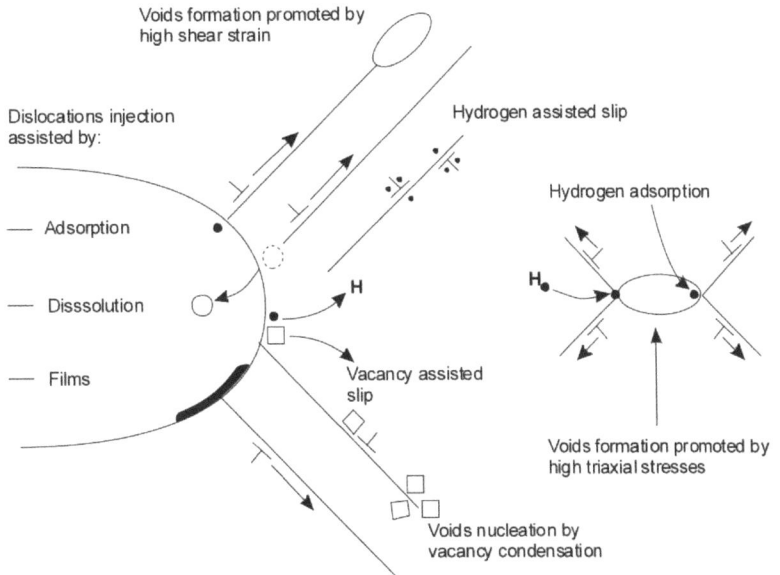

Figure 14. Processes resulting in hydrogen-induced cracking by localized slip and microvoid coalescence [129].

7. Hydrogen Embrittlement Characterization Techniques

To better understand the impact of hydrogen uptake in steels, it is necessary to establish experimental techniques to identify the presence and effect of hydrogen on steel microstructures and to provide valuable insights into the extent of its influence on the mechanical properties of the steels [130]. Microscopic observation, hydrogen permeation tests, thermal desorption analysis (TDA), and mechanical testing are the major categories of applied techniques in previous research [131]. Whilst the former three mainly aim to characterize the hydrogen content and its effect on microstructures, the latter aims to evaluate the influence of hydrogen adsorption on the mechanical properties of steels. The purpose of this section is to introduce mechanical testing methods, both established and new, that have provided valuable insights in HE research.

7.1. Tensile Tests

Hydrogen-induced embrittlement may cause a loss of load-bearing capacity, leading to premature fracture and reduced ultimate tensile strength. Tensile testing of hydrogen-charged specimens allows for the determination of parameters such as the reduction in area and fracture surface morphology, providing insights into the fracture mode and the extent of hydrogen-induced damage [132–134]. Tensile tests may be conducted utilizing either pre-charged hydrogen (ex situ) or the introduction of hydrogen during the straining (in situ). Whilst the ex situ test is more widely adopted due to its simpler instrumentation involving a separate charging unit to the tensile test setup, the in situ approach is generally considered more representative (particularly for simulating steel pipe service conditions characterized by consistent and prolonged hydrogen pressure [135]), ensuring continuous hydrogen presence by integrating a hydrogen charging mechanism into the test setup [131]. The conventional strain rate test (CSRT) and slow strain rate test (SSRT) are the most commonly applied tensile tests in previous studies for investigating the HE susceptibility of steels conducted in both in situ and ex situ conditions. While the range of strain rate for the CSRT method is reported in the literature to be 1–12 mm/min [132,136,137], this range for the SSRT is recounted as 0.001–0.12 mm/min [136,138,139]. This controlled and gradual deformation rate better mimics the actual stress conditions experienced by materials in practical application, offering more reliable insights into a material's response to hydrogen exposure and its potential for embrittlement [140]. The SSRT is standardized in ASTM G 129 [141]. Typical samples for tensile tests are shown in Figure 15a–c. Both smooth and notched geometries can be employed to differentiate the localized or universal effect of hydrogen-induced strain.

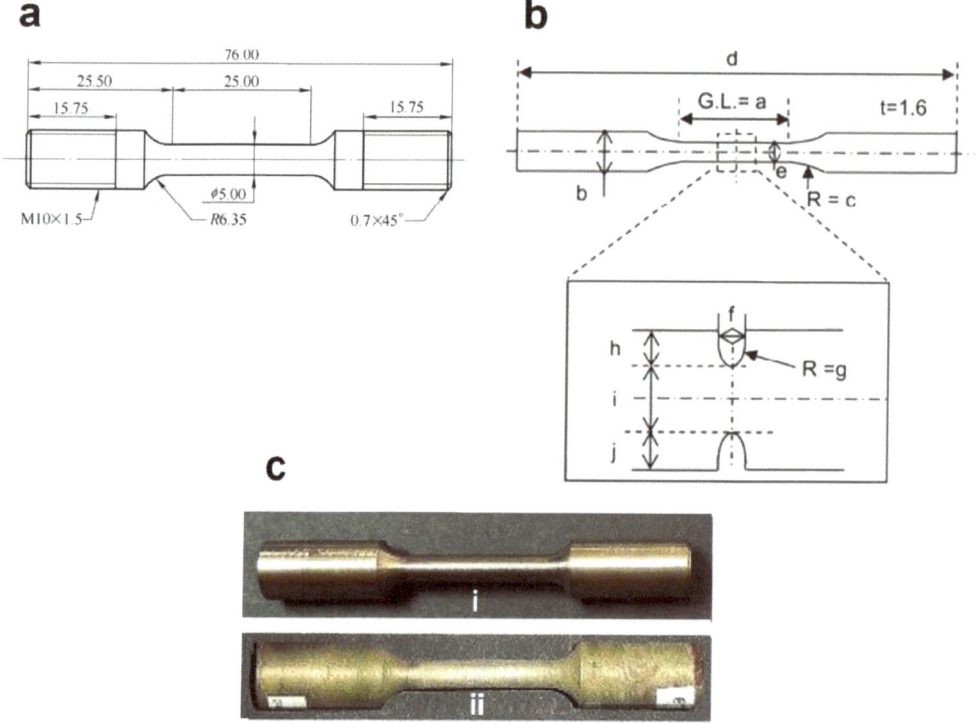

Figure 15. Tensile test specimens. (**a**) Typical drawing for smooth sample [138]. (**b**) Typical drawing for notched sample [136]. (**c**) Typical tensile specimen: (**i**) untreated and (**ii**) hydrogen embrittled [142].

The stress–displacement curves of X80 steel specimens after electrochemical hydrogen charging in a solution of 0.5 mol/L H_2SO_4 with a constant current density of 20 mA/cm^2 and tensile testing by the CSRT method are shown in Figure 16a. This suggests that with increased charging time, a modest reduction in tensile strength became apparent at a limited scale. Furthermore, the assessment of hydrogen embrittlement sensitivity involves the utilization of relative plasticity damages, characterized by elongation loss rate (I_δ) and area reduction rate (I_ψ), calculated by the following formula and shown in Figure 16b for the above-mentioned specimen:

$$\begin{cases} I_\delta = \dfrac{\delta_0 - \delta_H}{\delta_0} \times 100\% & (27) \\ I_\psi = \dfrac{\psi_0 - \psi_H}{\psi_0} \times 100\% & (28) \end{cases}$$

where δ_0, δ_H, ψ_0, and ψ_H are the elongation and reduction in area of the steel before and after hydrogen charging, respectively. Figure 16b indicates that as the duration of hydrogen charging increased, there was a tendency for the hydrogen content to rise, leading to a substantial decrease in the plasticity of the utilized X80 steel. Also, Takagi et al. [136]

investigated the differences between the critical HE conditions of steels obtained by the CSRT and SSRT under a constant load condition for a 1300 MPa-class JIS-SCM435 steel as the representative material. Consequently, the assessment of hydrogen embrittlement's critical conditions, as determined by applied stress and average diffusible hydrogen content (H_D), followed a sequence aligned with the SSRT and CSRT, starting with low stress and hydrogen content levels (Figure 17a). A comparison of the critical conditions derived from these two techniques was also conducted using the fracture initiation point's local stress and local diffusible hydrogen content index. It was observed that as the local diffusible hydrogen content increased, the local stress associated with critical conditions exhibited a decline (Figure 17b). In contrast, Hagihara [143,144] demonstrated that the critical condition of HE of TS 1300 MPa grade tempered martensitic steel obtained by the SSRT utilizing circumferential notched specimens yielded results nearly identical to the CSRT when assessing critical conditions based on local stress and local hydrogen distribution at the point of fracture initiation.

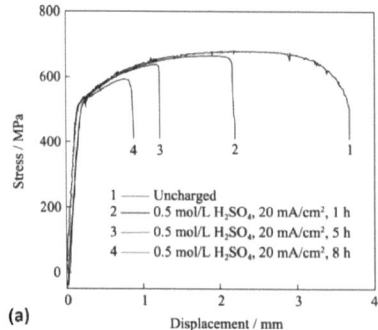
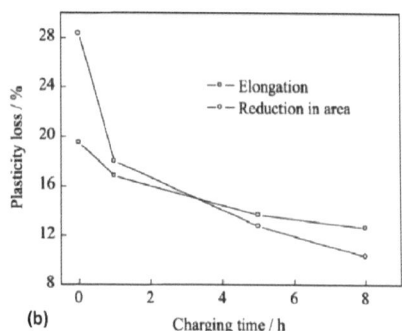

Figure 16. CSRT results for the X80 steel under different charging durations: (**a**) stress–displacement curves; (**b**) plasticity loss [132].

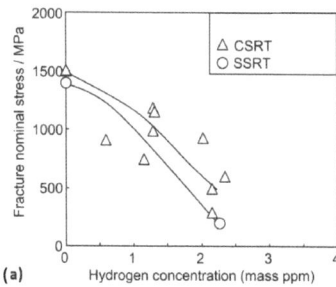
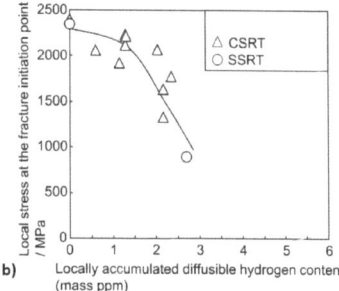

Figure 17. (**a**) Comparison of relationships between fracture nominal stress and hydrogen concentration of SCM435 steel obtained by CSRT and SSRT. (**b**) Comparison of relationships between local stress at fracture initiation points and local accumulated diffusible hydrogen concentration of SCM435 steel obtained by CSRT and SSRT [136].

Koyama et al. [145] carried out an interrupted CSRT on a high-strength ferrite/martensite dual-phase (DP) steel and compared the results with the non-interrupted and without charging samples to investigate the involved HE mechanisms. During the interrupted test, hydrogen was introduced prior to conducting the tensile test, the test was halted when the strain reached 6%, and then the load was removed. Following a period of 10 days in contact with air, the sample was reloaded until it fractured. The contrast between results with and without hydrogen charging distinctly illustrates the HE effect. It is evident that the yield and tensile strengths remained largely unchanged following hydrogen charging, while the elongation before fracture experienced a significant decline. Importantly, the interrupted test showed a partial recovery in elongation due to the desorption of hydrogen, suggesting that the degradation of tensile properties caused by hydrogen was significantly influenced by both HEDE- and HELP-assisted crack propagation within the crack growth regime (Figure 18).

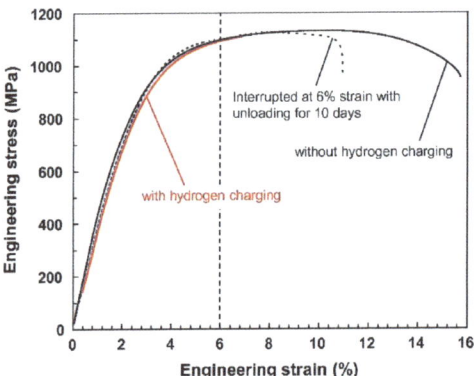

Figure 18. Engineered stress–strain curves of the specimens with and without charging and with interruption in the tensile test [145].

A drawback of the SSRT method was that upon surpassing the threshold stress (σ_{TH}), the specimens underwent prolonged extension, leading to extended failure times and operational inconvenience. To address this challenge, a novel approach termed linearly increasing stress testing (LIST) was introduced [146–148]. This technique shares some similarities with the SSRT but offers distinct advantages. In the LIST method, a sample is subjected to a gradually increasing stress until failure occurs. This is accomplished through controlled weight displacement facilitated by a motor-driven mechanism, as illustrated in Figure 19. One important difference between the SSRT and LIST is that the SSRT operates based on displacement control, whereas LIST operates on load control [148]. It is reported that the LIST and SSRT are basically the same up to the initiation of the crack, yielding identical values for σ_{TH}. The difference begins once the critical crack is reached. While it takes a relatively long time for the fracture of the SSRT sample to happen, the LIST sample fails at much shorter time due to experiencing plastic instability [149].

An earlier tensile testing method called the constant load test involves a notched or smooth specimen under an applied static load exposed to the environment (in situ). The constant load test was first introduced by Baldy [150] in the 1960s, and then became the NACE TM0177 [151] method A, also described in detail in ASTM E 1681 [152]. Typically, the assessment of HE susceptibility using this method relies on the time taken for failure to occur. Tensile test samples subjected to specific stress levels yield either a pass or fail outcome. By conducting tests on multiple specimens under different stress levels, it becomes possible to obtain an apparent threshold stress for HE [153]. Tensile tests can be performed either with constant-load or sustained-load (proof-ring or spring-loaded) devices as described in ASTM G49 [154]. While evaluating HE susceptibility through sustained-load test outcomes necessitates a visual inspection of specimens to identify crack presence, employing constant-load apparatus guarantees complete separation for materials prone to susceptibility. An issue frequently encountered in constant-load testing is the absence of a guarantee for sample failure, potentially leading to prolonged test durations. In such instances, a practical solution involves concluding the test after a specific duration (e.g., 100 h) has passed without the occurrence of specimen fracture [155].

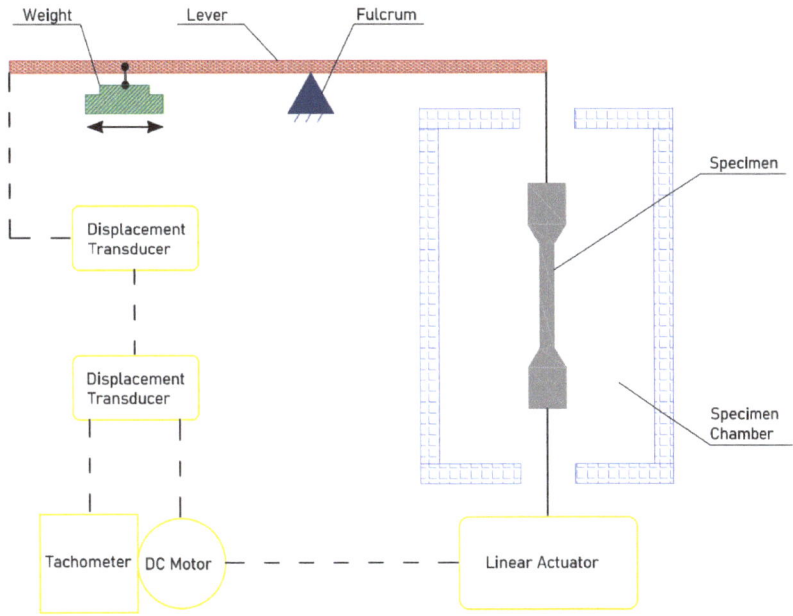

Figure 19. Schematic of the linear increasing stress testing (LIST) apparatus.

Al-Mansour et al. [156] investigated the HE susceptibility of API-X100 high-strength low-alloy steel using proof-ring constant-load testing with NACE TM-0177 solution A and generated an SCC threshold stress value of 46% of the yield strength. The specimens were loaded at stress values equivalent to 30% up to 80% of the material's yield strength, and time to failure (TTF) or no failure was recorded based on a maximum test duration of 720 h. The results are shown in Figure 20. The low threshold stress value of the material was attributed mainly to the X100 microstructure, having a banded structure providing higher hydrogen trapping site density in front of the crack tip than homogenous microstructures [157]. Similar research has been conducted on other HSLA steels, X60, X65, and X70, reporting threshold stress values of 60%, 69%, and 80% of the YS, respectively [157–159], while quenching and tempering treatment has been shown to increase the threshold stress by removing the banded structure and provoking a more homogenous one [157].

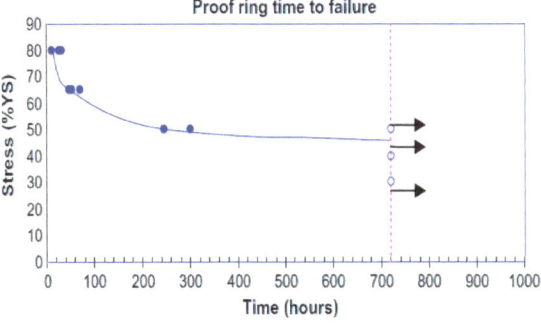

Figure 20. Proof-ring time to failure for the X100 in the NACE TM-0177 "A" solution (dark = fail, clear = pass; dashed line indicates the 720 h test duration) [156].

Using the SSRT and proof-ring testing, Li et al. [160] investigated the influence of a surface martensite layer on the HE of TWIP (austenitic high-Mn twinning-induced plasticity) steels in a wet H_2S environment. TWIP steels, owing to their mechanically induced austenitic twins and fcc structure of austenite phase, have an outstanding combination of both strength and ductility, and also high solubility and low diffusivity of hydrogen, making them good candidates for applications where high hardenability and formability is required at the same time as the high resistance to HE [161–164]. Two types of TWIP steels with different surface martensite microstructures were studied, Fe-16Mn-0.4C-2Mo (wt.%) (16Mn), with a surface layer containing ε-martensite, α′-martensite, and austenitic twins, and Fe-25Mn-0.4C-2Mo (wt.%) (25Mn), with a full α′-martensite surface layer. The results for the SSRT and proof-ring testing are shown in Figure 21a–d. It was seen that the strength reduction in 16Mn steel is approximately twice that of 25Mn steel due to the ε-martensite presence, which decreased hydrogen embrittlement resistance; removing surface martensite helped 16Mn steel but had little effect on 25Mn steel with only α′-martensite. The results from the proof-ring tests were also consistent with the tensile test.

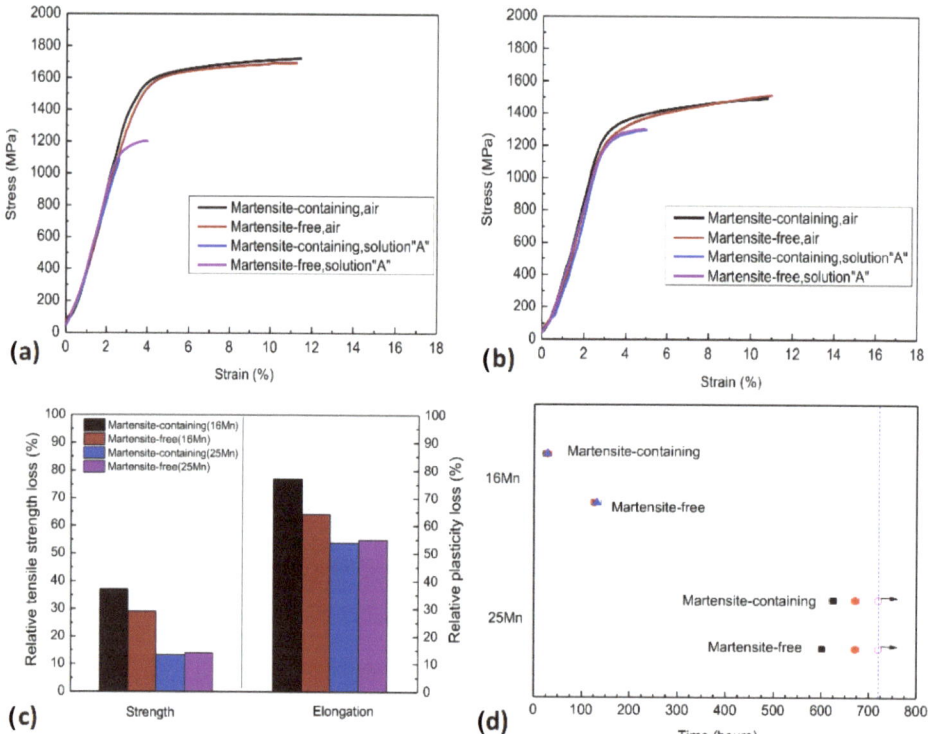

Figure 21. Influence of surface martensite layer on HE of 16Mn and 25Mn steels in different conditions in wet H_2S environment: (**a**,**b**) engineered tensile stress–strain curves; (**c**) relative tensile strength and ductility loss; (**d**) proof-ring time to failure in the NACE TM0177 "A" solution [160].

7.2. Double Cantilever Beam Test

The double cantilever beam (DCB) test is one of the most widely utilized tests to evaluate the resistance of steels to sulfide stress cracking (SSC), a particular form of HE. Due to its quantitative nature, high sensitivity, and minimal dependence on specimen surface finishing, the DCB test stands out as an exceptional quality control [165]. Initially proposed by Heady [166] in the early 1970s, the test is based on fracture mechanics, using the DCB specimen which is loaded to pre-define the critical stress intensity factor (K_{ISSC}).

Later, the test became the NACE TM0177 standard method D [163]. The standard DCB test specimen is shown in Figure 22 involving two beams (arms) separated by a slot. The specimen is then loaded either by inserting the wedge or by utilizing tensile equipment to produce an arm displacement which will create an initial stress intensity factor, K_{I0}, at the chevron notch. In air, K_{I0} is below K_{IC}, so no crack propagation will occur; however, when the specimen is put in the sour environment (i.e., NACE standard solution), the steel becomes embrittled and the crack will grow, leading the specimen to be progressively unloaded. The crack growth eventually stops when the applied stress intensity factor, K_I, matches the critical stress intensity factor of the steel in a corrosive environment, K_{ISSC}. The value of K_{ISSC} is calculated by the below equation:

$$K_{ISSC} = \frac{Pa\left(2\sqrt{3} + \frac{2.38h}{a}\right)\left(\frac{B}{B_n}\right)^{1/\sqrt{3}}}{Bh^{3/2}}$$

(29)

where:
K_{ISSC} = threshold stress intensity factor for SSC;
P = lift-off load;
a = crack length;
h = height of each arm;
B = DCB test specimen thickness;
B_n = web thickness.

It is noteworthy to mention that the DCB test is only designed to compare the resistance of diverse steels to the HE and that the K_{ISSC} is not an intrinsic material property [151], but also depends on test parameters such as specimen thickness and arm displacement [167–170]. In addition, the sensitivity of K_{ISSC} to other factors has also been studied [171–174], and the following factors are considered to have a greater impact on the value of the calculated K_{ISSC}: temperature, solution chemistry, specimen preparation, wedge introduction, etc. The test results for the two laboratories using the same material are compared by Szklarz [175] to investigate the effects of some of these factors. The test conditions were all the same in the two laboratories, except from the test vessel (12-liter capacity glass vessel in lab 1 and 6 or 10-liter depending on the number of tested specimens in lab 2), the use of a diffuser (lab 1 with diffuser and lab 2 without diffuser), and the opening of the specimens (with a hammer and chisel in lab 1 and with a tensile machine in lab 2). Figure 23a shows the results for the K_{ISSC} values in both laboratories. It can be seen that there was around 12% difference between the results, which may be influenced in part by the actual arm displacement utilized by the two laboratories (shown on Figure 23b).

A similar trend for the effect of arm displacement on the K_{ISSC} value was reported by Linne et al. [176], Sponseller [165], and Asahi et al. [173]. Also, Moderer et al. [177] investigated the influence of arm displacement, initial crack length, pre-cracking, and notch type (slot with a chevron or an electro-discharged machine slot (EDM)) on the K_{ISSC} values. The results showed minor sensitivity of K_{ISSC} to the notch type and pre-cracking, but a higher number of valid specimens were attained for EDM-notched specimens. Furthermore, shorter initial crack length and higher arm displacement led to a slight increase in K_{ISSC} values.

The DCB test is mainly designed to test the higher-strength materials in extreme sour environments, so there are limitations when applying this method to lower-strength steel grades (i.e., with SMYS values of $\leq$450 MPa) and mild sour environments due to crack growth beyond the acceptance criteria for a valid test and also the relaxation of stress at the crack tip because of arm bending. To overcome this issue, Maldonado et al. [178] conducted a large-scale DCB test (Figure 24) and obtained the K_{ISSC} value which met the requirement of the project.

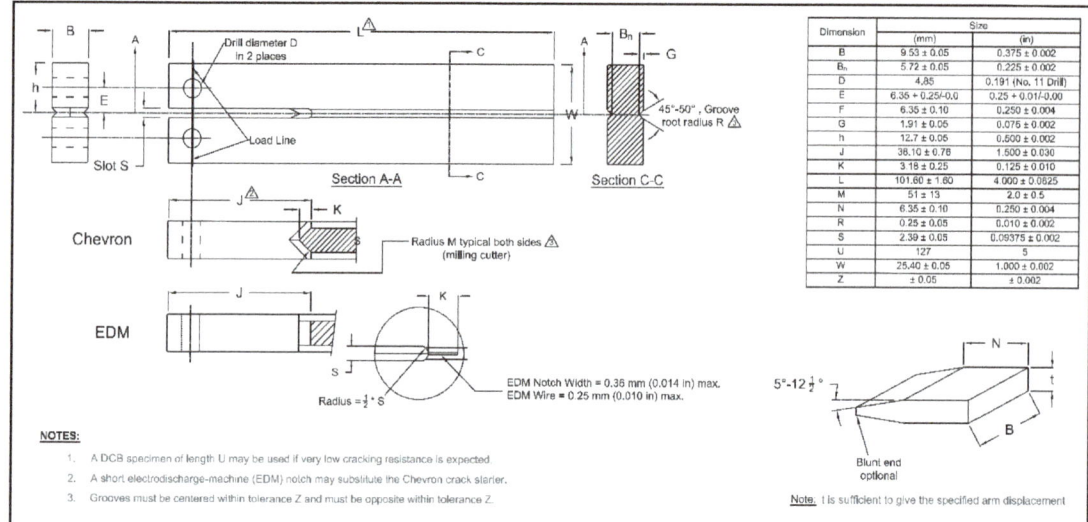

The table in the figure:

Dimension	Size (mm)	(in)
B	9.53 ± 0.05	0.375 ± 0.002
Bn	5.72 ± 0.05	0.225 ± 0.002
D	4.85	0.191 (No. 11 Drill)
E	6.35 + 0.25-0.0	0.25 + 0.01/-0.00
F	6.35 ± 0.10	0.250 ± 0.004
G	1.91 ± 0.05	0.075 ± 0.002
h	12.7 ± 0.05	0.500 ± 0.002
J	38.10 ± 0.76	1.500 ± 0.030
K	3.18 ± 0.25	0.125 ± 0.010
L	101.60 ± 1.60	4.000 ± 0.0625
M	51 ± 13	2.0 ± 0.5
N	6.35 ± 0.10	0.250 ± 0.004
R	0.25 ± 0.05	0.010 ± 0.002
S	2.39 ± 0.05	0.09375 ± 0.002
U	127	5
W	25.40 ± 0.05	1.000 ± 0.002
Z	± 0.05	± 0.002

Section A-A Section C-C

Chevron — Radius M typical both sides (milling cutter)

EDM

EDM Notch Width = 0.36 mm (0.014 in) max.
EDM Wire = 0.25 mm (0.010 in) max.
Radius = ½ · S

Blunt end

Note. t is sufficient to give the specified arm displacement

NOTES:
1. A DCB specimen of length U may be used if very low cracking resistance is expected.
2. A short electrodischarge-machine (EDM) notch may substitute the Chevron crack starter.
3. Grooves must be centered within tolerance Z and must be opposite within tolerance Z.

Figure 22. DCB test specimen.

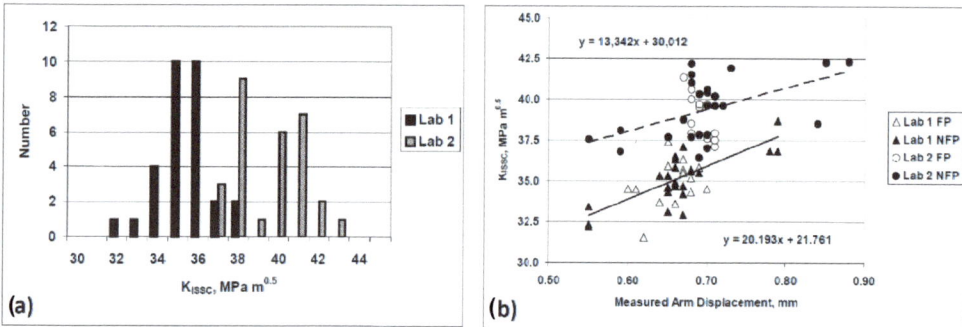

Figure 23. (**a**) Distribution of results by laboratory. (**b**) Arm displacement effect [175].

Figure 24. Large-scale DCB specimen from experiment by Maldonado [178] (the width and side notch dimensions were maintained within the NACE TM0177 limits, while all other dimensions were scaled up by a factor of 3.9x).

The stress field condition of the material plays a crucial role in influencing its hydrogen diffusion behavior and capture mechanism [179,180]. Xing et al. [181] investigated the relationship between the subsurface hydrogen content (C_0) and the threshold stress intensity factor of the hydrogen-induced cracking arrest (K_{HSC}) of X80 pipe steels through hydrogen permeation and DCB tests. For this purpose, samples were cut from the pipe steel and cathodically charged in a 0.5 mol/L H_2SO_4 and 0.2 g/L CH_4N_2S solution under different current densities, and crack propagation was monitored. As shown in Figure 25a, no crack propagation occurred when the current density was small (i.e., 1 or 3 mA/cm^2), but for higher current densities, the crack length increased with the increase in applied current density. The computed threshold stress intensity factor K_{HSC} of hydrogen-induced crack diminished as the applied current density rose, as shown in Figure 25b. The values for the subsurface hydrogen concentration (C_0) were also obtained from permeation tests with the same solution and current densities of the DCB test, and the relationship between C_0 and K_{HSC} satisfied the expression of K_{HSC} α-lnC_0, as illustrated in Figure 25c.

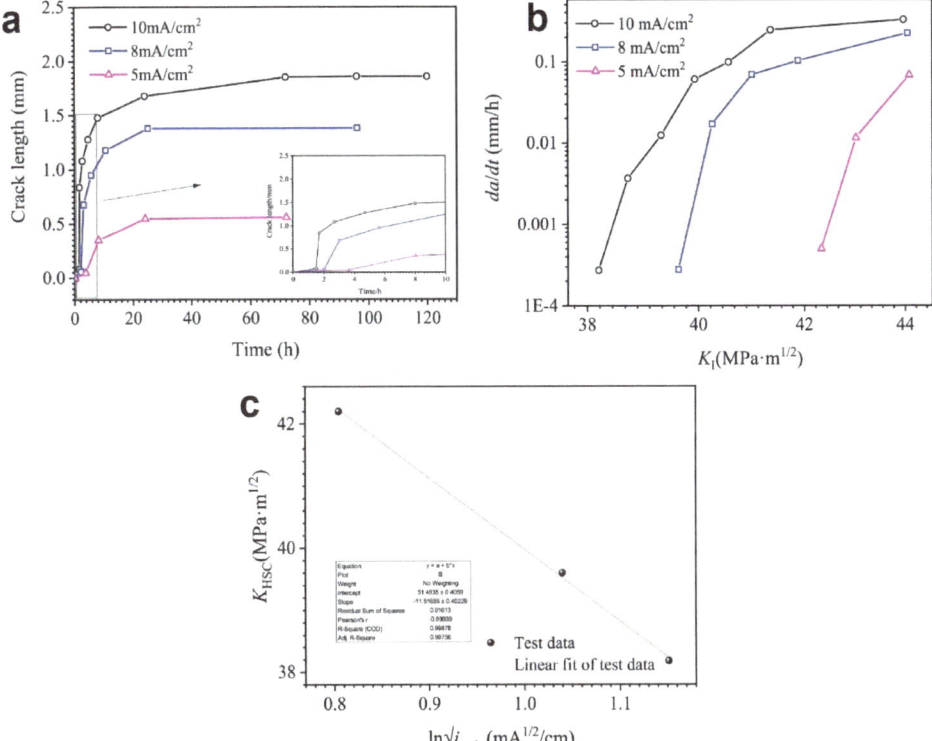

Figure 25. Hydrogen-induced cracking of X80 pipe steel under different current densities. (**a**) Variation curves of crack propagation length with time, (**b**) variation curves of crack propagation rate with stress intensity factor, and (**c**) relationship between K_{HSC} and subsurface absorbed hydrogen concentration C_0 [181].

7.3. Bent Beam Test

Bent beam tests are another kind of mechanical test utilized for the evaluation of HE susceptibility of carbon and low-alloy steels in the presence of a stress concentration, firstly introduced by Fraser [182] and later becoming the NACE method B [156]. Bent beam test specimens (Figure 26a) are loaded by test apparatus (Figure 26b) to varying particular deflections and then exposed to the test environment for a specific duration (i.e.,

720 h), and failure/no failure of the test will be assessed based on observations of cracks in the specimens. A statistically based pseudo-stress (S_c) for a 50% probability of failure is calculated to indicate the material's resistance to SCC. For a three-point bending test (such as the NACE method B), the deflection of the test specimen is calculated by the below formula:

$$D = \frac{Sl^2}{6Et} \tag{30}$$

where:

D = deflection;
S = nominal outer fiber pseudo-stress, typically in the range of 69 MPa from 22 to 24 HRC for carbon and low-alloy steel;
l = distance between centerlines of end supports;
E = elastic modulus;
t = thickness of test specimen.

a

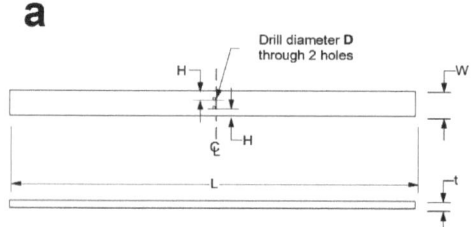

Dimension	Size	
	(mm)	(in)
L	67.3 ± 1.3	2.65 ± 0.050
t	1.52 ± 0.13	0.060 ± 0.0050
W	4.57 ± 0.13	0.180 ± 0.0050
H	1.58 ± 0.05	0.062 ± 0.002
D	0.71 ± 0.0013	0.028 ± 0.0005 (No. 70 Drill)

b

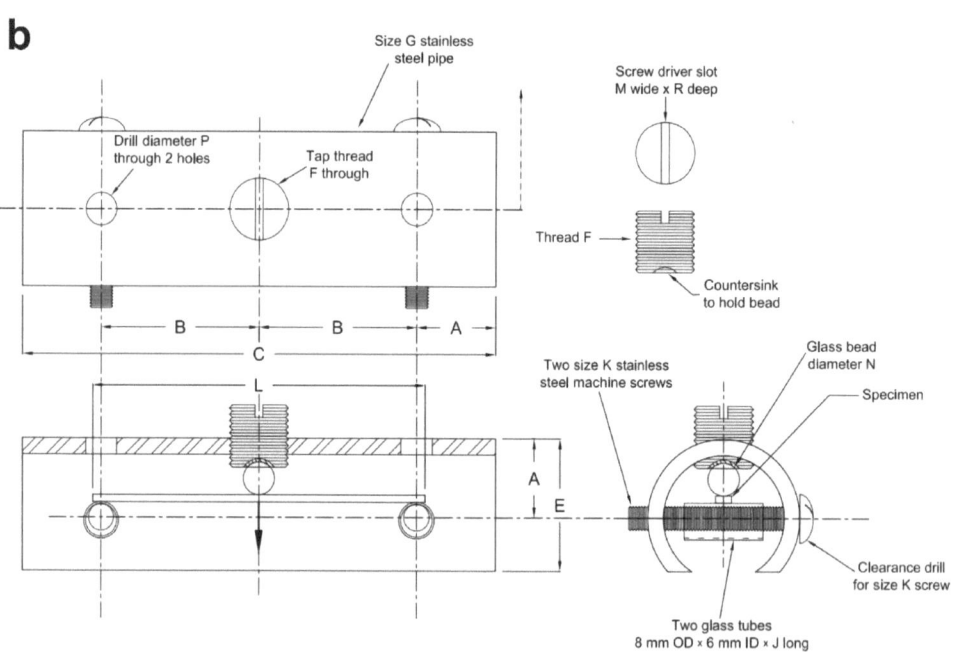

Figure 26. (**a**) Dimensional drawing of standard 3-point bent beam specimen. (**b**) Test fixture.

Then, the pseudo-stress S_c is calculated based on the below formula:

$$S_c = \frac{\frac{\sum S}{68.95\ Mpa} + 2\sum T}{n} \tag{31}$$

where:

T = the test result (i.e., +1 for passing and −1 for failure);

n = the total number of test specimens tested.

It should be noted that the computed pseudo-stress, lacking accuracy in reflecting the actual stress distribution, plastic deformation effects, and stress changes during crack growth, is unsuitable for determining threshold stress.

Delayed fracture strength (DFS), which is the maximum bending stress that does not cause failure of the specimen, is another parameter for the evaluation of HE properties of materials obtained through four-point bend experiments [136,137,182,183]. In this method, the specimen (Figure 27a) is cathodically pre-charged with hydrogen, then loaded by four-point bending for a defined duration (i.e., 5 or 100 h), counting the fracture time from the start of the loading. The critical HE is established as the maximum applied load among conditions of the specimen that remain unfractured after the test duration. This method has more in common with proof-ring testing rather than the three-point bent beam test. The delayed fracture limit stress as a function of diffusible hydrogen content for two ultra-high-strength steels is shown in Figure 27b,c. In both steels, the DFS decreased with increasing H_D; however, the V-added steel showed higher resistance to HE than the SCM435 steel at the same level of H_D.

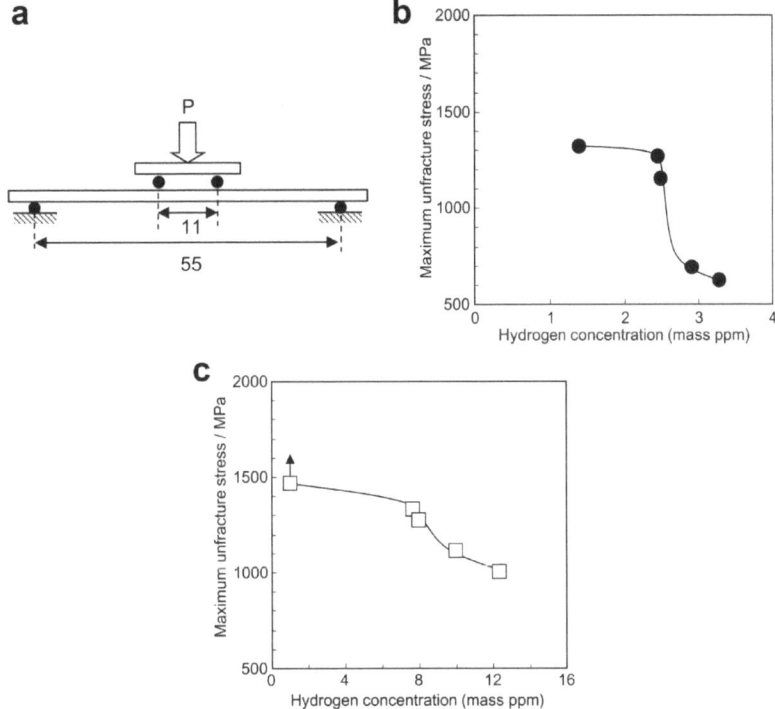

Figure 27. (a) Four-point bending specimen. (b,c) Diffusible hydrogen content–delayed fracture limit stress curves obtained by 4-point bending test for SCM435 and V-added ultra-high-strength steels, respectively [136].

Si et al. [184] applied a U-bend test to compare the HE resistance of two 1500 MPa martensitic steels, in which a hydraulic press was used to impose constant pressure downward to form a 180° bend on the samples that were fixed by bolts. The deformed samples were then immersed in 0.5 mol·L^{-1} HCL solution and time to fracture was recorded to compare the HE resistance of the two samples (Figure 28a–c). While sample #1 had obvious cracking after 10 h of the bending immersion test (Figure 29a), sample #2 cracked after 50 h, indicating that #2 had a better anti-HE effect than #1 under equivalent conditions, which was attributed to their different microstructures and precipitated phases, leading to different values of hydrogen desorption rates (Figure 29b).

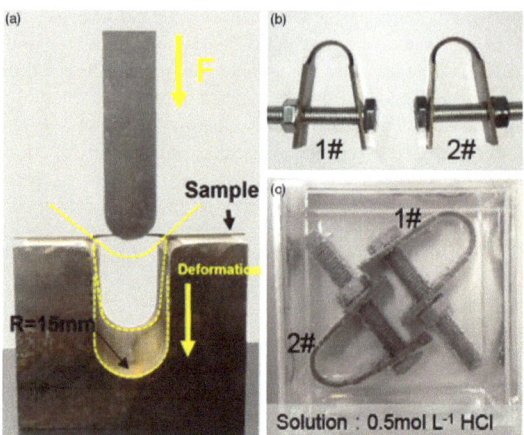

Figure 28. U-bend test: (**a**) test device, (**b**) sample pressed after forming, and (**c**) sample placed in 0.5 mol·L^{-1} HCl solution [184].

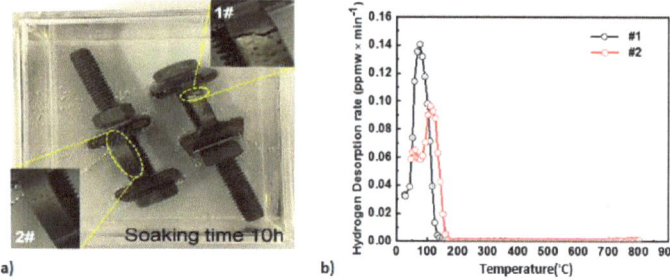

Figure 29. (**a**) Cracking of samples 1# and 2# after soaking for 10 h. (**b**) TDA spectra on #1 and #2 steels after hydrogen charging at a heating rate of 100 °C h^{-1} [184].

7.4. Fatigue Test

Fatigue testing is another important method for characterizing hydrogen embrittlement. The presence of hydrogen significantly reduces the fatigue life of steels, making them more susceptible to fatigue crack initiation and propagation. Hydrogen-enhanced crack growth occurs due to the acceleration of crack growth rates under the influence of hydrogen atoms [15,185,186]. Fatigue testing involves subjecting hydrogen-charged specimens to cyclic loading, typically using techniques such as rotating beam or axial fatigue tests. The resulting fatigue life curves and crack growth rates can be used to assess the influence of hydrogen on the material's fatigue behavior. If a significant amount of hydrogen enters into the material, it can have an adverse impact on the material's static fracture and fatigue properties. This could lead to an undesirable rise in the rate at which cracks develop within the material, a phenomenon commonly denoted as hydrogen-affected fatigue crack growth rate (HAFCGR) [16]. Fatigue crack growth rate (FCGR) testing is usually performed on compact tension (CT) specimens. The recommended dimensions for CT specimens are described in the ASTM E647 standard [187] (Figure 30a). FCGR tests can be performed both ex situ and in situ; however, the in situ test has the advantage of reflecting the real condition of the material under the working environment. A typical test setup for in situ FCGR testing is shown in Figure 30b. The typical crack growth rate behavior in materials is characterized by ΔK-da/dN plots, which identify three domains: stage I (threshold domain), stage II (linear or Paris domain), and stage III (final fracture), as shown in Figure 31. While stage III is linked to unstable crack growth and failure, both stage I and particularly stage II (Paris domain) can be influenced by hydrogen presence [188]. Determining hydrogen's impact on crack growth is challenging due to its dependence on various factors like the material, load frequency, temperature, pressure, or cathodic potential. The Paris law provides a quantitative description of the stage II fatigue crack growth domain [189]:

$$\frac{da}{dN} = C.\Delta K^m \tag{32}$$

where:
a = crack length;
N = number of the cycles;
ΔK = variation in the stress intensity factor encountered by the material throughout fatigue cycles;
C and m = constants that depend on the material and the testing conditions.

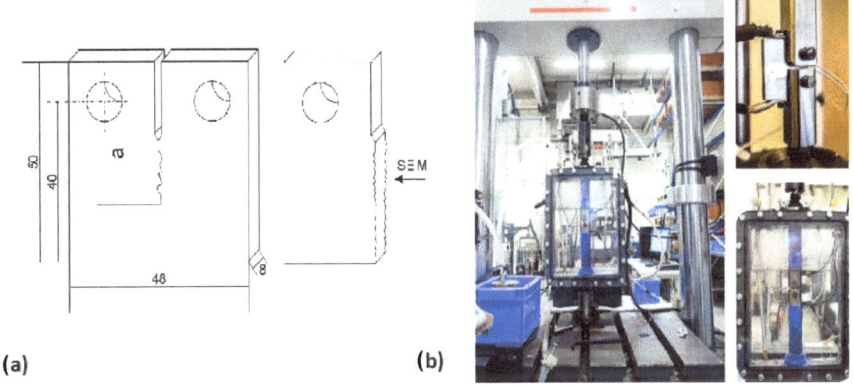

(a) (b)

Figure 30. (**a**) Typical geometry of the CT specimen and schematic representation of the fracture surface. (**b**) Setup for in situ electrochemically charged fatigue crack growth rate test [16].

Meng et al. [190] investigated the impact of hydrogen on mechanical properties of X-80 pipeline steel in natural gas/hydrogen mixtures with 5.0, 10.0, 20.0, and 50.0 vol% hydrogen at a pressure of 12 MPa using FCGR testing. As shown in Figure 32a,b, it was concluded that the quantity of introduced hydrogen is a crucial factor in the HE of X80 steel, and the rate of fatigue crack growth was notably accelerated as hydrogen levels increased. Also, the fatigue lifespan of the X80 steel pipeline experienced a significant decline due to the introduction of hydrogen. In a nitrogen gas environment, the fatigue life was 24,431 cycles, whereas in a 5% hydrogen blend, it reduced to 2130 cycles.

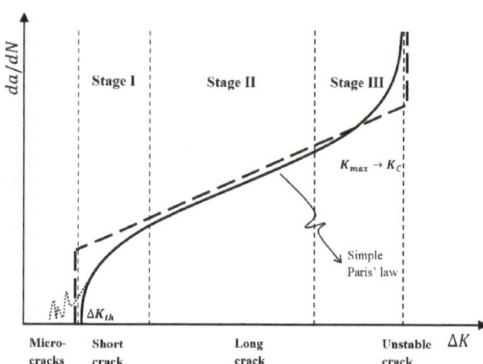

Figure 31. Schematic diagram of a normal fatigue cracking process [188].

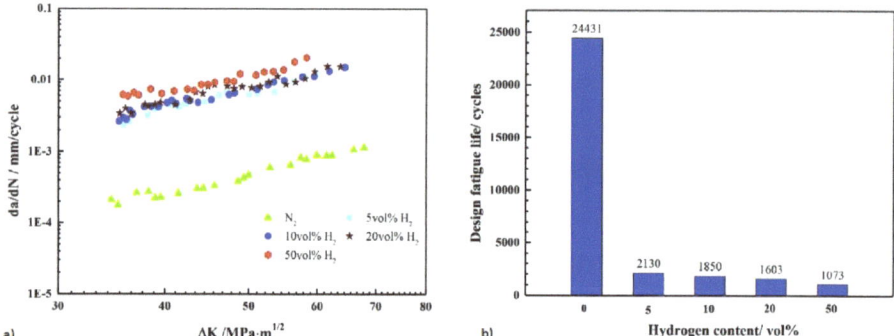

Figure 32. Hydrogen effects on X80 pipeline steel mechanical properties. (**a**) da/dN versus ΔK curves in nitrogen gas and hydrogen blends. (**b**) The fatigue life of the example pipeline [190].

8. Conclusions

The interaction between hydrogen and metals is a highly intricate issue. This problem encompasses two aspects: firstly, the interaction of hydrogen with the metal surface, and secondly, the diffusion of hydrogen once it has entered the metal. Based on the aforementioned content, we can draw the following conclusions:

1. The interaction of hydrogen with metals involves three steps: physisorption, chemisorption, and diffusion into the steel.
2. The diffusion of hydrogen into metals is influenced by various factors. The impact of hydrogen traps is particularly significant. Point defects, line defects, plane defects, and volumetric defects in metal–hydrogen can all serve as hydrogen traps. These traps can be classified as reversible and irreversible based on their binding energy, but a case-by-case analysis is also pertinent. For instance, some traps might transition from being irreversible hydrogen traps at room temperature to reversible traps at elevated temperatures due to the temperature dependency of trap binding energy. Notably, though grain boundaries can expedite hydrogen diffusion due to the short-circuit effect, an excessively high density of grain boundaries can decrease the rate of hydrogen diffusion, as nodal points can capture hydrogen. Additionally, hydrogen can promote the clustering of vacancies in metals, leading to the formation of voids. As a consequence, molecular hydrogen can develop within these voids. This could result in severe degradation of the metal, which should be avoided.
3. The microstructure of metals also significantly influences hydrogen diffusion. For instance, cementite is considered one of the microstructures that enables rapid hydro-

Materials **2024**, *17*, 965

gen diffusion. Apart from microstructures, the influence of phase boundaries is also pivotal, like the martensite/ferrite, martensite/austenite, and ferrite/austenite phase boundaries which act as reversible hydrogen traps.

4. It is evident that studying the behavior of hydrogen diffusion in metals should take into consideration all potential influencing factors. Therefore, a practical approach to understanding the impact of hydrogen traps on hydrogen permeation involves conducting specific permeation experiments for a given material and focusing on measuring and characterizing various types of hydrogen traps rather than concentrating on a single influencing factor. Moreover, the limitations posed by the present characterization methods demand the development of more innovative test techniques. Such a comprehensive investigation offers a more holistic perspective on hydrogen diffusion mechanisms and is instrumental in devising effective strategies to address and mitigate hydrogen-induced problems in materials.

Author Contributions: Q.L.: writing initial draft, H.G.: writing initial draft, V.J.: writing initial draft, T.A.: review and edit, Z.F.: review, edit, supervision and M.A.I.: review, edit. All authors have read and agreed to the published version of the manuscript.

Funding: This work was funded by Natural Sciences and Engineering Research Council of Canada (NSERC), grant # ALLRP 570445-2021.

Conflicts of Interest: Author Tahrim Alam was employed by the company Enbridge Gas Inc. The remaining authors declare that the research was conducted in the absence of any commercial or financial relationships that could be construed as a potential conflict of interest.

Nomenclature

List of Symbols	Definition
HE	hydrogen embrittlement
HIC	hydrogen-induced cracking
physisorption	physical adsorption
chemisorption	chemical adsorption
fcc	face-centered cubic
hcp	hexagonal close-packed
bcc	body-centered cubic
O sites	octahedral sites
T sites	tetrahedral sites
HMT	hydrogen microprint technique
SEM	scanning electron microscopy
TEM	transmission electron microscopy
$AgBr$	silver bromide
$NaNO_2$	sodium nitrite
$Na_2S_2O_3$	sodium thiosulfate
NH_4SCN	ammonium thiocyanate
CH_4N_2S	thiourea
CH_3COOH	acetic acid
CH_3COONa	sodium acetate
PSU	power supply unit
Pt	platinum
SSRT	slow strain rate tension test
CGHAZ	coarse-grained heat-affected zone
J	diffusion flux
D	diffusion coefficient
$\partial c/\partial x$	concentration gradient
$\partial c/\partial t$	the change in concentration over time
$\partial^2 c/\partial x^2$	the second derivative of the concentration with respect to the distance
c_{0R} (mol·m^{-3})	the hydrogen concentration on the entry side (also referred to as charging side) in the hydrogen permeation test

i_t (μA)	transient current in the hydrogen permeation test
i_{ss} (μA)	the steady-state current in the hydrogen permeation test
t_{lag} (s)	the time elapsed from the beginning of the permeation test until the ratio of transient-to-steady-state current (i_t / i_{ss}) reaches 0.63
t_b (s)	the time calculated by extrapolating the linear portion of the rising hydrogen permeation current transient
D_{eff} (m$^2 \cdot$s^{-1})	the effective diffusion coefficient in the hydrogen permeation test
J_{SS} (mol$\cdot$m^{-2}s^{-1})	permeation flux at steady state in the hydrogen permeation test
i_0 (μA)	background current in the hydrogen permeation test
N_T (m^{-3})	total density of hydrogen trap
D_l (m$^2 \cdot$s^{-1}A)	lattice diffusion coefficient of hydrogen
E_b (J$\cdot$mol^{-1})	binding energy
N_L (m^{-3})	the density of the interstitial sites in the steel
N_{ir} (m^{-3})	the density of irreversible traps in the steel
N_r (m^{-3})	the density of reversible traps in the steel
ΔH (J$\cdot$mol^{-1})	dissolution enthalpy of hydrogen into steel
f_{H_2} (MPa)	fugacity of hydrogen
LP	laser peening
HEDE	hydrogen-enhanced decohesion mechanism
CTOD	critical crack tip opening displacement
HELP	hydrogen-enhanced localized plasticity
AIDE	adsorption-induced dislocation emission
HEMP	hydrogen-enhanced macroscopic plasticity
TDA	thermal desorption analysis
CSRT	conventional strain rate test
I_δ	elongation loss rate
I_ψ	area reduction rate
δ_0	elongation before charging
δ_H	elongation after charging
ψ_0	area reduction before charging
ψ_H	area reduction after charging
σ_{TH}	threshold stress
LIST	linearly increasing stress testing
SCC	stress corrosion cracking
TWIP	twinning-induced plasticity
DCB	double cantilever beam
SSC	sulfide stress cracking
K_{ISSC}	threshold stress intensity factor for SSC
K_{I0}	initial stress intensity factor
K_{IC}	critical stress intensity factor
a	DCB and fatigue test—specimen crack length
P	DCB test—lift-off load
h	DCB test—specimen height of each arm
B	DCB test—specimen thickness
B_n	DCB test—specimen web thickness
EDM	electro-discharged machine
H_2SO_4	sulfuric acid
D	bent beam test—deflection
S	bent beam test—nominal outer fiber pseudo-stress
l	bent beam test—distance between centerlines of end supports
E	elastic modulus
t	bent beam test—thickness of test specimen
T	bent beam test result
n	bent beam test—total number of specimens tested
DFS	delayed fracture strength
HCl	hydrochloric acid
FGCR	fatigue crack growth rate
HAFCGR	hydrogen-affected fatigue crack growth rate

CT	compact tension
N	fatigue test—number of the cycles
ΔK	fatigue test variation in the stress intensity factor

References

1. Johnson, W.H., II. On Some Remarkable Changes Produced in Iron and Steel by the Action of Hydrogen and Acids. *Proc. R. Soc. Lond.* **1875**, *23*, 168–179. [CrossRef]
2. Gangloff, R.P.; Somerday, B.P. *Gaseous Hydrogen Embrittlement of Materials in Energy Technologies: Mechanisms, Modelling and Future Developments*; Elsevier: Amsterdam, The Netherlands, 2012; ISBN 0857095374.
3. Alvaro, A.; Jensen, I.T.; Kheradmand, N.; Løvvik, O.M.; Olden, V. Hydrogen Embrittlement in Nickel, Visited by First Principles Modeling, Cohesive Zone Simulation and Nanomechanical Testing. *Int. J. Hydrogen Energy* **2015**, *40*, 16892–16900. [CrossRef]
4. Lu, X.; Wang, D. Effect of Hydrogen on Deformation Behavior of Alloy 725 Revealed by In-Situ Bi-Crystalline Micropillar Compression Test. *J. Mater. Sci. Technol.* **2021**, *67*, 243–253. [CrossRef]
5. Bhadeshia, H.K.D.H. Prevention of Hydrogen Embrittlement in Steels. *ISIJ Int.* **2016**, *56*, 24–36. [CrossRef]
6. Deng, Y.; Barnoush, A. Hydrogen Embrittlement Revealed via Novel in Situ Fracture Experiments Using Notched Micro-Cantilever Specimens. *Acta Mater.* **2018**, *142*, 236–247. [CrossRef]
7. Koyama, M.; Akiyama, E.; Lee, Y.-K.; Raabe, D.; Tsuzaki, K. Overview of Hydrogen Embrittlement in High-Mn Steels. *Int. J. Hydrogen Energy* **2017**, *42*, 12706–12723. [CrossRef]
8. Martin, M.L.; Dadfarnia, M.; Nagao, A.; Wang, S.; Sofronis, P. Enumeration of the Hydrogen-Enhanced Localized Plasticity Mechanism for Hydrogen Embrittlement in Structural Materials. *Acta Mater.* **2019**, *165*, 734–750. [CrossRef]
9. Li, X.; Zhang, J.; Fu, Q.; Song, X.; Shen, S.; Li, Q. A Comparative Study of Hydrogen Embrittlement of 20SiMn2CrNiMo, PSB1080 and PH13-8Mo High Strength Steels. *Mater. Sci. Eng. A* **2018**, *724*, 518–528. [CrossRef]
10. Li, X.; Zhang, J.; Shen, S.; Wang, Y.; Song, X. Effect of Tempering Temperature and Inclusions on Hydrogen-Assisted Fracture Behaviors of a Low Alloy Steel. *Mater. Sci. Eng. A* **2017**, *682*, 359–369. [CrossRef]
11. Neeraj, T.; Srinivasan, R.; Li, J. Hydrogen Embrittlement of Ferritic Steels: Observations on Deformation Microstructure, Nanoscale Dimples and Failure by Nanovoiding. *Acta Mater.* **2012**, *60*, 5160–5171. [CrossRef]
12. Zhu, X.; Li, W.; Hsu, T.Y.; Zhou, S.; Wang, L.; Jin, X. Improved Resistance to Hydrogen Embrittlement in a High-Strength Steel by Quenching–Partitioning–Tempering Treatment. *Scr. Mater.* **2015**, *97*, 21–24. [CrossRef]
13. Zhou, P.; Li, W.; Zhu, X.; Li, Y.; Jin, X.; Chen, J. Graphene Containing Composite Coatings as a Protective Coatings against Hydrogen Embrittlement in Quenching & Partitioning High Strength Steel. *J. Electrochem. Soc.* **2016**, *163*, D160.
14. Zhao, T.; Liu, Z.; Xu, X.; Li, Y.; Du, C.; Liu, X. Interaction between Hydrogen and Cyclic Stress and Its Role in Fatigue Damage Mechanism. *Corros. Sci.* **2019**, *157*, 146–156. [CrossRef]
15. Ronevich, J.A.; Somerday, B.P.; San Marchi, C.W. Effects of Microstructure Banding on Hydrogen Assisted Fatigue Crack Growth in X65 Pipeline Steels. *Int. J. Fatigue* **2016**, *82*, 497–504. [CrossRef]
16. Alvaro, A.; Wan, D.; Olden, V.; Barnoush, A. Hydrogen Enhanced Fatigue Crack Growth Rates in a Ferritic Fe-3 Wt% Si Alloy and a X70 Pipeline Steel. *Eng. Fract. Mech.* **2019**, *219*, 106641. [CrossRef]
17. Ogawa, Y.; Matsunaga, H.; Yamabe, J.; Yoshikawa, M.; Matsuoka, S. Fatigue Limit of Carbon and CrMo Steels as a Small Fatigue Crack Threshold in High-Pressure Hydrogen Gas. *Int. J. Hydrogen Energy* **2018**, *43*, 20133–20142. [CrossRef]
18. Wang, R. Effects of Hydrogen on the Fracture Toughness of a X70 Pipeline Steel. *Corros. Sci.* **2009**, *51*, 2803–2810. [CrossRef]
19. Chatzidouros, E.V.; Traidia, A.; Devarapalli, R.S.; Pantelis, D.I.; Steriotis, T.A.; Jouiad, M. Effect of Hydrogen on Fracture Toughness Properties of a Pipeline Steel under Simulated Sour Service Conditions. *Int. J. Hydrogen Energy* **2018**, *43*, 5747–5759. [CrossRef]
20. Song, Y.; Chai, M.; Yang, B.; Han, Z.; Ai, S.; Liu, Y.; Cheng, G.; Li, Y. Investigation of the Influence of Pre-Charged Hydrogen on Fracture Toughness of as-Received 2.25Cr1Mo0.25V Steel and Weld. *Materials* **2018**, *11*, 1068. [CrossRef]
21. Bhuiyan, M.S.; Toda, H.; Shimizu, K.; Su, H.; Uesugi, K.; Takeuchi, A.; Watanabe, Y. The Role of Hydrogen on the Local Fracture Toughness Properties of 7XXX Aluminum Alloys. *Metall. Mater. Trans. A* **2018**, *49*, 5368–5381. [CrossRef]
22. Pallaspuro, S.; Yu, H.; Kisko, A.; Porter, D.; Zhang, Z. Fracture Toughness of Hydrogen Charged As-Quenched Ultra-High-Strength Steels at Low Temperatures. *Mater. Sci. Eng. A* **2017**, *688*, 190–201. [CrossRef]
23. Yamabe, J.; Yoshikawa, M.; Matsunaga, H.; Matsuoka, S. Effects of Hydrogen Pressure, Test Frequency and Test Temperature on Fatigue Crack Growth Properties of Low-Carbon Steel in Gaseous Hydrogen. *Procedia Struct. Integr.* **2016**, *2*, 525–532. [CrossRef]
24. Staehle, R.W. *Fundamental Aspects of Stress Corrosion Cracking*; National Association of Corrosion Engineers: Houston, TX, USA, 1969; p. 3.
25. Kittel, C.; McEuen, P. *Introduction to Solid State Physics*; John Wiley & Sons: Hoboken, NJ, USA, 2018; ISBN 1119454166.
26. Smialowski, M. *Hydrogen in Steel: Effect of Hydrogen on Iron and Steel during Production, Fabrication, and Use*; Elsevier: Amsterdam, The Netherlands, 2014; ISBN 1483213714.
27. Smialowski, M. Chapter 1—Metals and hydrogen. In *Hydrogen in Steel*; Smialowski, M., Ed.; Pergamon: Oxford, UK, 1962; pp. 1–24; ISBN 978-0-08-009697-1.
28. Tien, C. Chapter 1—Introduction. In *Introduction to Adsorption*; Tien, C., Ed.; Elsevier: Amsterdam, The Netherlands, 2019; pp. 1–6; ISBN 978-0-12-816446-4.

29. Atkins, P.W.; De Paula, J. *Atkins' Physical Chemistry*; Oxford University Press: Oxford, UK, 2014; ISBN 019969740X.
30. Christmann, K. Interaction of Hydrogen with Solid Surfaces. *Surf. Sci. Rep.* **1988**, *9*, 1–163. [CrossRef]
31. Liu, Q.; Atrens, A. A Critical Review of the Influence of Hydrogen on the Mechanical Properties of Medium-Strength Steels. *Corros. Rev.* **2013**, *31*, 85–103. [CrossRef]
32. Lu, X.; Wang, D.; Johnsen, R. Hydrogen Diffusion and Trapping in Nickel-Based Alloy 625: An Electrochemical Permeation Study. *Electrochim. Acta* **2022**, *421*, 140477. [CrossRef]
33. Martin, F.; Feaugas, X.; Oudriss, A.; Tanguy, D.; Briottet, L.; Kittel, J. State of Hydrogen in Matter: Fundamental Ad/Absorption, Trapping and Transport Mechanisms. In *Mechanics-Microstructure-Corrosion Coupling*; Elsevier: Amsterdam, The Netherlands, 2019; pp. 171–197.
34. Popov, B.N.; Lee, J.-W.; Djukic, M.B. Hydrogen Permeation and Hydrogen-Induced Cracking. In *Handbook of Environmental Degradation of Materials*; Elsevier: Amsterdam, The Netherlands, 2018; pp. 133–162.
35. Liu, Q.; Atrens, A.D.; Shi, Z.; Verbeken, K.; Atrens, A. Determination of the Hydrogen Fugacity during Electrolytic Charging of Steel. *Corros. Sci.* **2014**, *87*, 239–258. [CrossRef]
36. Protopopoff, E.; Marcus, P. *Surface Effects on Hydrogen Entry into Metals*; Marcel Dekker: New York, NY, USA, 2002.
37. Li, X.; Ma, X.; Zhang, J.; Akiyama, E.; Wang, Y.; Song, X. Review of Hydrogen Embrittlement in Metals: Hydrogen Diffusion, Hydrogen Characterization, Hydrogen Embrittlement Mechanism and Prevention. *Acta Metall. Sin. Engl. Lett.* **2020**, *33*, 759–773. [CrossRef]
38. Kirchheim, R.; Pundt, A. 25—Hydrogen in Metals. In *Physical Metallurgy*, 5th ed.; Laughlin, D.E., Hono, K., Eds.; Elsevier: Oxford, UK, 2014; pp. 2597–2705; ISBN 978-0-444-53770-6.
39. Flynn, C.P.; Stoneham, A.M. Quantum Theory of Diffusion with Application to Light Interstitials in Metals. *Phys. Rev. B* **1970**, *1*, 3966. [CrossRef]
40. Paxton, A.T. From Quantum Mechanics to Physical Metallurgy of Steels. *Mater. Sci. Technol.* **2014**, *30*, 1063–1070. [CrossRef]
41. Gillan, M.J. Quantum Transition State Theory and the Diffusion of Hydrogen in Metals. *J. Less Common Met.* **1991**, *172–174*, 529–537. [CrossRef]
42. Díaz, A.; Alegre, J.M.; Cuesta, I.I. A Review on Diffusion Modelling in Hydrogen Related Failures of Metals. *Eng. Fail. Anal.* **2016**, *66*, 577–595. [CrossRef]
43. Fukai, Y. Phase Diagrams and Statistical Thermodynamics of Binary M-H Systems. In *The Metal-Hydrogen System: Basic Bulk Properties*; Fukai, Y., Ed.; Springer: Berlin/Heidelberg, Germany, 1993; pp. 1–42; ISBN 978-3-662-02801-8.
44. Ovejero-García, J. Hydrogen Microprint Technique in the Study of Hydrogen in Steels. *J. Mater. Sci.* **1985**, *20*, 2623–2629. [CrossRef]
45. Jack, T.A. Investigation of Hydrogen Induced Cracking Susceptibility of API 5L X65 Pipeline Steels. Master's Thesis, University of Saskatchewan, Saskatoon, SK, Canada, 2021.
46. Thomas, A.; Szpunar, J.A. Hydrogen Diffusion and Trapping in X70 Pipeline Steel. *Int. J. Hydrogen Energy* **2020**, *45*, 2390–2404. [CrossRef]
47. Ronevich, J.A.; Speer, J.G.; Krauss, G.; Matlock, D.K. Improvement of the Hydrogen Microprint Technique on AHSS Steels. *Metallogr. Microstruct. Anal.* **2012**, *1*, 79–84. [CrossRef]
48. Devanathan, M.A.V.; Stachurski, Z. The Adsorption and Diffusion of Electrolytic Hydrogen in Palladium. *Proc. R. Soc. Lond. A Math. Phys. Sci.* **1962**, *270*, 90–102.
49. Zhao, W.; Zhang, T.; Zhao, Y.; Sun, J.; Wang, Y. Hydrogen Permeation and Embrittlement Susceptibility of X80 Welded Joint under High-Pressure Coal Gas Environment. *Corros. Sci.* **2016**, *111*, 84–97. [CrossRef]
50. *ASTM G148-97*; Standard Practice for Evaluation of Hydrogen Uptake, Permeation, and Transport in Metals by an Electrochemical Technique. ASTM International: West Conshohocken, PA, USA, 2011.
51. Ajito, S.; Hojo, T.; Koyama, M.; Akiyama, E. Effects of Ammonium Thiocyanate and PH of Aqueous Solutions on Hydrogen Absorption into Iron under Cathodic Polarization. *ISIJ Int.* **2021**, *61*, 1209–1214. [CrossRef]
52. Fuji, H.; Fujishiro, T.; Hara, T. Effect of Ammonium Thiocyanate on Hydrogen Entry Behavior of Low Alloy Steel under Galvanostatic Cathode Charging. *ISIJ Int.* **2021**, *61*, 1151–1158. [CrossRef]
53. Mohtadi-Bonab, M.A.; Szpunar, J.A.; Razavi-Tousi, S.S. A Comparative Study of Hydrogen Induced Cracking Behavior in API 5L X60 and X70 Pipeline Steels. *Eng. Fail. Anal.* **2013**, *33*, 163–175. [CrossRef]
54. Fallahmohammadi, E.; Bolzoni, F.; Lazzari, L. Measurement of Lattice and Apparent Diffusion Coefficient of Hydrogen in X65 and F22 Pipeline Steels. *Int. J. Hydrogen Energy* **2013**, *38*, 2531–2543. [CrossRef]
55. Koren, E.; Hagen, C.M.H.; Wang, D.; Lu, X.; Johnsen, R.; Yamabe, J. Experimental Comparison of Gaseous and Electrochemical Hydrogen Charging in X65 Pipeline Steel Using the Permeation Technique. *Corros. Sci.* **2023**, *215*, 111025. [CrossRef]
56. Zhang, S.; Li, J.; An, T.; Zheng, S.; Yang, K.; Lv, L.; Xie, C.; Chen, L.; Zhang, L. Investigating the Influence Mechanism of Hydrogen Partial Pressure on Fracture Toughness and Fatigue Life by In-Situ Hydrogen Permeation. *Int. J. Hydrogen Energy* **2021**, *46*, 20621–20629. [CrossRef]
57. Wang, C.; Zhang, J.; Liu, C.; Hu, Q.; Zhang, R.; Xu, X.; Yang, H.; Ning, Y.; Li, Y. Study on Hydrogen Embrittlement Susceptibility of X80 Steel through In-Situ Gaseous Hydrogen Permeation and Slow Strain Rate Tensile Tests. *Int. J. Hydrogen Energy* **2023**, *48*, 243–256. [CrossRef]

58. Cheng, Y.F. Analysis of Electrochemical Hydrogen Permeation through X-65 Pipeline Steel and Its Implications on Pipeline Stress Corrosion Cracking. *Int. J. Hydrogen Energy* **2007**, *32*, 1269–1276. [CrossRef]
59. Turnbull, A. 4—Hydrogen Diffusion and Trapping in Metals. In *Gaseous Hydrogen Embrittlement of Materials in Energy Technologies*; Gangloff, R.P., Somerday, B.P., Eds.; Woodhead Publishing: Sawston, UK, 2012; Volume 1, pp. 89–128; ISBN 978-0-85709-536-7.
60. Zakroczymski, T. Adaptation of the Electrochemical Permeation Technique for Studying Entry, Transport and Trapping of Hydrogen in Metals. *Electrochim. Acta* **2006**, *51*, 2261–2266. [CrossRef]
61. Doyle, D.M.; Palumbo, G.; Aust, K.T.; El-Sherik, A.M.; Erb, U. The Influence of Intercrystalline Defects on Hydrogen Activity and Transport in Nickel. *Acta Metall. Et. Mater.* **1995**, *43*, 3027–3033. [CrossRef]
62. Rudomilova, D.; Prošek, T.; Salvetr, P.; Knaislová, A.; Novák, P.; Kodým, R.; Schimo-Aichhorn, G.; Muhr, A.; Duchaczek, H.; Luckeneder, G. The Effect of Microstructure on Hydrogen Permeability of High Strength Steels. *Mater. Corros.* **2020**, *71*, 909–917. [CrossRef]
63. Song, Y.; Han, Z.; Chai, M.; Yang, B.; Liu, Y.; Cheng, G.; Li, Y.; Ai, S. Effect of Cementite on the Hydrogen Diffusion/Trap Characteristics of 2.25 Cr-1Mo-0.25 V Steel with and without Annealing. *Materials* **2018**, *11*, 788. [CrossRef]
64. Dong, C.F.; Liu, Z.Y.; Li, X.G.; Cheng, Y.F. Effects of Hydrogen-Charging on the Susceptibility of X100 Pipeline Steel to Hydrogen-Induced Cracking. *Int. J. Hydrogen Energy* **2009**, *34*, 9879–9884. [CrossRef]
65. Haq, A.J.; Muzaka, K.; Dunne, D.P.; Calka, A.; Pereloma, E.V. Effect of Microstructure and Composition on Hydrogen Permeation in X70 Pipeline Steels. *Int. J. Hydrogen Energy* **2013**, *38*, 2544–2556. [CrossRef]
66. Zhou, P.; Li, W.; Zhao, H.; Jin, X. Role of Microstructure on Electrochemical Hydrogen Permeation Properties in Advanced High Strength Steels. *Int. J. Hydrogen Energy* **2018**, *43*, 10905–10914. [CrossRef]
67. San Marchi, C.; Somerday, B.P.; Robinson, S.L. Permeability, Solubility and Diffusivity of Hydrogen Isotopes in Stainless Steels at High Gas Pressures. *Int. J. Hydrogen Energy* **2007**, *32*, 100–116. [CrossRef]
68. JO'M, B.; Subramanyan, P.K. The Equivalent Pressure of Molecular Hydrogen in Cavities within Metals in Terms of the Overpotential Developed during the Evolution of Hydrogen. *Electrochim. Acta* **1971**, *16*, 2169–2179.
69. Sandler, S.I. *Chemical, Biochemical, and Engineering Thermodynamics*; John Wiley & Sons: Hoboken, NJ, USA, 2017; ISBN 047050479X.
70. Venezuela, J.; Tapia-Bastidas, C.; Zhou, Q.; Depover, T.; Verbeken, K.; Gray, E.; Liu, Q.; Liu, Q.; Zhang, M.; Atrens, A. Determination of the Equivalent Hydrogen Fugacity during Electrochemical Charging of 3.5 NiCrMoV Steel. *Corros. Sci.* **2018**, *132*, 90–106. [CrossRef]
71. Atrens, A.; Mezzanotte, D.; Fiore, N.F.; Genshaw, M.A. Electrochemical Studies of Hydrogen Diffusion and Permeability in Ni. *Corros. Sci.* **1980**, *20*, 673–684. [CrossRef]
72. Liu, Q.; Gray, E.; Venezuela, J.; Zhou, Q.; Tapia-Bastidas, C.; Zhang, M.; Atrens, A. Equivalent Hydrogen Fugacity during Electrochemical Charging of 980DP Steel Determined by Thermal Desorption Spectroscopy. *Adv. Eng. Mater.* **2018**, *20*, 1700469. [CrossRef]
73. Polfus, J.M.; Løvvik, O.M.; Bredesen, R.; Peters, T. Hydrogen Induced Vacancy Clustering and Void Formation Mechanisms at Grain Boundaries in Palladium. *Acta Mater.* **2020**, *195*, 708–719. [CrossRef]
74. Lee, S.-M.; Lee, J.-Y. The Trapping and Transport Phenomena of Hydrogen in Nickel. *Metall. Trans. A* **1986**, *17*, 181–187. [CrossRef]
75. Chu, W. *Hydrogen Embrittlement and Stress Corrosion: Fundamentals*; Science Press: Beijing, China, 2013; ISBN 9787030388841.
76. Frappart, S.; Feaugas, X.; Creus, J.; Thebault, F.; Delattre, L.; Marchebois, H. Study of the Hydrogen Diffusion and Segregation into Fe–C–Mo Martensitic HSLA Steel Using Electrochemical Permeation Test. *J. Phys. Chem. Solids* **2010**, *71*, 1467–1479. [CrossRef]
77. Pressouyre, G.M. Trap Theory of Hydrogen Embrittlement. *Acta Metall.* **1980**, *28*, 895–911. [CrossRef]
78. Mohtadi-Bonab, M.A.; Eskandari, M.; Szpunar, J.A. Texture, Local Misorientation, Grain Boundary and Recrystallization Fraction in Pipeline Steels Related to Hydrogen Induced Cracking. *Mater. Sci. Eng. A* **2015**, *620*, 97–106. [CrossRef]
79. Pressouyre, G.M.; Bernstein, I.M. An Example of the Effect of Hydrogen Trapping on Hydrogen Embrittlement. *Metall. Trans. A* **1981**, *12*, 835–844. [CrossRef]
80. Fu, Y.; Li, T.; Yan, Y.-B.; Wang, X.-Y.; Zhu, M.-L.; Xuan, F.-Z. A First Principles Study on H-Atom Interaction with Bcc Metals. *Int. J. Hydrogen Energy* **2023**, *48*, 9911–9920. [CrossRef]
81. Beck, W.; Bockris, J.O.M.; Genshaw, M.A.; Subramanyan, P.K. Diffusivity and Solubility of Hydrogen as a Function of Composition in Fe-Ni Alloys. *Metall. Trans.* **1971**, *2*, 883–888. [CrossRef]
82. Fukuda, K.; Tojo, A.; Matsumoto, R. Evaluating Solubility and Diffusion Coefficient of Hydrogen in Martensitic Steel Using Computational Mechanics. *Mater. Trans.* **2020**, *61*, 1287–1293. [CrossRef]
83. Pressouyre, G.M.; Bernstein, I.M. A Quantitative Analysis of Hydrogen Trapping. *Metall. Trans. A* **1978**, *9*, 1571–1580. [CrossRef]
84. Zaw, A.K.; Chernov, I.I.; Staltsov, M.S.; Kalin, B.A.; Efimov, V.S. Hydrogen Retention by Vanadium-Titanium Alloys. *Inorg. Mater. Appl. Res.* **2015**, *6*, 138–142. [CrossRef]
85. Paul, A.; Laurila, T.; Vuorinen, V.; Divinski, S.V. Short-Circuit Diffusion. In *Thermodynamics, Diffusion and the Kirkendall Effect in Solids*; Paul, A., Laurila, T., Vuorinen, V., Divinski, S.V., Eds.; Springer International Publishing: Cham, Switzerland, 2014; pp. 429–491; ISBN 978-3-319-07461-0.
86. Oudriss, A.; Creus, J.; Bouhattate, J.; Savall, C.; Peraudeau, B.; Feaugas, X. The Diffusion and Trapping of Hydrogen along the Grain Boundaries in Polycrystalline Nickel. *Scr. Mater.* **2012**, *66*, 37–40. [CrossRef]
87. Ichimura, M.; Sasajima, Y.; Imabayashi, M. Grain Boundary Effect on Diffusion of Hydrogen in Pure Aluminum. *Mater. Trans. JIM* **1991**, *32*, 1109–1114. [CrossRef]

88. Choo, W.Y.; Lee, J.Y. Effect of Cold Working on the Hydrogen Trapping Phenomena in Pure Iron. *Metall. Trans. A* **1983**, *14*, 1299–1305. [CrossRef]
89. Hull, D.; Bacon, D.J. Chapter 2—Observation of Dislocations. In *Introduction to Dislocations*, 5th ed.; Hull, D., Bacon, D.J., Eds.; Butterworth-Heinemann: Oxford, UK, 2011; pp. 21–41; ISBN 978-0-08-096672-4.
90. Song, Y.; Huang, S.; Sheng, J.; Agyenim-Boateng, E.; Jiang, Y.; Liu, Q.; Zhu, M. Improvement of Hydrogen Embrittlement Resistance of 2205 Duplex Stainless Steel by Laser Peening. *Int. J. Hydrogen Energy* **2023**, *48*, 18930–18945. [CrossRef]
91. Kurkela, M.; Latanision, R.M. The Effect of Plastic Deformation on the Transport of Hydrogen in Nickel. *Scr. Metall.* **1979**, *13*, 927–932. [CrossRef]
92. Chêne, J.; Brass, A.M. Hydrogen Transport by Mobile Dislocations in Nickel Base Superalloy Single Crystals. *Scr. Mater.* **1999**, *40*, 537–542. [CrossRef]
93. Mohtadi-Bonab, M.A.; Masoumi, M. Different Aspects of Hydrogen Diffusion Behavior in Pipeline Steel. *J. Mater. Res. Technol.* **2023**, *24*, 4762–4783. [CrossRef]
94. Park, G.T.; Koh, S.U.; Jung, H.G.; Kim, K.Y. Effect of Microstructure on the Hydrogen Trapping Efficiency and Hydrogen Induced Cracking of Linepipe Steel. *Corros. Sci.* **2008**, *50*, 1865–1871. [CrossRef]
95. Turnbull, A.; Hutchings, R.B. Analysis of Hydrogen Atom Transport in a Two-Phase Alloy. *Mater. Sci. Eng. A* **1994**, *177*, 161–171. [CrossRef]
96. Van den Eeckhout, E.; Laureys, A.; Van Ingelgem, Y.; Verbeken, K. Hydrogen Permeation through Deformed and Heat-Treated Armco Pure Iron. *Mater. Sci. Technol.* **2017**, *33*, 1515–1523. [CrossRef]
97. Pfeil, L.B. The Effect of Occluded Hydrogen on the Tensile Strength of Iron. *Proc. R. Soc. Lond. Ser. A Contain. Pap. A Math. Phys. Character* **1926**, *112*, 182–195.
98. Troiano, A.R. The Role of Hydrogen and Other Interstitials in the Mechanical Behavior of Metals. *Trans. ASM* **1960**, *52*, 54–81. [CrossRef]
99. Wipf, H. *Hydrogen in Metals*; Springer: Berlin/Heidelberg, Germany, 1997; Volume 3.
100. Sun, B.; Wang, D.; Lu, X.; Wan, D.; Ponge, D.; Zhang, X. Current Challenges and Opportunities toward Understanding Hydrogen Embrittlement Mechanisms in Advanced High-Strength Steels: A Review. *Acta Metall. Sin. Engl. Lett.* **2021**, *34*, 741–754. [CrossRef]
101. Martínez-Pañeda, E.; del Busto, S.; Niordson, C.F.; Betegón, C. Strain Gradient Plasticity Modeling of Hydrogen Diffusion to the Crack Tip. *Int. J. Hydrogen Energy* **2016**, *41*, 10265–10274. [CrossRef]
102. Katzarov, I.H.; Paxton, A.T. Hydrogen Embrittlement II. Analysis of Hydrogen-Enhanced Decohesion across (111) Planes in α-Fe. *Phys. Rev. Mater.* **2017**, *1*, 033603. [CrossRef]
103. Lynch, S. Hydrogen Embrittlement Phenomena and Mechanisms. *Corros. Rev.* **2012**, *30*, 105–123. [CrossRef]
104. Dwivedi, S.K.; Vishwakarma, M. Hydrogen Embrittlement in Different Materials: A Review. *Int. J. Hydrogen Energy* **2018**, *43*, 21603–21616. [CrossRef]
105. Choudhary, S.; Vishwakarma, M.; Dwivedi, S.K. Evaluation and Prevention of Hydrogen Embrittlement by NDT Methods: A Review. *Mater. Proc.* **2021**, *6*, 18.
106. Kappes, M.; Iannuzzi, M.; Carranza, R.M. Hydrogen Embrittlement of Magnesium and Magnesium Alloys: A Review. *J. Electrochem. Soc.* **2013**, *160*, C168. [CrossRef]
107. Zapffe, C.A.; Sims, C.E. Hydrogen Embrittlement, Internal Stress and Defects in Steel. *Trans. Aime* **1941**, *145*, 225–271.
108. Beachem, C.D. A New Model for Hydrogen-Assisted Cracking (Hydrogen "Embrittlement"). *Metall. Mater. Trans. B* **1972**, *3*, 441–455. [CrossRef]
109. Pradhan, A.; Vishwakarma, M.; Dwivedi, S.K. A Review: The Impact of Hydrogen Embrittlement on the Fatigue Strength of High Strength Steel. *Mater. Today Proc.* **2020**, *26*, 3015–3019. [CrossRef]
110. Song, J.; Curtin, W.A. Atomic Mechanism and Prediction of Hydrogen Embrittlement in Iron. *Nat. Mater.* **2013**, *12*, 145–151. [CrossRef]
111. Sofronis, P.; Robertson, I.M. Transmission Electron Microscopy Observations and Micromechanical/Continuum Models for the Effect of Hydrogen on the Mechanical Behaviour of Metals. *Philos. Mag. A* **2002**, *82*, 3405–3413. [CrossRef]
112. Robertson, I.M. The Effect of Hydrogen on Dislocation Dynamics. *Eng. Fract. Mech.* **2001**, *68*, 671–692. [CrossRef]
113. Robertson, I.M.; Sofronis, P.; Nagao, A.; Martin, M.L.; Wang, S.; Gross, D.W.; Nygren, K.E. Hydrogen Embrittlement Understood. *Metall. Mater. Trans. A* **2015**, *46*, 2323–2341. [CrossRef]
114. Hirth, J.P. Effects of Hydrogen on the Properties of Iron and Steel. *Metall. Trans. A* **1980**, *11*, 861–890. [CrossRef]
115. Matsui, H.; Kimura, H.; Kimura, A. The Orientation Dependence of the Yield and Flow Stress of High Purity Iron Single Crystals Doped with Hydrogen. In *Strength of Metals and Alloys*; Elsevier: Amsterdam, The Netherlands, 1979; pp. 977–982.
116. Li, X.; Zhang, J.; Akiyama, E.; Fu, Q.; Li, Q. Hydrogen Embrittlement Behavior of Inconel 718 Alloy at Room Temperature. *J. Mater. Sci. Technol.* **2019**, *35*, 499–502. [CrossRef]
117. Xie, D.; Li, S.; Li, M.; Wang, Z.; Gumbsch, P.; Sun, J.; Ma, E.; Li, J.; Shan, Z. Hydrogenated Vacancies Lock Dislocations in Aluminium. *Nat. Commun.* **2016**, *7*, 13341. [CrossRef] [PubMed]
118. Wu, X.; Zhang, H.; Yang, M.; Jia, W.; Qiu, Y.; Lan, L. From the Perspective of New Technology of Blending Hydrogen into Natural Gas Pipelines Transmission: Mechanism, Experimental Study, and Suggestions for Further Work of Hydrogen Embrittlement in High-Strength Pipeline Steels. *Int. J. Hydrogen Energy* **2022**, *47*, 8071–8090. [CrossRef]

119. Lynch, S.P. Hydrogen Embrittlement and Liquid-Metal Embrittlement in Nickel Single Crystals. *Scr. Metall.* **1979**, *13*, 1051–1056. [CrossRef]
120. Lynch, S.P. Mechanisms of Hydrogen Assisted Cracking—A Review. In Proceedings of the International Conference on Hydrogen Effects on Material Behaviour and Corrosion Deformation Interactions, Moran, WY, USA, 22–26 September 2002; pp. 449–466.
121. Westlake, D.G. *Generalized Model for Hydrogen Embrittlement*; Argonne National Laboratory: Lemont, IL, USA, 1969.
122. Birnbaum, H.K. *Mechanisms of Hydrogen Related Fracture of Metals/Hydrogen Effects on Materials Behavior*; Moody, N.R., Thompson, A.W., Eds.; TMS: Warrendale, PA, USA, 1990; pp. 639–658.
123. Birnbaum, H.K.; Robertson, I.M.; Sofronis, P.; Teter, D. Mechanisms of Hydrogen Related Fracture—A Review. In Proceedings of the Second International Conference on Corrosion-Deformation Interactions. CDI'96, Nice, France, 24–26 September 1996; pp. 172–195.
124. Uhlig, H.H. *Evaluation of Stress-Corrosion Cracking Mechanisms*; Academic Press: Cambridge, MA, USA, 1969.
125. Song, E.J.; Bhadeshia, H.; Suh, D.-W. Effect of Hydrogen on the Surface Energy of Ferrite and Austenite. *Corros. Sci.* **2013**, *77*, 379–384. [CrossRef]
126. Zhou, C.; Huang, Q.; Guo, Q.; Zheng, J.; Chen, X.; Zhu, J.; Zhang, L. Sulphide Stress Cracking Behaviour of the Dissimilar Metal Welded Joint of X60 Pipeline Steel and Inconel 625 Alloy. *Corros. Sci.* **2016**, *110*, 242–252. [CrossRef]
127. Liu, Q.; Zhou, Q.; Venezuela, J.; Zhang, M.; Atrens, A. Evaluation of the Influence of Hydrogen on Some Commercial DP, Q&P and TWIP Advanced High-Strength Steels during Automobile Service. *Eng. Fail. Anal.* **2018**, *94*, 249–273.
128. Dwivedi, S.K.; Vishwakarma, M. Effect of Hydrogen in Advanced High Strength Steel Materials. *Int. J. Hydrogen Energy* **2019**, *44*, 28007–28030. [CrossRef]
129. Ćwiek, J.; Michalska-Ćwiek, J. Evaluation of Hydrogen Degradation of High-Strength Weldable Steels. *J. Achiev. Mater. Manuf. Eng.* **2010**, *42*, 103–110.
130. Koyama, M.; Rohwerder, M.; Tasan, C.C.; Bashir, A.; Akiyama, E.; Takai, K.; Raabe, D.; Tsuzaki, K. Recent Progress in Microstructural Hydrogen Mapping in Steels: Quantification, Kinetic Analysis, and Multi-Scale Characterisation. *Mater. Sci. Technol.* **2017**, *33*, 1481–1496. [CrossRef]
131. Li, H.; Niu, R.; Li, W.; Lu, H.; Cairney, J.; Chen, Y.-S. Hydrogen in Pipeline Steels: Recent Advances in Characterization and Embrittlement Mitigation. *J. Nat. Gas Sci. Eng.* **2022**, *105*, 104709. [CrossRef]
132. Dong, C.; Xiao, K.; Liu, Z.; Yang, W.; Li, X. Hydrogen Induced Cracking of X80 Pipeline Steel. *Int. J. Miner. Metall. Mater.* **2010**, *17*, 579–586. [CrossRef]
133. Elhoud, A.M.; Renton, N.C.; Deans, W.F. Hydrogen Embrittlement of Super Duplex Stainless Steel in Acid Solution. *Int. J. Hydrogen Energy* **2010**, *35*, 6455–6464. [CrossRef]
134. Liu, Y.; Wang, M.; Liu, G. Effect of Hydrogen on Ductility of High Strength 3Ni–Cr–Mo–V Steels. *Mater. Sci. Eng. A* **2014**, *594*, 40–47. [CrossRef]
135. Li, L.; Song, B.; Cai, Z.; Liu, Z.; Cui, X. Effect of Vanadium Content on Hydrogen Diffusion Behaviors and Hydrogen Induced Ductility Loss of X80 Pipeline Steel. *Mater. Sci. Eng. A* **2019**, *742*, 712–721. [CrossRef]
136. Takagi, S.; Hagihara, Y.; Hojo, T.; Urushihara, W.; Kawasaki, K. Comparison of Hydrogen Embrittlement Resistance of High Strength Steel Sheets Evaluated by Several Methods. *ISIJ Int.* **2016**, *56*, 685–692. [CrossRef]
137. Hojo, T.; Kobayashi, J.; Sugimoto, K.; Nagasaka, A.; Akiyama, E. Effects of Alloying Elements Addition on Delayed Fracture Properties of Ultra High-Strength TRIP-Aided Martensitic Steels. *Metals* **2019**, *10*, 6. [CrossRef]
138. Zhong, Y.; Zhou, C.; Chen, S.; Wang, R. Effects of Temperature and Pressure on Stress. Corrosion Cracking Behavior of 310S Stainless Steel in Chloride Solution. *Chin. J. Mech. Eng.* **2017**, *30*, 200–206. [CrossRef]
139. Natividad, C.; Salazar, M.; Contreras, A.; Albiter, A.; Pérez, R.; Gonzalez-Rodriguez, J.G. Sulfide Stress Cracking Susceptibility of Welded X-60 and X-65 Pipeline Steels. *Corrosion* **2006**, *62*, 375–382. [CrossRef]
140. Parkins, R.N. *Development of Strain-Rate Testing and Its Implications*; ASTM International: Conshohocken, PA, USA, 1979.
141. *ASTM G129-21*; Standard Practice for Slow Strain Rate Testing to Evaluate the Susceptibility of Metallic Materials to Environmentally Assisted Cracking. ASTM International: West Conshohocken, PA, USA, 2021.
142. Basukumar, H.K.; Arun, K. V Experimental Investigation on Hydrogen Embrittlement of EN47 Spring Steel. *IOP Conf. Ser. Mater. Sci. Eng.* **2021**, *1126*, 012077. [CrossRef]
143. Hagihara, Y. Evaluation of Delayed Fracture Characteristics of High-Strength Bolt Steels by CSRT. *ISIJ Int.* **2012**, *52*, 292–297. [CrossRef]
144. Hagihara, Y.; Ito, C.; Hisamori, N.; Suzuki, H.; Takai, K.; Akiyama, E. Evaluation of Delayed Fracture Characteristics of High Strength Steel Based on CSRT Method. *Tetsu Hagane/J. Iron Steel Inst. Jpn.* **2008**, *94*, 215–221. [CrossRef]
145. Koyama, M.; Tasan, C.C.; Akiyama, E.; Tsuzaki, K.; Raabe, D. Hydrogen-Assisted Decohesion and Localized Plasticity in Dual-Phase Steel. *Acta Mater.* **2014**, *70*, 174–187. [CrossRef]
146. Atrens, A.; Dannhäuser, G.; Bäro, G. Stress-Corrosion-Cracking of Zircaloy-4 Cladding Tubes: Part 1. Threshold in the Presence of Iodine. *J. Nucl. Mater.* **1984**, *126*, 91–102. [CrossRef]
147. Atrens, A.; Oehlert, A. Linearly-Increasing-Stress Testing of Carbon Steel in 4 N NaNO3 and in Bayer Liquor. *J. Mater. Sci.* **1998**, *33*, 783–788. [CrossRef]
148. Atrens, A.; Brosnan, C.C.; Ramamurthy, S.; Oehlert, A.; Smith, I.O. Linearly Increasing Stress Test (LIST) for SCC Research. *Meas. Sci. Technol.* **1993**, *4*, 1281. [CrossRef]

149. Winzer, N.; Atrens, A.; Dietzel, W.; Song, G.; Kainer, K.U. Comparison of the Linearly Increasing Stress Test and the Constant Extension Rate Test in the Evaluation of Transgranular Stress Corrosion Cracking of Magnesium. *Mater. Sci. Eng. A* **2008**, *472*, 97–106. [CrossRef]
150. Baldy, M.F. Sulfide Stress Cracking of Steels for API Grade N-80 Tubular Products. *Corrosion* **1961**, *17*, 509t–513t. [CrossRef]
151. NACE International. *Laboratory Testing of Metals for Resistance to Specific Forms of Environmental Cracking in H2S Environments: Standard Test Method*; NACE International: Houston, TX, USA, 1996; ISBN 157590036X.
152. *ASTM E1681-03*; Standard Test Method for Determining Threshold Stress Intensity Factor for Environment-Assisted Cracking of Metallic Materials. ASTM International: West Conshohocken, PA, USA, 2008.
153. Greer, J.B. *Results of Interlaboratory Sulfide Stress Cracking Using the NACE t-1f-9 Proposed Test Method*; Exxon Production Research Co.: Houston, TX, USA, 1977.
154. *ASTM G49-85*; Standard Practice for Preparation and Use of Direct Tension Stress-Corrosion Test. ASTM International: West Conshohocken, PA, USA, 2011.
155. Venezuela, J.; Liu, Q.; Zhang, M.; Zhou, Q.; Atrens, A. A Review of Hydrogen Embrittlement of Martensitic Advanced High-Strength Steels. *Corros. Rev.* **2016**, *34*, 153–186. [CrossRef]
156. Al-Mansour, M.; Alfantazi, A.M.; El-Boujdaini, M. Sulfide Stress Cracking Resistance of API-X100 High Strength Low Alloy Steel. *Mater. Des.* **2009**, *30*, 4088–4094. [CrossRef]
157. Carneiro, R.A.; Ratnapuli, R.C.; Lins, V.d.F.C. The Influence of Chemical Composition and Microstructure of API Linepipe Steels on Hydrogen Induced Cracking and Sulfide Stress Corrosion Cracking. *Mater. Sci. Eng. A* **2003**, *357*, 104–110. [CrossRef]
158. Pontremoli, M.; Buzzichelli, G.; De Vito, A.; Bufalini, P. Composition, Microstructure and Properties of Pipeline Steels with High HIC and SSCC Resistance. In Proceedings of the XXII International Metallurgy Congress: Innovation for Quality (L'Innovazione per la Qualita), Bologna, Italy, 17–19 May 1988; Volume 2, pp. 1543–1561.
159. Ume, K.; Taira, T.; Hyodo, T. *Initiation and Propagation Morphology of Sulfide Stress Corrosion Cracking at Welds on Linepipe Steels*; Ume, K., Taira, T., Hyodo, T., Kobayashi, Y., Eds.; Nippon Kokan K. K.: Tokyo, Japan, 1996; pp. 405–414.
160. Li, S.; Chen, C.; Liu, Y.; Yu, H.; Wang, X. Influence of Surface Martensite Layer on Hydrogen Embrittlement of Fe-Mn-C-Mo Steels in Wet H2S Environment. *Int. J. Hydrogen Energy* **2018**, *43*, 16728–16736. [CrossRef]
161. Seol, J.-B.; Jung, J.E.; Jang, Y.W.; Park, C.G. Influence of Carbon Content on the Microstructure, Martensitic Transformation and Mechanical Properties in Austenite/ε-Martensite Dual-Phase Fe–Mn–C Steels. *Acta Mater.* **2013**, *61*, 558–578. [CrossRef]
162. Wei, Y.; Li, Y.; Zhu, L.; Liu, Y.; Lei, X.; Wang, G.; Wu, Y.; Mi, Z.; Liu, J.; Wang, H. Evading the Strength–Ductility Trade-off Dilemma in Steel through Gradient Hierarchical Nanotwins. *Nat. Commun.* **2014**, *5*, 3580. [CrossRef]
163. So, K.H.; Kim, J.S.; Chun, Y.S.; Park, K.-T.; Lee, Y.-K.; Lee, C.S. Hydrogen Delayed Fracture Properties and Internal Hydrogen Behavior of a Fe–18Mn–1.5 Al–0.6 C TWIP Steel. *ISIJ Int.* **2009**, *49*, 1952–1959. [CrossRef]
164. Shao, C.; Hui, W.; Zhang, Y.; Zhao, X.; Weng, Y. Effect of Intercritical Annealing Time on Hydrogen Embrittlement of Warm-Rolled Medium Mn Steel. *Mater. Sci. Eng. A* **2018**, *726*, 320–331. [CrossRef]
165. Sponseller, D.L.; Sponseller, T.E. The Double Cantilever Beam (DCB) Test at Forty. *BHM Berg-Und Hüttenmännische Monatshefte* **2016**, *1*, 19–26. [CrossRef]
166. Heady, R.B. Evaluation of Sulfide Corrosion Cracking Resistance in Low Alloy Steels. *Corrosion* **1977**, *33*, 98–107. [CrossRef]
167. Sponseller, D.L. Interlaboratory Testing of Seven Alloys for SSC Resistance by the Double Cantilever Beam Method. *Corrosion* **1992**, *48*, 159–171. [CrossRef]
168. Guntz, G.C.; Linne, C.; Puissochet, F.; Orlans-Joliet, B.; Popperling, R.; Fliethmann, J.M. DCB Test: A Review of Critical Parameters. In Proceedings of the NACE Corrosion, San Antonio, TX, USA, 25–30 April 1999; p. NACE-99606.
169. Szklarz, K.E. Understanding the Size Effect in NACE TM0177 Method D (DCB) Testing and Implications for Users. In Proceedings of the NACE Corrosion, Houston, TX, USA, 11–16 March 2001; p. NACE-01074.
170. Ernst, H.; Villasante, J.; Pachao, N. Specimen Geometry Effects on Crack Growth in Sour Environments. In Proceedings of the NACE Corrosion, Denver, CO, USA, 7–11 April 2002; p. NACE-02060.
171. Morales, C.R.; Pérez, T.E.; Fitzsimons, G.L. Sulfide Stress Cracking: Some Osbervations about The Dcb Test. In Proceedings of the NACE Corrosion, New Orleans, LA, USA, 9–14 March 1997; p. NACE-97052.
172. Asahi, H.; Sogo, Y.; Higashiyama, H. Effects of Test Conditions on KISSC Values Influenced by SSC Susceptibility of Materials. In *Corrosion/1987, Paper No. 290*; Houston, TX, USA, 1987.
173. Asahi, H. Sulfide stress cracking resistance evaluation methods for steels used in oil field environments-features and problems. In *CORROSION 91/29*; Asahi, H., Tsukano, Y., Ueno, M., Eds.; NACE: Houston, TX, USA, 1991.
174. Echaniz, G.; Morales, C.; Perez, T. *The Effect of Microstructure on the K {Sub ISSC} Low Alloy Carbon Steels*; NACE International: Houston, TX, USA, 1998.
175. Szklarz, K.E.; Sutter, P.; Leyer, J.; Orlans-Joliet, B. NACE TM0177 Method D Test Procedure (DCB) Learnings from Comparative Laboratory Testing. In Proceedings of the NACE Corrosion, San Diego, CA, USA, 16–20 March 2003; p. NACE-03104.
176. Linne, C.P.; Blanchard, F.; Puissochet, F.; Orlans-Joliet, B.J.; Hamilton, R.S. *Heavy Wall Casing in C110 Grade for Sour Service*; NACE International: Houston, TX, USA, 1998.
177. Moderer, L.; Holzer, C.; Mori, G.; Klösch, G.; Klarner, J.; Sonnleitner, R. Einfluss Unterschiedlicher Testparameter Auf Den NACE Standard TM0177 DCB-Sauergastest. *BHM Berg-Und Hüttenmännische Monatshefte* **2013**, *158*, 348–354. [CrossRef]

178. Maldonado, J.G.; Dent, P.; MacDonald, J.; Fowler, C. Sour Service Fitness for Purpose Testing of Linepipe Using Full Ring, 4-Point BB and Large DCBs. In Proceedings of the NACE Corrosion, Phoenix, AZ, USA, 15–19 April 2018; p. NACE-2018.
179. Sun, Y.; Cheng, Y.F. Hydrogen Permeation and Distribution at a High-Strength X80 Steel Weld under Stressing Conditions and the Implication on Pipeline Failure. *Int. J. Hydrogen Energy* **2021**, *46*, 23100–23112. [CrossRef]
180. Zhao, W.; Zhang, T.; He, Z.; Sun, J.; Wang, Y. Determination of the Critical Plastic Strain-Induced Stress of X80 Steel through an Electrochemical Hydrogen Permeation Method. *Electrochim. Acta* **2016**, *214*, 336–344. [CrossRef]
181. Xing, Y.; Yang, Z.; Zhao, Q.; Qiao, L.; Yang, J.; Zhang, L. Stress-Based Hydrogen Damage Model of X80 Pipeline Steel and Its Damage Risk Assessment under Cathodic Interference. *Int. J. Press. Vessel. Pip.* **2023**, *203*, 104947. [CrossRef]
182. Fraser, J.P.; Eldredge, G.G.; Treseder, R.S. Laboratory and Field Methods for Quantitative Study of Sulfide Corrosion Cracking. *Corrosion* **1958**, *14*, 37–43. [CrossRef]
183. Hojo, T.; Zhou, Y.; Kobayashi, J.; Sugimoto, K.; Takemoto, Y.; Nagasaka, A.; Koyama, M.; Ajito, S.; Akiyama, E. Effects of Thermomechanical Processing on Hydrogen Embrittlement Properties of Ultrahigh-Strength TRIP-Aided Bainitic Ferrite Steels. *Metals* **2022**, *12*, 269. [CrossRef]
184. Si, Y.; Tang, Y.; Xu, Z.; Yu, S.; Zhou, X.; Li, K.; Cao, P.; Ma, Y. Hydrogen Embrittlement and Microstructure Characterization of 1500 MPa Martensitic Steel. *Steel Res. Int.* **2022**, *93*, 2200295. [CrossRef]
185. Briottet, L.; Moro, I.; Lemoine, P. Quantifying the Hydrogen Embrittlement of Pipeline Steels for Safety Considerations. *Int. J. Hydrogen Energy* **2012**, *37*, 17616–17623. [CrossRef]
186. Ronevich, J.A.; D'Elia, C.R.; Hill, M.R. Fatigue Crack Growth Rates of X100 Steel Welds in High Pressure Hydrogen Gas Considering Residual Stress Effects. *Eng. Fract. Mech.* **2018**, *194*, 42–51. [CrossRef]
187. *ASTM E 647*; Standard Test Method for Measurement of Fatigue Crack Growth Rates. American Society of Testing and Materials: Philadelphia, PA, USA, 2001.
188. Cheng, A.; Chen, N.-Z. Fatigue Crack Growth Modelling for Pipeline Carbon Steels under Gaseous Hydrogen Conditions. *Int. J. Fatigue* **2017**, *96*, 152–161. [CrossRef]
189. Paris, P.; Erdogan, F. A Critical Analysis of Crack Propagation Laws. *J. Basic Eng.* **1963**, *85*, 528–533. [CrossRef]
190. Meng, B.; Gu, C.; Zhang, L.; Zhou, C.; Li, X.; Zhao, Y.; Zheng, J.; Chen, X.; Han, Y. Hydrogen Effects on X80 Pipeline Steel in High-Pressure Natural Gas/Hydrogen Mixtures. *Int. J. Hydrogen Energy* **2017**, *42*, 7404–7412. [CrossRef]

Article

Effects of C and Al Alloying on Constitutive Model Parameters and Hot Deformation Behavior of Medium-Mn Steels

Guangshun Guo [1], Mingming Wang [1,*], Hongchao Ji [1,2], Xiaoyan Zhang [3], Dongdong Li [1], Chenyang Wei [4] and Fucheng Zhang [3]

1 College of Mechanical Engineering, North China University of Science and Technology, Tangshan 063210, China
2 School of Materials Science and Engineering, Zhejiang University, Hangzhou 310030, China
3 College of Metallurgy and Energy, North China University of Science and Technology, Tangshan 063210, China
4 State Key Laboratory of Metastable Materials Science and Technology, Yanshan University, Qinhuangdao 066004, China
* Correspondence: wangmingming@ncst.edu.cn

Citation: Guo, G.; Wang, M.; Ji, H.; Zhang, X.; Li, D.; Wei, C.; Zhang, F. Effects of C and Al Alloying on Constitutive Model Parameters and Hot Deformation Behavior of Medium-Mn Steels. *Materials* **2024**, *17*, 732. https://doi.org/10.3390/ma17030732

Academic Editor: Davide Palumbo

Received: 19 December 2023
Revised: 17 January 2024
Accepted: 27 January 2024
Published: 3 February 2024

Abstract: Single-pass isothermal hot compression tests on four medium-Mn steels with different C and Al contents were conducted using a Gleeble-3500 thermal simulation machine at varying deformation temperatures (900–1150 °C) and strain rates (0.01–5 s^{-1}). Based on friction correction theory, the friction of the test stress–strain data was corrected. On this basis, the Arrhenius constitutive model of experimental steels considering Al content and strain compensation and hot processing maps of different experimental steels at a strain of 0.9 were established. Moreover, the effects of C and Al contents on constitutive model parameters and hot processing performance were analyzed. The results revealed that the increase in C content changed the trend of the thermal deformation activation energy Q with the true strain. The Q value of 2C7Mn3Al increased by about 50 KJ/mol compared with 7Mn3Al at a true strain greater than 0.4. In contrast, increasing the Al content from 0 to 1.14 wt.% decreased the activation energy of thermal deformation in the true strain range of 0.4–0.9. Continuing to increase to 3.30 wt.% increased the Q of 7Mn3Al over 7Mn by about 65 KJ/mol over the full strain range. In comparison, 7Mn1Al exhibited the best hot processing performance under the deformation temperature of 975–1125 °C and strain rate of 0.2–5 s^{-1}. This is due to the addition of C element reduces the δ-ferrite volume fraction, which leads to the precipitation of κ-carbides and causes the formation of microcracks; an increase in Al content from 0 to 1.14 wt.% reduces the austenite stability and improves the hot workability, but a continued increase in the content up to 3.30 wt.% results in the emergence of δ-ferrite in the microstructure, which slows down the austenite DRX and not conducive to the hot processing performance.

Keywords: medium-Mn steel; hot deformation behavior; constitutive model; processing map

1. Introduction

Environmental protection, weight reduction, and improved safety stand as the primary requirements within the automotive industry. One of the means to achieve lightweight automobiles is the utilization of advanced high-strength steels. As an important component in third-generation advanced high-strength steels, medium-Mn steel has attracted considerable attention owing to its multiphase, multi-scale, and metastable characteristics [1,2]. The ultra-fine dual-phase microstructure of medium-Mn steel, obtained through intercritical annealing treatment, is accompanied by a large amount of retained austenite [3–6]. Transformation-induced plasticity (TRIP) is a major strengthening and toughening mechanism for medium-Mn steels. During plastic deformation, austenite grains possessing appropriate stability transform into martensite, thereby delaying crack nucleation and micropore aggregation. Consequently, medium-Mn steels exhibit higher mechanical properties [7]. The microstructure heredity during the preparation of medium-Mn steel affects its final microstructure and mechanical properties. Therefore, modifying the initial microstructure before annealing by optimizing the hot deformation process parameters is a crucial method for controlling the mechanical properties of medium-Mn steels. However, current research mainly focuses on the relationship between microstructure mechanical properties, the austenite reverse phase transformation mechanism, and the stability of austenite medium-Mn steels [5,8–10]. Studies on flow stress, dynamic recovery (DRV), and dynamic recrystallization during hot deformation and the effects of alloying elements on constitutive model parameters and hot-forming properties of steels remain to be explored further.

Upon introducing the Zener–Hollomon parameter, Wan et al. [11] established the dynamic recrystallization (DRX) flow stress model for Fe-Mn-C-Al steel and investigated the optimal hot working parameters to achieve a uniform fine microstructure. The occurrence of dynamic recrystallization behavior in the steel became challenging with an increase in the Z parameters. Li et al. [12] studied the DRX characteristics of Fe-8Mn-6Al-0.2C steel and analyzed its microstructure evolution qualitatively. Their findings indicated that elevated temperatures and reduced strain rates can effectively promote the growth of austenite and α-ferrite, respectively. Liu et al. [13] analyzed the deformation microstructure of Fe-Mn-C-Al dual-phase steel during the DRX process and further verified its microstructure characteristics by combining it with the hot processing map. Based on the DRX behavior and microstructure evolution of medium-Mn steel, Sun et al. [14] established a strain-compensated constitutive model and the grain size evolution model. Their findings suggested that the flow stress–strain curve showed evident positive deformation temperature and negative strain rate sensitivity. The recrystallized grain size gradually increased with the deformation temperature and strain rate, and the size distribution range was expanded considerably. Li et al. [15] delineated how δ-ferrite and austenite affected the flow stress characteristics of Fe-Mn-C-Al lightweight steel during the early and late stages of compression–deformation, respectively. Li et al. established the austenite DRX kinetic model without considering DRV and δ-ferrite DRX. They also highlighted that the growth of austenite grains at low strain rates promoted the transformation of δ-ferrite from a banded structure to an island structure. In addition, various scholars have explored the effects of alloying elements on the hot workability of medium-Mn steels. The increase in Ti content reduces the deformation activation energy, decreases the original austenite grain size, accelerates the occurrence of the DRX process, and promotes the formation of fine lath martensite microstructure [16]. Krbat'a et al. [17] investigated the high-temperature plastic behavior of low manganese Si-containing medium-Mn steel. They analyzed the effects of deformation temperature on peak stress via flow curve and discussed the phase transformation of the steel from the perspective of DRV and DRX.

The aforementioned research mainly examined the effects of hot deformation process parameters and alloy elements on the flow stress and microstructure, but the application of medium-Mn steels from sheet to automobile parts involves a complex hot forming process. Therefore, it is necessary to study the effects of C and Al content on the constitutive model parameters, hot processing map, and microstructure after deformation of medium-Mn steels and obtain the optimal forming process interval and variation rule of medium-Mn steels with different element content, so as to provide theoretical guidance for the numerical simulation and practical application of medium-Mn steels as auto parts. In this study, the high-temperature plastic deformation behavior of medium-Mn steels was studied by single-pass hot compression tests. The flow stress, under varying deformation conditions, was friction-corrected, and a constitutive model of the steels considering the content of Al was established. The effects of C and Al contents on the constitutive model parameters and hot deformation behavior of medium-Mn steel were studied. The relationship between element content and microstructure was analyzed by selecting deformed samples with different hot working conditions in combination with hot processing maps.

2. Materials and Methods

Steels were smelted in a vacuum furnace and cast into Φ200 mm ingots. These ingots were heated to 1200 °C for 2 h and subsequently hot forged into 50 mm × 50 mm billets. Square-shaped specimens, with 15 mm × 15 mm × 15 mm dimensions, were processed from the forging billet. After the samples were ground, their chemical composition was tested using SparkCCD 7000 optical emission spectrometer (OES, NCS Testing Technology Co., Ltd., Beijing, China). Table 1 displays the chemical composition of four kinds of 7Mn steels with varying C and Al contents. Cylindrical samples, with dimensions Φ10 mm × 15 mm, were processed longitudinally from the forging billet using wire-cut electrical discharge machining (Suzhou New Spark Machine Tool Co., Ltd., Suzhou, China), and the circumference and end face of specimens were polished to ensure that the surface roughness was less than 0.4 μm. Isothermal hot compression tests were conducted on a Gleeble-3500 thermal simulator at six temperatures (900, 950, 1000, 1050, 1100, and 1150 °C) and four strain rates (0.01, 0.1, 1, and 5 s^{-1}). Samples were heated to 600 °C at a rate of 10 °C/s, then heated to 1200 °C at a rate of 1 °C/s and held for 180 s to homogenize the microstructure. Subsequently, they were cooled to the deformation temperature at a rate of 10 °C/s and held for 30 s to eliminate the temperature gradient before compression tests to ensure homogenization of the entire sample temperature. Samples were compressed according to the predetermined strain of 60%. Immediate water quenching was performed after compression to retain the high-temperature deformed microstructure. The hot deformation process is shown in Figure 1. Deformed samples were cut through the center parallel to the compression axis for microstructure observations. After the center sections were ground, polished, and etched in a 4% nitric acid alcohol solution (volume fraction) for 10 s, the microstructures were observed using a Scios scanning electron microscope (SEM, Thermo Fisher Scientific (CHINA) Co., Ltd., Shanghai, China) at 20 kV and 0.4 nA.

Table 1. Chemical composition of 7Mn steels (wt.%).

Steels	C	Mn	Al	Cr	Ni	Mo
7Mn	0.01	6.19	0.00	0.16	0.59	0.03
7Mn1Al	0.02	6.53	1.14	0.13	0.11	0.06
7Mn3Al	0.03	6.78	3.30	0.12	0.04	0.13
2C7Mn3Al	0.20	6.58	3.46	0.12	0.03	0.14

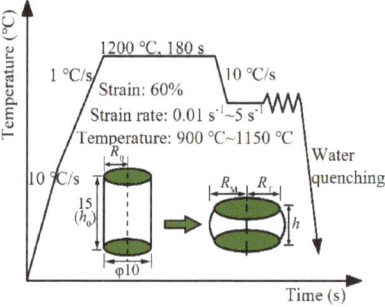

Figure 1. Schematic diagram of the isothermal hot compression test process.

3. Results

3.1. Thermodynamic Calculation

Based on the chemical composition of experimental steels in Table 1, the volume fraction of each phase in the equilibrium microstructure at various temperatures was calculated using the General Steel module of JMatPro 7.0 software. The start and end temperatures are 20 °C and 1600 °C, respectively, and the step is 10 °C. Figure 2 shows the results of the calculations. A comparison between 7Mn and 7Mn1Al reveals that with an increase in Al content from 0 to 1.14 wt.%, the phase diagram shapes remain similar, but the transformation temperatures alter. The complete austenitizing temperatures (A_3) rise from 708 °C to 812 °C, and the temperatures for the same mass fraction of the two phases increase from 645 °C to 694 °C. This change reduces the high-temperature processing range for the steels. At room temperature, the microstructure of both steels comprises ferrite and austenite, displaying minimal differences in phase content, as depicted in Figure 2a,b. Upon increasing the Al content to 3.3 wt.%, notable changes occur in the phase diagram compared with the previous compositions. Above 1400 °C, the liquid phase transforms into δ-ferrite, but an immediate transformation into austenite does not occur. At 1329 °C, some δ-ferrite transitions into austenite, reaching its maximum content around 1060 °C. Subsequently, austenite starts transforming into α-ferrite, as indicated in Figure 2c. Notably, δ-ferrite persists, owing to the stabilizing effect of the Al element through solid solution strengthening, enabling its retention at room temperature [18]. Comparing 7Mn3Al and 2C7Mn3Al, it becomes evident that with an increase in C content to 0.2 wt.%, the liquid phase initially transforms into δ-ferrite at high temperatures. At 1455 °C, the δ-ferrite content peaks at 84.4 wt.%, then transitions into austenite. At approximately 1220 °C, the austenite content reaches a maximum of 62.6 wt.%. Carbides begin precipitating around 690 °C, as depicted in Figure 2d.

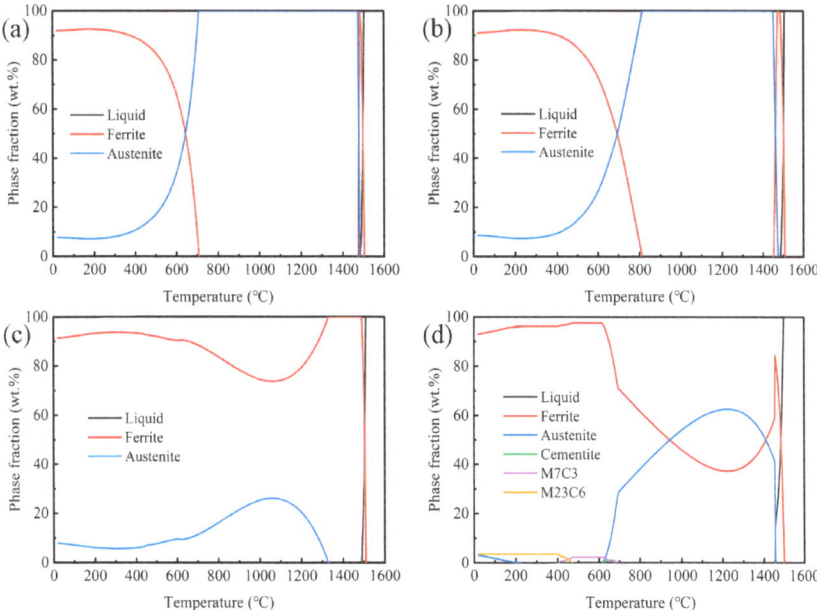

Figure 2. Equilibrium phase diagrams of steels: (**a**) 7Mn, (**b**) 7Mn1Al, (**c**) 7Mn3Al, and (**d**) 2C7Mn3Al.

An increase in Al content reduces the austenite content at the selected deformation temperature range, whereas an increase in C content elevates the austenite content within the same temperature range. This aligns with the established understanding that Al and C elements tend to stabilize ferrite and austenite, respectively. Specifically, in the temperature range of 900 °C to 1150 °C, the microstructure of 7Mn and 7Mn1Al consists solely of single-phase austenite. Conversely, 7Mn3Al and 2C7Mn3Al exhibit a dual-phase microstructure of austenite and ferrite within this range, albeit with differing phase proportions. The former primarily comprises ferrite, whereas the latter is predominantly composed of austenite.

3.2. Hot Deformation Flow Stress Curve and Friction-Correction

During the hot compression process, the sample undergoes friction from the equipment's indenter, causing metal accumulation along the flow direction. This phenomenon, known as "bulging", leads to a certain deviation between the experimental and actual values, as depicted in Figure 1. To enhance accuracy, the high-temperature flow stress is adjusted or corrected to refine test data.

In general, the following relationship between the stress before and after friction correction is expressed as follows [19]:

$$\frac{\sigma_0}{\sigma} = \frac{8bR}{H}\left\{\left[\frac{1}{12}+\left(\frac{H}{Rb}\right)^2\right]^{\frac{3}{2}}-\left(\frac{H}{Rb}\right)^3-\frac{me^{-\frac{b}{2}}}{24\sqrt{3}\left(e^{-\frac{b}{2}}-1\right)}\right\} \tag{1}$$

where σ_0 is the original flow stress data (MPa); σ is the flow stress after friction-correction (MPa); b is the drum belly parameter; R and H are the instantaneous radius and height in the hot compression process (mm), respectively; m is the friction factor; and e is the natural constant.

b, R, H, and m can be calculated by Equations (2)–(5), respectively:

$$b = 4 \times \frac{R_M - R_T}{R_f} \times \frac{h}{h_0 - h} \tag{2}$$

$$R = R_0 e^{\frac{\varepsilon}{2}} \tag{3}$$

$$H = h_0 e^{-\varepsilon} \tag{4}$$

$$m = \frac{(R_f/H)b}{\left(4/\sqrt{3}\right)-\left(2b/3\sqrt{3}\right)} \tag{5}$$

where R_M is the radius of the bulging area after compression of the sample (mm); R_T is the radius of the end face of the sample after compression (mm); R_f is the average radius of the sample after compression (mm); h is the height of the sample after compression (mm); h_0 is the original height of the sample (mm); R_0 is the original radius of the sample (mm); and ε is the strain of the sample.

$$R_T = \sqrt{3 \times \frac{h_0}{h} \times R_0^2 - 2R_M^2} \tag{6}$$

$$R_f = R_0 \sqrt{\frac{h_0}{h}} \tag{7}$$

The true stress-true strain curves of 7Mn3Al before and after correction at varying temperatures are shown in Figure 3. Post-friction-correction, the curves exhibit general consistency with the pre-correction curves at lower strain levels. However, as the strain increases, the disparity between the two curves at the same deformation temperature progressively grows. This occurrence arises from the initial stages of hot compression–deformation, where the contact area between the sample's end face and the indenter is minimal, resulting in minimal frictional influence on metal flow. Consequently, the two curves remain largely similar. However, with increasing strain, the end face area enlarges, intensifying friction between the surfaces, thereby augmenting the metal's flow resistance. Ultimately, this contributes to a more pronounced difference between the pre and post-correction curves.

At the onset of deformation, dislocation multiplication and dislocation glide predominantly govern the process, leading to pronounced work hardening in the initial stages.

Consequently, the flow stress experiences rapid escalation with increasing strain, reaching a peak swiftly. Subsequent to this initial deformation-strengthening phase driven by work hardening, the nucleation and growth of dynamically recrystallized grains play a crucial role. This phase allows for the dissipation of deformation energy and the absorption of defects, such as dislocations and sub-grain boundaries formed during the deformation process. As a result, a softening effect emerges, diminishing the rate of work hardening and leading to a gradual decline in flow stress [20,21]. Eventually, a dynamic equilibrium is attained when the mechanisms of work hardening and dynamic softening reach a balance. At this stage, the flow stress tends to stabilize [22].

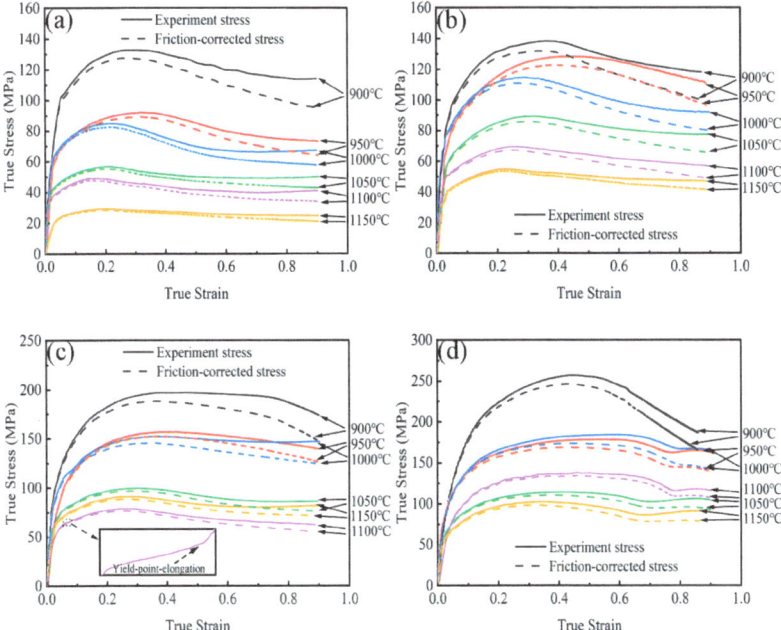

Figure 3. True stress–true strain curves of 7Mn3Al at various deformation parameters: (**a**) 0.01 s^{-1}, (**b**) 0.1 s^{-1}, (**c**) 1 s^{-1}, and (**d**) 5 s^{-1}.

The flow stress curve is significantly influenced by deformation temperature and strain rate. Elevated temperature increases the thermal vibration amplitude of metal atoms, thereby reducing dislocation motion resistance. Consequently, an increased number of slip systems become activated, promoting grain boundary migration and enhancing DRX. Simultaneously, higher deformation temperatures augment the driving force for vacancy atom diffusion, dislocation climb, and cross slip, favoring the occurrence of softening mechanisms. In contrast, higher strain rates result in shorter deformation, grain boundary migration, and dislocation reconstruction times. This shorter duration delays the onset of dynamic recrystallization, leading to abbreviated dynamic recrystallization times and predominantly manifesting DRV, resulting in a lesser softening effect [23]. Conversely, lower strain rates provide more time for DRX and DRV during deformation, intensifying the material's softening effect. Consequently, the material exhibits a lower flow stress value under slower strain rate conditions. Hence, flow stress curves demonstrate two distinct types: the DRX type curve, featuring pronounced softening effects at lower strain rates, and the DRV type curve, displaying less obvious softening effects at higher strain rates.

DRX flow stress curves can be categorized into three types: single-peak, multi-transient steady-state, and multi-peak [24]. As depicted in Figure 3d, at a lower temperature of 900 °C and a higher strain rate of 5 s^{-1}, the curve exhibits a single peak characteristic, with the flow stress declining after reaching its maximum value. This characteristic curve pattern is commonly observed in metals with high stacking fault energy [25]. Conversely, the occurrence of stress fluctuation with multiple peaks, known as multi-peak behavior, can be observed at lower strain rates, such as at a temperature of 900 °C and a strain rate of 0.01 s^{-1}, illustrated in Figure 3a. This behavior is attributable to the presence of multiple independent DRX cycles [24].

A yield-point elongation is observable on the flow stress curve at the initial stage of deformation under the deformation conditions of 1100 °C and a strain rate of 1 s^{-1}, as highlighted in the enlarged section of Figure 3c. This phenomenon arises from the uneven distribution of ferrite and austenite during deformation [26]. Due to the differing yield strengths of the two phases, deformation initiates in the softer ferrite phase. As deformation progresses, a multitude of dislocations proliferates and becomes entangled within the ferrite, resulting in work hardening. Consequently, the strain is subsequently transferred to the surrounding austenite, leading to the observation of two yield points—an occurrence known as yield-point elongation [26,27].

3.3. Establishment and Analysis of Constitutive Model

Metal materials exhibit complex rheological properties due to the influence of deformation temperature, strain rate, and strain during hot deformation. To describe the relationship between different factors, the constitutive relationship of steel was established using the Zener–Hollomon parameter [28]. The corresponding equations are as follows:

$$\dot{\varepsilon} = A f(\sigma) \exp\left(-\frac{Q}{RT}\right) \tag{8}$$

$$Z = \dot{\varepsilon} \exp\left(\frac{Q}{RT}\right) \tag{9}$$

where $\dot{\varepsilon}$ is the strain rate (s^{-1}); A is the material constant; $f(\sigma)$ is the function of stress; Q is the activation energy (J/mol); R is the molar gas constant (8.314 J·mol^{-1}·K^{-1}); and T is the thermodynamic temperature (K). The Zener–Hollomon parameter is generally considered a function of flow stress and is divided into three equations according to different stress levels [29]:

$$f(\sigma) = \begin{cases} \sigma^{n_1} & \alpha\sigma < 0.8 \ (\text{power function}) \\ \exp(\beta\sigma) & \alpha\sigma > 1.2 \ (\text{exponential function}) \\ [\sinh(\alpha\sigma)]^n & \text{for all } \sigma \ (\text{hyperbolic sine function}) \end{cases} \tag{10}$$

where n_1, n, α, and β are material constants, $\alpha = \beta/n_1$.

The hyperbolic sine function that describes the entire stress level is used as the constitutive model to describe the variation in the flow stress with temperature and strain rate during the hot deformation of steels [30]:

$$Z = \dot{\varepsilon} \exp\left(\frac{Q}{RT}\right) = A[\sinh(\alpha\sigma)]^n \tag{11}$$

Substituting Equation (10) into Equation (8), the logarithm on both sides is taken, and the partial derivative is calculated to obtain:

$$n_1 = \frac{\partial \ln \dot{\varepsilon}}{\partial \ln \sigma} \tag{12}$$

$$\beta = \frac{\partial \ln \dot{\varepsilon}}{\partial \sigma} \tag{13}$$

When the temperature T or $\dot{\varepsilon}$ are constant, the logarithm of two sides of Equation (11) and the partial derivative can be obtained:

$$n = \left[\frac{\partial \ln \dot{\varepsilon}}{\partial \ln[\sinh(\alpha\sigma)]} \right]_T \tag{14}$$

$$\frac{Q}{nR} = \left[\frac{\partial \ln[\sinh(\alpha\sigma)]}{\partial(1/T)} \right]_{\dot{\varepsilon}} \tag{15}$$

Four parameters (n_1, β, n, and Q/nR) of the peak stress constitutive model can be obtained by linear regression of $\ln\dot{\varepsilon} - \ln\sigma$, $\ln\dot{\varepsilon} - \sigma$, $\ln\dot{\varepsilon} - \ln[\sinh(\alpha\sigma)]$, and $\ln[\sinh(\alpha\sigma)] - 1/T$ under different hot working conditions to obtain the average slope value, and then α can be calculated by the equation of $\alpha = \beta/n_1$.

The parameters α, n, and Q/nR are introduced into Equation (11) under constant deformation temperature T and strain rate $\dot{\varepsilon}$ to determine A. Upon leveraging the properties of the hyperbolic sine function and the correlation between peak stress and deformation temperature, the Arrhenius peak stress constitutive model for the steel can be established through Equations (8) and (10). However, this model solely accounts for the impact of deformation temperature and strain rate on deformation resistance. Studies have revealed that under low-temperature conditions, deformation resistance is significantly influenced by strain [14,31]. Consequently, to accurately predict flow stress across the entire deformation range, α, n, Q, and A are fitted as polynomial functions of strain using a strain-compensation method. Figure 4 illustrates the relationship between the four parameters of the constitutive model and the true strain ε.

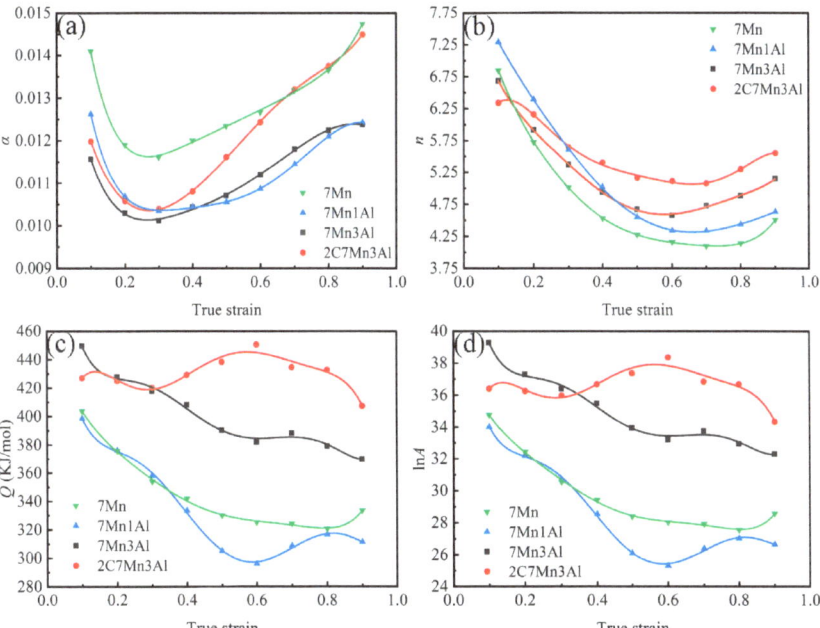

Figure 4. Relationship between (**a**) α, (**b**) n, (**c**) Q, (**d**) $\ln A$ and true strain of steels.

The stress level parameter α varies with material and deformation conditions [32]. Figure 4a shows that the stress level parameters of the four steels decrease to the lowest point and then increase gradually with an increase in strain. Comparing 7Mn, 7Mn1Al, and 7Mn3Al reveals that the addition of the Al element reduces the stress levels under the same strain in the absence of the C element. However, 7Mn1Al and 7Mn3Al exhibit alternating changes. Specifically, 7Mn1Al steel exhibits higher stress levels when the strain is less than 0.4, while 7Mn3Al demonstrates higher stress levels when the strain exceeds 0.4. Further comparison between 7Mn3Al and 2C7Mn3Al demonstrates that under the same Al content, the stress level parameters increase with a rise in C content under equivalent strains. This difference becomes progressively evident, particularly when the strain exceeds 0.3.

The stress index, represented by n, initially decreases and then increases with an increase in strain. Additionally, the stress index exhibits an upward trend with an increase in Al content. Specifically, the stress index of 7Mn1Al steel surpasses that of 7Mn3Al when the strain is below 0.4 but falls below that of 7Mn3Al when the strain exceeds 0.4. Furthermore, as the C content increases, the stress index of 2C7Mn3Al demonstrates higher values compared with the three other steels when the strain exceeds 0.3.

The thermal deformation activation energy Q characterizes the energy barrier that dislocations must surpass during thermal deformation, signifying the processing performance of the alloy. Figure 4c illustrates that, except for 2C7Mn3Al, the hot deformation activation energy of other steels decreases with an increase in deformation level. Specifically, for 7Mn1Al, the activation energy decreases initially, then increases, and subsequently decreases again, peaking at 0.6 strain. Comparing the thermal deformation activation energy Q between 7Mn and 7Mn3Al, as the Al content increases from 0 to 3 wt.%, the thermal activation energy of the steel increases under the same strain. Similarly, comparing the thermal deformation activation energy Q between 7Mn3Al and 2C7Mn3Al, the alteration in activation energy demonstrates a similar trend when the strain is below 0.3. However, as strain increases, the former exhibits a declining trend, while the latter initially rises and subsequently declines. Based on this analysis, an increase in Al and C content can enhance the thermal deformation activation energy under specific strain conditions.

The material's constitutive model parameters, considering the Al content and strain-compensation, are derived by taking a matrix of the material model parameters across various aluminum contents and multiple strain conditions, followed by nonlinear surface fitting. The fitting outcomes for different parameters are depicted in Figure 5.

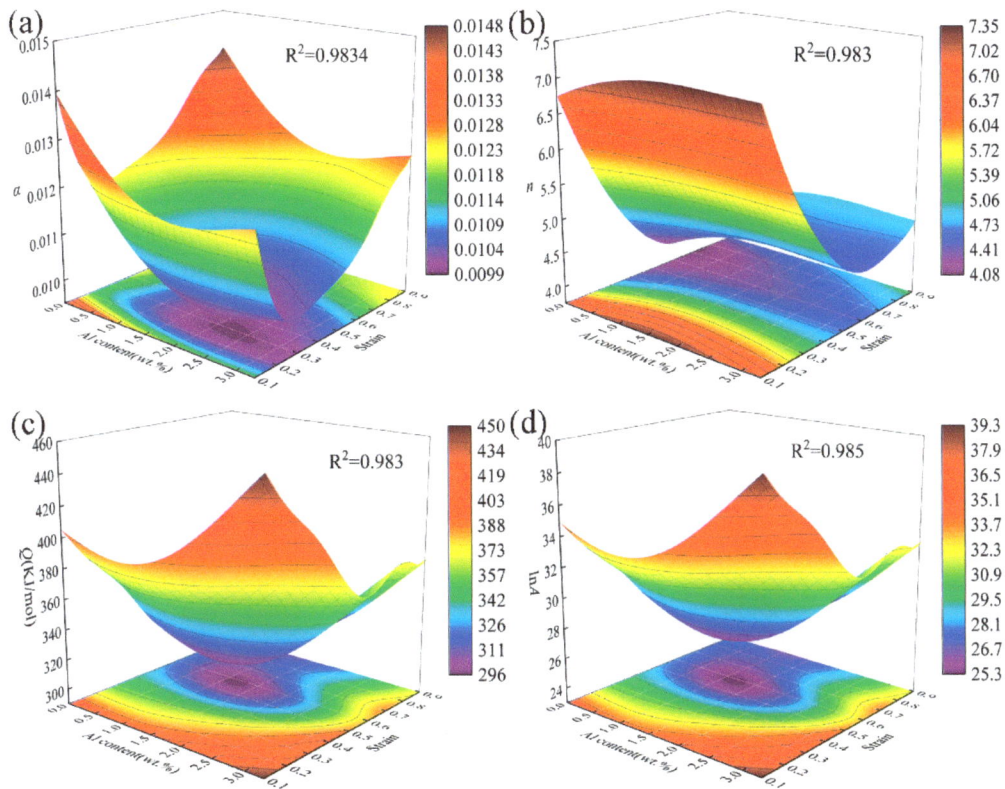

Figure 5. Related parameter fitting diagram: (**a**) α, (**b**) n, (**c**) Q, and (**d**) $\ln A$.

The connection between the constitutive model parameters and strain can be articulated through a polynomial function encompassing the aluminum content and strain, as demonstrated in Equation (16). Table 2 provides the coefficients of the fitting polynomials for various parameters.

$$G = k_0 + a_1 w + a_2 w^2 + a_3 w^3 + a_4 w^4 + b_1 \varepsilon + b_2 \varepsilon^2 + b_3 \varepsilon^3 + b_4 \varepsilon^4 + b_5 \varepsilon^5 + b_6 \varepsilon^6 + b_7 \varepsilon^7 \quad (16)$$

where G represents different material parameters (α, n, Q, and $\ln A$); k_0, $a_1 \ldots a_4$, and $b_1 \ldots b_7$ represent polynomial coefficients; w represents the content of Al element (wt.%); ε is strain.

Table 2. Constitutive model-related constants of Al-containing medium-Mn steels.

Coefficients	G			
	α	n	Q (KJ/mol)	lnA
k_0	1.69×10^{-2}	7.64	4.02×10^5	3.50×10^1
a_1	-1.73×10^{-3}	4.80×10^{-1}	-1.89×10^4	-1.93
a_2	1.09×10^{-4}	-1.16×10^{-1}	1.75×10^3	0.22
a_3	2.04×10^{-4}	-1.37×10^{-2}	6.79×10^3	0.64
a_4	-3.71×10^{-5}	5.09×10^{-3}	-1.20×10^3	-0.11
b_1	-2.32×10^{-2}	-6.23	6.38×10^5	4.53×10^1
b_2	-1.77×10^{-1}	-5.00×10^1	-1.08×10^7	-8.36×10^2
b_3	1.62	3.11×10^2	6.34×10^7	5.03×10^3
b_4	-5.00	-8.82×10^2	-1.91×10^8	-1.54×10^4
b_5	7.59	1.36×10^3	3.02×10^8	2.48×10^4
b_6	-5.68	-1.06×10^3	-2.40×10^8	-1.98×10^4
b_7	1.68	3.29×10^2	7.47×10^7	6.21×10^3

Upon analyzing the trends in material parameters, it is possible to determine the flow stress under specific strain conditions. The Arrhenius constitutive model for medium-Mn steels with varying aluminum content under different strain conditions can be formulated as follows:

$$\sigma = \frac{1}{\alpha} \ln \left\{ \left(\frac{Z}{A} \right)^{\frac{1}{n}} + \left[\left(\frac{Z}{A} \right)^{\frac{2}{n}} + 1 \right]^{\frac{1}{2}} \right\} \quad (17)$$

To validate the accuracy of the Arrhenius constitutive model for medium-Mn steels with different Al contents, the predicted outcomes under different Al contents and deformation conditions were compared with experimental results. Figure 6 displays the fitting results of both sets under different conditions. To further assess the model's precision, the correlation coefficient R and the average absolute relative error AARE (%) are utilized to characterize the model's performance.

$$R = \frac{\sum\limits_{i=1}^{N} (\sigma_{Ei} - \overline{\sigma}_E)(\sigma_{Pi} - \overline{\sigma}_{Pi})}{\sqrt{\sum\limits_{i=1}^{N} (\sigma_{Ei} - \overline{\sigma}_E)^2 \sum\limits_{i=1}^{N} (\sigma_{Pi} - \overline{\sigma}_{Pi})^2}} \quad (18)$$

$$AARE = \frac{1}{N} \sum\limits_{i=1}^{N} \left| \frac{\sigma_{Ei} - \sigma_{Pi}}{\sigma_{Ei}} \right| \quad (19)$$

where N is the total number of data; σ_{Ei} and $\overline{\sigma}_E$ represent the test value and its average value (MPa), respectively; and σ_{Pi} and $\overline{\sigma}_{Pi}$ represent the predicted value and its average value (MPa), respectively.

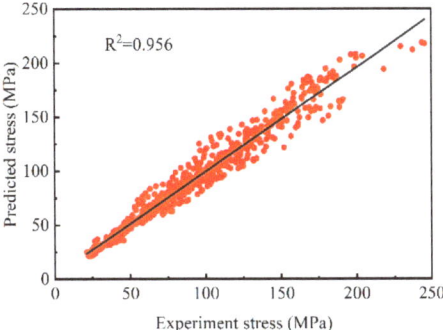

Figure 6. Comparison of predicted values and experimental values under different conditions.

Figure 6 depicts that the predicted value and the experimental value are basically near the straight line, indicating that the stress values of the two are basically the same under different conditions. Calculation shows that $R = 0.956$, AARE = 9.781%, less than 10%.

3.4. Establishment and Analysis of Hot Processing Map of Steels

Hot processing maps and instability maps play remarkable roles in investigating constitutive behavior and optimizing process parameters during material thermal deformation processes. Utilizing the DMM, the deformation process is perceived as a nonlinear energy dissipation system. This dissipation primarily comprises plastic deformation energy dissipation (G) and microstructure evolution energy dissipation (J) [30]. Consequently, the total energy (P) of the external input system can be mathematically expressed as follows:

$$P = G + J = \int_0^\sigma \dot{\varepsilon}\,\mathrm{d}\sigma + \int_0^{\dot{\varepsilon}} \sigma\,\mathrm{d}\dot{\varepsilon} \tag{20}$$

The two energy ratios can be expressed by the strain rate sensitivity index m:

$$m = \left(\frac{\partial J}{\partial G}\right)_{\varepsilon,T} = \frac{\partial P}{\partial G}\frac{\partial J}{\partial P} = \frac{\sigma \mathrm{d}\dot{\varepsilon}}{\dot{\varepsilon}\mathrm{d}\sigma} = \left[\frac{\partial \ln \sigma}{\partial \ln \dot{\varepsilon}}\right]_{\varepsilon,T} \tag{21}$$

When the energy dissipation process conforms to an ideal linear dissipation pattern, the value of J reaches its maximum, and m is 1. The power dissipation efficiency η is commonly employed to illustrate the correlation between microstructure evolution and energy dissipation, as follows:

$$\eta = \frac{J}{J_{\max}} = \frac{2m}{m+1} \tag{22}$$

The power dissipation efficiency is dependent on the deformation temperature and strain rate. Across a specified strain level, the variation in power dissipation efficiency concerning temperature (T) and strain rate constitutes the power dissipation diagram. Regions where the η value exceeds 0.3 signify stable microstructure alterations, indicating good plasticity. However, the instability zone may also exhibit high η values; hence, relying solely on the power dissipation coefficient η for analyzing hot working performance might not be adequate. To assess the material's behavior comprehensively, the plastic instability criterion to construct a hot processing map for further analysis must be integrated.

The Prasad instability criterion, founded on the DMM and Ziegler's plastic flow theory, serves to evaluate rheological instability in hot compression scenarios. The specific form of this criterion is expressed as follows:

$$\xi_\mathrm{p}(\dot{\varepsilon}, T) = \frac{\partial \ln\left(\frac{m}{m+1}\right)}{\partial \ln \dot{\varepsilon}} + m < 0 \tag{23}$$

where ζ_p is a function of strain rate and deformation temperature, the negative value indicates that the material has flow instability and is not suitable for processing within the corresponding process parameters.

The hot processing map is generated upon overlaying the power dissipation map and the rheological instability map. Figure 7 illustrates the hot processing map for various steels under a strain of 0.9. In this figure, the color-altered region indicates a high-power dissipation efficiency (η value), while the blue area signifies instability. A comparison between 7Mn and 7Mn1Al reveals that increasing the Al content from 0 to 1.14 wt.% expands the instability zone, primarily concentrated within the temperature range of 900–950 °C and 1025–1125 °C, under strain rates from 0.1–5 s^{-1}. Moreover, the broader high-power dissipation zone transforms from a narrow strip to a block-shaped area. Although the area expands, the peak value decreases from 0.45 to 0.42. Conversely, when the Al content increases from 1.14 wt.% to 3.30 wt.%, the instability zone in the temperature range of 900–950 °C widens to 900–1025 °C. Additionally, the higher power dissipation zone shifts to a narrow strip. However, the peak value of 7Mn steel increases from 0.45 to 0.57. This observation suggests that within these corresponding hot working conditions, DRX occurs readily, reaching maximum levels at the peak value of η, leading to material flow softening. Consequently, incorporating an appropriate amount of Al enhances hot workability.

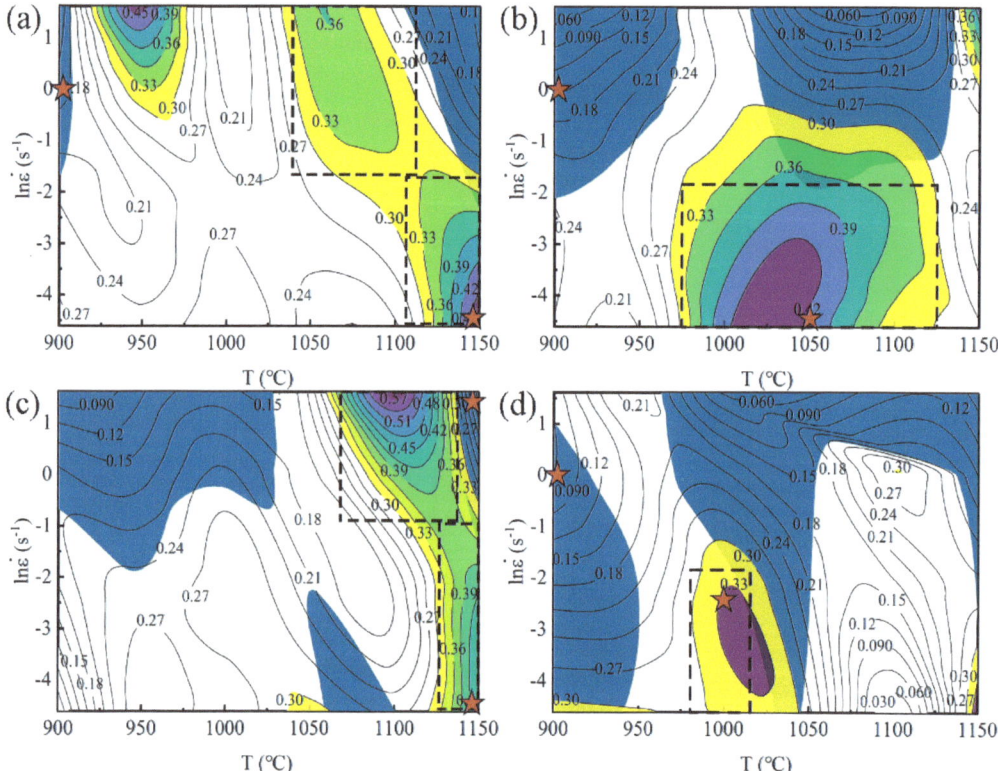

Figure 7. Hot processing map of the steels at 0.9 strain: (**a**) 7Mn, (**b**) 7Mn1Al, (**c**) 7Mn3Al, and (**d**) 2C7Mn3Al. The color-altered region indicates a high-power dissipation efficiency (η value), while the blue area signifies instability. Red stars represent the process conditions of samples for microstructure characterization.

Upon comparing 7Mn3Al and 2C7Mn3Al, the low-temperature instability zone (within the deformation temperature range of 1060 °C to 1140 °C and strain rates of 0.4 s^{-1} to 5 s^{-1}) shifts from high to low strain rates. Additionally, the area of the instability zone decreases as the C content escalates from 0.03 wt.% to 0.2 wt.%. In contrast to 7Mn3Al, the region where the power dissipation coefficient of 2C7Mn3Al exceeds 0.3 shrinks, and the peak value decreases from 0.57 to 0.33. This reduction suggests an environment less conducive to dynamic recrystallization. Simultaneously, within the temperature range of 1075–1125 °C and strain rates from 0.01–0.1 s^{-1}, although the instability parameter surpasses zero, the power dissipation coefficient remains below 0.15. This result indicates that the potential presence of an adiabatic shear band within the sample's microstructure formed under these conditions causes localized plastic deformation during material deformation. The localized heat generated might not dissipate quickly enough to the surrounding area, resulting in decreased local flow stress. The formation of the shear band consumes substantial energy, leading to a low power dissipation coefficient, accompanied by remarkable local shear deformation. Typically, this region may exhibit crack formation. Once cracks emerge, they tend to propagate along the direction of the shear band, resulting in transgranular cracking during the deformation process. Consequently, the increase in C content appears to worsen thermal processing performance. This finding aligns with the observed trend of hot deformation activation energy Q concerning the C and Al element content under a strain of 0.9.

The optimum hot working range for steels is represented by the black dashed wire-frames in Figure 7. For 7Mn, the deformation temperature ranges from 1040 °C to 1110 °C, with a strain rate of 0.2 s^{-1} to 5 s^{-1}. Additionally, the deformation temperature ranges from 1105 °C to 1150 °C, with a strain rate of 0.01 s^{-1} to 0.2 s^{-1}. As for 7Mn1Al steel, the deformation temperature lies between 975 °C and 1125 °C, with a strain rate of 0.2 s^{-1} to 5 s^{-1}. In the case of 7Mn3Al, the deformation temperature ranges from 1060 °C to 1140 °C, with a strain rate of 0.4 s^{-1} to 5 s^{-1}, and from 1125 °C to 1150 °C, with a strain rate of 0.01 s^{-1} to 0.4 s^{-1}. Finally, the deformation temperature for 2C7Mn3Al is between 980 °C and 1120 °C, and the strain rate varies from 0.01 s^{-1} to 0.2 s^{-1}. The comparison shows that the optimum hot working interval area of 7Mn1Al is the largest, and the distribution is continuous. Hence, its hot working performance is the best.

3.5. Microstructure Evolution

The microstructures of various steels subjected to different hot working conditions are depicted in Figure 8. Across all steels, the formation of martensitic microstructures during the cooling phase is evident. Figure 8a,b present the microstructures of 7Mn under varied hot working conditions, predominantly comprising martensite and retained austenite. However, the morphology and size of martensite differ based on distinct deformation processes. At 1150 °C and 0.01 s^{-1}, martensite predominantly displays a lath-shaped structure. In contrast, at 900 °C and 1 s^{-1}, two distinct forms—lath and block—are observed, with the former notably smaller in size compared with the latter.

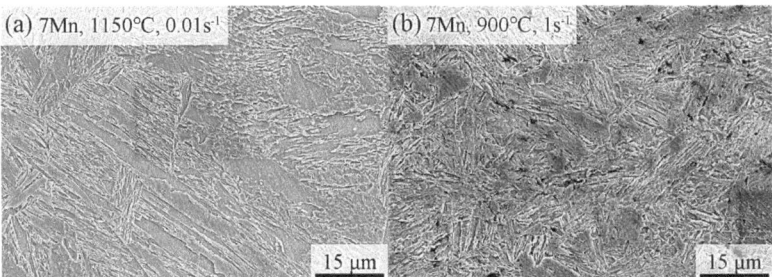

Figure 8. *Cont.*

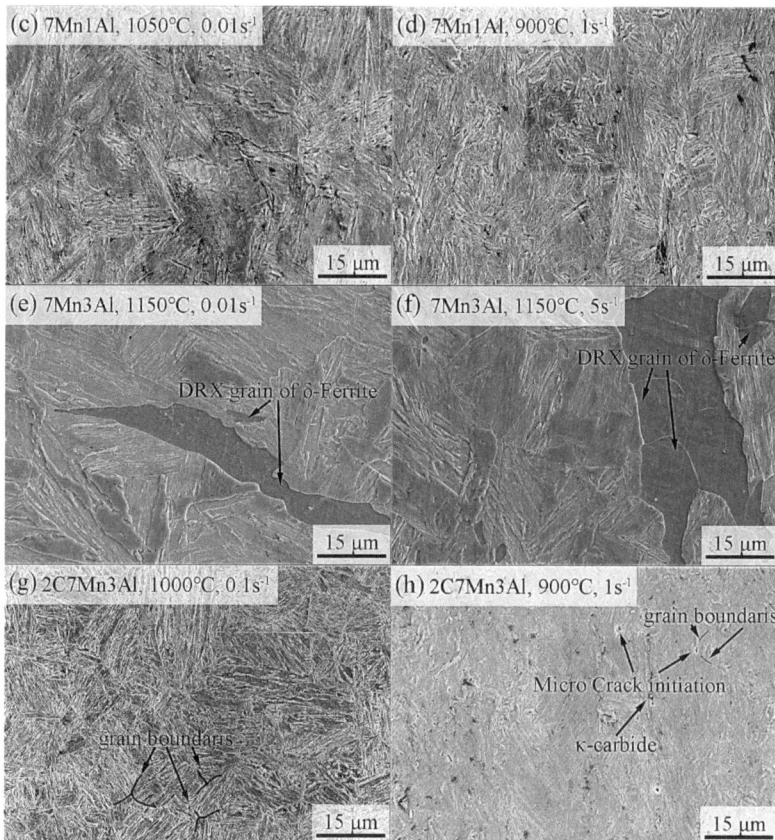

Figure 8. Microstructures of the steels under different hot working conditions: (**a**) 1150 °C, 0.01 s^{-1} and (**b**) 900 °C, 1 s^{-1} of 7Mn, (**c**) 1050 °C, 0.01 s^{-1} and (**d**) 900 °C, 1 s^{-1} of 7Mn1Al, (**e**) 1150 °C, 0.01 s^{-1} and (**f**) 1150 °C, 5 s^{-1} of 7Mn3Al, and (**g**) 1000 °C, 0.1 s^{-1} and (**h**) 900 °C, 1 s^{-1} of 2C7Mn3Al.

In Figure 8c,d, the microstructure of 7Mn1Al predominantly reveals martensite. This observation indicates that the increase in Al content from 0 to 1.14 wt.% diminishes the stability of austenite, causing a complete transformation into martensite during the cooling phase. Concurrently, during hot deformation, the less stable austenite necessitates a lower critical strain to initiate dislocation slip, enhancing its propensity for plastic deformation. As a result, this enhancement potentially augments the hot-forming performance.

In Figure 8e,f, the microstructures of 7Mn3Al under different hot working conditions are depicted. Notably, distinct banded δ-ferrite is visibly apparent in the microstructure, consistent with the phase composition calculated at high temperatures using JMatPro. At this temperature, DRX is more prominently observed in δ-ferrite rather than in austenite. This disparity arises because the critical stress and strain required for DRX initiation in austenite exceed those in δ-ferrite [33]. In addition, compared with the sample with a strain rate of 5 s^{-1}, inhomogeneous DRX grains were observed in the sample with a strain rate of 0.01 s^{-1}, as shown in Figure 8e,f. This difference may be attributable to the lower strain rate, causing an earlier onset and prolonged duration of dynamic recrystallization, facilitating the comprehensive growth of DRX grains.

Upon comparison with 7Mn3Al, an evident reduction in the volume of δ-ferrite is observed in Figure 8g,h, when the carbon content increases from 0.03 wt.% to 0.2 wt.%. Additionally, the size of martensite laths notably diminishes in comparison to 7Mn3Al.

Notably, carbides become apparent in the steel microstructure. Studies suggest that κ-carbide, composed of Fe, C, Al, and Mn, located at grain boundaries, may promote cracks, leading to quasi-cleavage or cleavage fracture modes [33]. Research indicates that carbides in medium-Mn steel might not favor microstructural stability [34]. The appearance of cracks dissipates a substantial amount of energy through plastic deformation, resulting in a lower η value in the corresponding region on the hot processing map under specific hot working conditions.

In summary, incorporating a specific quantity of Al can reduce the stability of austenite, thereby enhancing hot workability. However, an escalation in Al content can induce the appearance of δ-ferrite in the microstructure, consequently retarding the dynamic recrystallization of austenite and compromising hot workability. The addition of C serves to impede δ-ferrite formation, yet it triggers the formation of κ-carbide within the matrix. This occurrence renders the material susceptible to micro-crack formation at the precipitates, leading to intergranular fractures.

4. Conclusions

In this study, the hot deformation behavior of four types of medium-Mn steels with varying C and Al contents was examined through isothermal hot compression tests. The investigation focused on understanding the effects of elemental content on constitutive model parameters, hot workability, and microstructure. The findings are summarized as follows:

1. The relationship between the equilibrium phase content of the four steel variants and temperature was computed using JMatPro. At elevated temperatures, 7Mn and 7Mn1Al exhibit a single-phase structure of austenite, wherein the presence of Al elevates the complete austenitizing temperature. Conversely, 7Mn3Al and 2C7Mn3Al show a microstructure at high temperatures characterized by a mix of ferrite and austenite.

2. The flow stress of steel exhibits sensitivity to deformation temperature and strain rate, tending to rise with decreasing deformation temperature and increasing strain rate. The introduction of Al promotes ferrite formation in 7Mn3Al and 2C7Mn3Al at elevated temperatures. Uneven strain distribution between ferrite and austenite contributes to the elongation of the yield point in the initial stage of the flow stress curve. An Arrhenius constitutive model, incorporating Al content and strain compensation, was developed to predict the flow stress behavior in steel accurately.

3. The hot processing maps of the steels under 0.9 strain were established, and the 7Mn1Al exhibits the best hot workability. The optimum hot working interval is deformation temperature 975–1125 °C, strain rate 0.2–5 s^{-1}.

4. The rise in Al content from 0 to 1.14 wt.% diminished the stability of austenite, enhancing the hot processing capabilities of steel. However, when the content reached 3.3 wt.%, the emergence of δ-ferrite in the microstructure hindered austenite DRX, negatively affecting hot workability. The addition of C reduces the volume of δ-ferrite but results in κ-carbide precipitation, leading to micro-crack formation, which adversely affects processing.

Author Contributions: Conceptualization, M.W.; Data curation, G.G., H.J., X.Z., D.L., C.W. and F.Z.; Investigation, G.G., C.W. and F.Z.; Methodology, M.W.; Software, G.G., X.Z. and D.L.; Validation., M.W.; Writing—original draft, G.G.; Writing—review and editing, M.W.; Visualization, H.J., D.L., C.W. and F.Z. All authors have read and agreed to the published version of the manuscript.

Funding: This research was funded by the Natural Science Foundation of Hebei Province China (No. E2021209012 and E2021209026), Science and Technology Project of Hebei Education Department (No. QN2022054 and KJZX202203), the National Key R&D Program of China (No. 2022YFB3705200), Tangshan Basic Research Science and Technology Program (No. 21130209D).

Institutional Review Board Statement: Not applicable.

Informed Consent Statement: Not applicable.

Data Availability Statement: Data are contained within the article.

Acknowledgments: The authors thank the support of isothermal hot compression tests from the State Key Laboratory of Metastable Materials Science and Technology, Yanshan University.

Conflicts of Interest: The authors declare no conflict of interest.

References

1. Zhao, X.; Shen, Y.; Qiu, L.; Liu, Y.; Sun, X.; Zuo, L. Effects of Intercritical Annealing Temperature on Mechanical Properties of Fe-7.9Mn-0.14Si-0.05Al-0.07C Steel. *Materials* **2014**, *7*, 7891–7906. [CrossRef] [PubMed]
2. Lv, C.; Wang, Y.; Huang, X.; Zhang, L.; Fu, Q.; Wu, M.; Zhang, J. The Novel Combination of Strength and Ductility in 0.4C-7Mn-3.2Al Medium Manganese Steel by Intercritical Annealing. *Steel Res. Int.* **2019**, *90*, 1900228. [CrossRef]
3. Yang, D.P.; Wu, D.; Yi, H.L. Reverse transformation from martensite into austenite in a medium-Mn steel. *Scr. Mater.* **2019**, *161*, 1–5. [CrossRef]
4. Han, J.; Lee, S.-J.; Lee, C.-Y.; Lee, S.; Jo, S.Y.; Lee, Y.-K. The size effect of initial martensite constituents on the microstructure and tensile properties of intercritically annealed Fe-9Mn-0.05C steel. *Mater. Sci. Eng. A* **2015**, *633*, 9–16. [CrossRef]
5. Shao, C.; Hui, W.; Zhang, Y.; Zhao, X.; Weng, Y. Microstructure and mechanical properties of hot-rolled medium-Mn steel containing 3% aluminum. *Mater. Sci. Eng. A* **2017**, *682*, 45–53. [CrossRef]
6. Arlazarov, A.; Gouné, M.; Bouaziz, O.; Hazotte, A.; Petitgand, G.; Barges, P. Evolution of microstructure and mechanical properties of medium Mn steels during double annealing. *Mater. Sci. Eng. A* **2012**, *542*, 31–39. [CrossRef]
7. Kozłowska, A.; Grzegorczyk, B.; Morawiec, M.; Grajcar, A. Explanation of the PLC Effect in Advanced High-Strength Medium-Mn Steels. A Review. *Materials* **2019**, *12*, 4175. [CrossRef]
8. Leták, R.; Jirková, H.; Kučerová, L.; Jeníček, Š.; Volák, J. Effect of Forming and Heat Treatment Parameters on the Mechanical Properties of Medium Manganese Steel with 5% Mn. *Materials* **2023**, *16*, 4340. [CrossRef] [PubMed]
9. Li, Y.; Li, W.; Min, N.; Liu, W.; Jin, X. Effects of hot/cold deformation on the microstructures and mechanical properties of ultra-low carbon medium manganese quenching-partitioning-tempering steels. *Acta Mater.* **2017**, *139*, 96–108. [CrossRef]
10. Wang, M.; Liang, X.; Ren, W.; Tong, S.; Sun, X. Effect of Mn Content on the Toughness and Plasticity of Hot-Rolled High-Carbon Medium Manganese Steel. *Materials* **2023**, *16*, 2299. [CrossRef] [PubMed]
11. Wan, P.; Kang, T.; Li, F.; Gao, P.; Zhang, L.; Zhao, Z. Dynamic recrystallization behavior and microstructure evolution of low-density high-strength Fe–Mn–Al–C steel. *J. Mater. Res. Technol.* **2021**, *15*, 1059–1068. [CrossRef]
12. Li, X.; Song, R.B.; Kang, T.; Zhou, N.P. Hot Deformation and Dynamic Recrystallization Behavior of Fe-8Mn-6Al-0.2C Steel. *Mater. Sci. Forum* **2017**, *898*, 797–802. [CrossRef]
13. Liu, D.; Ding, H.; Cai, M.; Han, D. Hot Deformation Behavior and Processing Map of a Fe-11Mn-10Al-0.9C Duplex Low-Density Steel Susceptible to κ-Carbides. *J. Mater. Eng. Perform.* **2019**, *28*, 5116–5126. [CrossRef]
14. Sun, X.; Wang, B.; Duan, Y.; Liu, Q.; Xu, X.; Wang, S.; Yang, X.; Feng, X. Kinetics and microstructure evolution of dynamic recrystallization of medium-Mn steel during hot working. *J. Mater. Res. Technol.* **2023**, *23*, 5631–5643. [CrossRef]
15. Li, Y.-P.; Song, R.-B.; Wen, E.-D.; Yang, F.-Q. Hot Deformation and Dynamic Recrystallization Behavior of Austenite-Based Low-Density Fe–Mn–Al–C Steel. *Acta Metall. Sin.* **2016**, *29*, 441–449. [CrossRef]
16. Dong, Y.; Zhang, B.; Du, Y.; Du, L.-X.; Misra, R.D.K. The Significant Impact of Ti Addition on the Hot Deformation Behavior of Medium-Manganese Microalloyed Steel. *Steel Res. Int.* **2021**, *92*, 2100074. [CrossRef]
17. Krbaťa, M.; Barényi, I.; Eckert, M.; Križan, D.; Kaar, S.; Breznická, A. Hot deformation analysis of lean medium-manganese 0.2C3Mn1.5Si steel suitable for quenching et partitioning process. *Met. Mater.* **2021**, *59*, 379–390. [CrossRef]
18. Chen, S.; Rana, R.; Haldar, A.; Ray, R.K. Current state of Fe-Mn-Al-C low density steels. *Prog. Mater. Sci.* **2017**, *89*, 345–391. [CrossRef]
19. Najafizadeh, R.E. A new method for evaluation of friction in bulk metal forming. *J. Mater. Process. Technol.* **2004**, *152*, 136–143.
20. Chen, X.-M.; Lin, Y.C.; Wen, D.-X.; Zhang, J.-L.; He, M. Dynamic recrystallization behavior of a typical nickel-based superalloy during hot deformation. *Mater. Des.* **2014**, *57*, 568–577. [CrossRef]
21. Lin, Y.C.; Wu, X.-Y.; Chen, X.-M.; Chen, J.; Wen, D.-X.; Zhang, J.-L.; Li, L.-T. EBSD study of a hot deformed nickel-based superalloy. *J. Alloys Compd.* **2015**, *640*, 101–113. [CrossRef]
22. Su, L.; Liu, H.; Jing, L.; Yu, Z.; Wang, W.; Zhou, L. Flow stress characteristics and microstructure evolution during hot compression of Nb-47Ti alloy. *J. Alloys Compd.* **2019**, *797*, 735–743. [CrossRef]
23. Sun, B.; Aydin, H.; Fazeli, F.; Yue, S. Microstructure Evolution of a Medium Manganese Steel During Thermomechanical Processing. *Metall. Mater. Trans. A* **2016**, *47*, 1782–1791. [CrossRef]
24. Saadatkia, S.; Mirzadeh, H.; Cabrera, J.-M. Hot deformation behavior, dynamic recrystallization, and physically-based constitutive modeling of plain carbon steels. *Mater. Sci. Eng. A* **2015**, *636*, 196–202. [CrossRef]
25. Mirzadeh, H.; Parsa, M.H.; Ohadi, D. Hot deformation behavior of austenitic stainless steel for a wide range of initial grain size. *Mater. Sci. Eng. A* **2013**, *569*, 54–60. [CrossRef]

26. Liu, D.; Ding, H.; Hu, X.; Han, D.; Cai, M. Dynamic recrystallization and precipitation behaviors during hot deformation of a κ-carbide-bearing multiphase Fe–11Mn–10Al–0.9C lightweight steel. *Mater. Sci. Eng. A* **2020**, *772*, 138682. [CrossRef]
27. Cizek, P. The microstructure evolution and softening processes during high-temperature deformation of a 21Cr–10Ni–3Mo duplex stainless steel. *Acta Mater.* **2016**, *106*, 129–143. [CrossRef]
28. McQueen, H.J.; Yue, S.; Ryan, N.D.; Fry, E. Hot working characteristics of steels in austenitic state. *J. Mater. Process. Technol.* **1995**, *53*, 293–310. [CrossRef]
29. Mirzadeh, H.; Cabrera, J.M.; Najafizadeh, A. Constitutive relationships for hot deformation of austenite. *Acta Mater.* **2011**, *59*, 6441–6448. [CrossRef]
30. Yan, N.; Di, H.-S.; Huang, H.-Q.; Misra, R.D.K.; Deng, Y.-G. Hot Deformation Behavior and Processing Maps of a Medium Manganese TRIP Steel. *Acta Metall. Sin.* **2018**, *32*, 1021–1031. [CrossRef]
31. Zhang, Y.; Chai, Z.; Volinsky, A.A.; Tian, B.; Sun, H.; Liu, P.; Liu, Y. Processing maps for the Cu-Cr-Zr-Y alloy hot deformation behavior. *Mater. Sci. Eng. A* **2016**, *662*, 320–329. [CrossRef]
32. McQueen, H.J.; Ryan, N.D. Constitutive analysis in hot working. *Mater. Sci. Eng. A* **2002**, *322*, 43–63. [CrossRef]
33. Liu, G.; Wang, J.; Ji, Y.; Hao, R.; Li, H.; Li, Y.; Jiang, Z. Hot Deformation Behavior and Microstructure Evolution of Fe-5Mn-3Al-0.1C High-Strength Lightweight Steel for Automobiles. *Materials* **2021**, *14*, 2478. [CrossRef] [PubMed]
34. Jeong, J.; Lee, C.-Y.; Park, I.-J.; Lee, Y.-K. Isothermal precipitation behavior of κ-carbide in the Fe–9Mn–6Al–0.15C lightweight steel with a multiphase microstructure. *J. Alloys Compd.* **2013**, *574*, 299–304. [CrossRef]

Article

Prediction of Crack Width in RC Piles Exposed to Local Corrosion in Chloride Environment

Wei Shao [1], Xiaoqing He [2], Danda Shi [1] and Wenjin Zhu [3,*]

[1] College of Ocean Science and Engineering, Shanghai Maritime University, Shanghai 201306, China; weishao@shmtu.edu.cn (W.S.); ddshi@shmtu.edu.cn (D.S.)
[2] Department of Hydraulic Engineering, School of Civil Engineering, Tongji University, Shanghai 200092, China; hexiaoqing0719@163.com
[3] School of Civil and Ocean Engineering, Jiangsu Ocean University, Lianyungang 222005, China
* Correspondence: zhucius@jou.edu.cn

Abstract: A novel prediction model for crack development of reinforced concrete (RC) piles with localized chloride corrosion in the marine environment is proposed. A discrete method is used to solve the corrosion pit radius model and a crack extension model is developed to investigate the initiation and extension of cracks. The maximum corrosion degree of the reinforced concrete pile is predicted according to the limit crack criterion, and finally, a sensitivity analysis is carried out on the important parameters of crack extension. The results show that the radius of the corrosion pit, the depth corrosion pit, and the cross-sectional area loss of reinforcement gradually increase as the corrosion level increases. The loss of the local reinforcement section at crack initiation increases with the increase in the ratio of concrete cover to initial diameter and increases with the increase in the pitting factor. The required pit depth for reinforcement cracking increases with the increase in the ratio of concrete cover thickness to diameter. The loss of the cross-sectional area of reinforcement and the radius of the corrosion pit increase with the increase in the initial diameter of reinforcement. Increasing the pitting factor can reduce the pit depth and make the crack width develop faster before reaching the limit crack width. Increasing the concrete cover thickness can provide an improvement in the propagation of cracks. A comparative analysis shows that the localized corrosion pattern is more in conformity with marine engineering practice.

Keywords: localized corrosion; corrosion pit radius; steel section loss; crack extension; maximum corrosion level

check for updates

Citation: Shao, W.; He, X.; Shi, D.; Zhu, W. Prediction of Crack Width in RC Piles Exposed to Local Corrosion in Chloride Environment. *Materials* **2023**, *16*, 6403. https://doi.org/ 10.3390/ma16196403

Academic Editor: Ming Liu

Received: 4 August 2023
Revised: 19 September 2023
Accepted: 20 September 2023
Published: 26 September 2023

1. Introduction

In recent years, the corrosion of reinforced concrete (RC) structures in the marine environment has received great concern. RC structures can be affected by chloride, air, seawater splash, wind, and other external factors during service, which can cause corrosion of reinforced concrete structures. Corrosion can produce some corrosion products on the surface of RC structures, which can cause the volume to increase and create tensile stress around the reinforced concrete. When the maximum tensile strength exceeds the tensile stress of the concrete, the cracking of the concrete cover initiates. Concrete cracking may cause early intrusion of local chloride ions and corrosion of reinforcement. Concrete cracking may cause early localized chloride ion ingress and reinforcement corrosion, corrosion-induced cracks penetrate the concrete cover and provide a way for corrosive media such as chloride ions and oxygen to rapidly enter the reinforcement [1,2]. The cracks start to spread outwards from the steel–concrete interface and finally, lead to the complete cracking of the concrete cover. Concrete cracking leads to changes in the load-bearing performance of RC structures, causing structural instability, and in severe cases, may lead to major accidents. Furthermore, the development of cracks can provide effective information on the corrosion of RC structures, enabling us to monitor the health status of concrete structures in a timely manner. In order to comprehensively predict the durability of reinforced concrete structures under local chloride ion erosion, it is necessary to conduct in-depth research on the mechanisms of corrosion pit development and crack propagation. Therefore, it is very meaningful to study the development of cracks for the durability life prediction and repair of RC structures.

The mechanism of damage and crack expansion in RC structures caused by corrosion and expansion of reinforcement has always been a hot topic of research in this field. In the actual marine environment, chloride generally penetrates into concrete structures in a unidirectional manner, and corrosion may start from the outermost layer of steel bars, causing uneven corrosion on the surface of steel bars [3]. Local invasion of chloride ions causes random corrosion pits of varying shapes and sizes to form on the surface of steel bars. However, the development of cracks is related to the growth of corrosion pits. It is a rare idea to combine the development of corrosion pits with the development of cracks to study the durability of RC piles under local chloride ion erosion.

In recent years, many scholars have studied the changes in the load-bearing performance of RC structures after local corrosion [4–10]. Only a small number of studies have conducted detailed research on the development of corrosion pits. Kim and Kim [11] estimated the maximum corrosion depth of local corrosion of reinforcement considering chloride ion diffusion. Li et al. [12] conducted tests on pit corrosion of reinforcement in chloride ion corroded concrete and analyzed the relationship between pit depth, residual sectional area, and area of the smaller planar principal moment of inertia and corrosion grade. Darowicki et al. [13] investigated the localized corrosion of stainless steel using optical surface measurement technology and deduced the cumulative distribution function of pitting corrosion. Kioumarsi et al. [14] studied the influence of corrosion pit disturbance on the damage probability of reinforced concrete beams after localized corrosion. These studies have provided us with a better understanding of the generation and development of corrosion pits and also provided a new entry point for the durability research of reinforced concrete structures under localized corrosion conditions.

In addition, some scholars have also conducted research on the cracking and crack development of reinforced concrete. Vidal et al. [15] predicted the crack width of corrosion in reinforced concrete structures and proposed a relationship between reinforcement section loss and cracking. Zhang et al. [16] measured the crack initiation, maximum load, fracture energy, and other parameters of short-cut basalt fiber concrete through tests and studied the crack resistance of concrete. Zhang et al. [17] presented a new model to predict steel corrosion from corrosion cracks based on the average steel section loss parameter. Li et al. [18] proposed a theoretical model for the corrosion crack width of reinforced concrete structures. Zhao et al. [19] presented a concrete cracking model to investigate stress and cracks in concrete cover caused by reinforcing steel corrosion. Shao et al. [20] proposed an analysis model for the durability life of RC piles with chloride local corrosion considering the impact of the crack growth stage on the structure. Guzmán and Gálvez [21] used embedded cohesive cracking finite element analysis to study the effect of localized corrosion on the cracking of concrete cover, deriving different cracking patterns. These studies can help us predict the cracking time of RC structures and have a clear understanding of the development of cracks.

Although the above studies have studied corrosion pits and concrete cracking and crack development, few scholars have combined the growth and crack development of corrosion pits to study the durability life of RC piles. This article aims to combine the growth of corrosion pits with the development of crack width, establish the relationship between the depth and degree of corrosion pits and crack development, and provide new analytical ideas for the development of cracks in locally corroded RC piles. In this paper, a novel prediction model for crack development of RC piles with localized chloride corrosion in the marine environment is proposed. The maximum corrosion degree of the RC pile is predicted according to the limit crack criterion. A discrete method is used to solve the corrosion pit radius model, then crack initiation and extension are investigated based on the crack extension model, and finally, sensitivity analyses are conducted on the important parameters of crack extension.

2. Localized Corrosion Modeling

2.1. Geometry Model

RC piles exposed to the marine environment are easily corroded by chloride ions. The corrosion of RC structures is generally divided into uniform corrosion and local corrosion, and local corrosion is the key factor for predicting the durability life of RC piles. When reinforced concrete is corroded, corrosion pits are created on the surface of the reinforcement. The location and shape of these corrosion pits are formed randomly, and they are related to factors such as the environment and the level of corrosion. According to existing research, the shapes of corrosion pits may be semi-circular, circular, elliptical, and cup-shaped. In order to facilitate the calculation of the radius of corrosion pits and establish a relationship between local steel section loss and crack width, this paper assumes that the corrosion pits are semi-circular. Considering the uneven corrosion on the surface of the steel bar with a single pit at the top of the steel bar, it is assumed that all corrosion is limited to one pit [22].

2.2. Modeling of Corrosion Pits

As the corrosion process proceeds, corrosion products can be produced on the surface of the reinforcement, and corrosion products can expand on the surface of the reinforcement. In this paper, the distribution of corrosion pits on the surface of reinforcement is regarded as a simple geometric distribution. As shown in Figure 1, it is assumed that the center C of the corrosion crater is directly above the center A of the reinforcement. The intersection points of the corrosion pit and reinforcement are B and D, respectively, the intersection of AC and BD is I, and the radius of the corrosion pit is R'. According to Figure 1, the shadow area is the corrosion pit area, which is composed of two semicircles with different radii. The pit area is the sum of the areas of zone 1 and zone 2.

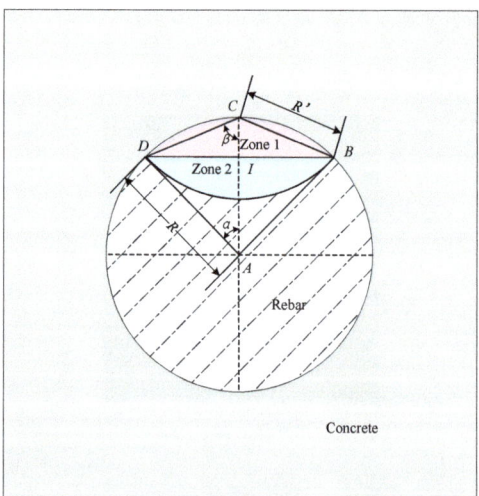

Figure 1. Area loss of reinforcement.

In the corrosion pit model, it is assumed that

$$IB = ID. \tag{1}$$

The area of zone 1 can be expressed as

$$A_1 = A_{S_{ABD}} - A_{T_{ABD}}. \tag{2}$$

The area of triangle ABD obtained from geometric knowledge can be expressed as

$$A_{T_{ABD}} = IB \times IA. \tag{3}$$

IA can be expressed as

$$IA = \sqrt{OB^2 - BI^2} = \sqrt{R^2 - R'^2}. \tag{4}$$

Therefore, the area of the triangle ABD is expressed as

$$A_{T_{ABD}} = R' \times \sqrt{R^2 - R'^2}. \tag{5}$$

The expression for the area of the sector ABD can be expressed as

$$A_{S_{ABD}} = \frac{2\pi R^2 \alpha}{360}. \tag{6}$$

The angle α can be expressed as

$$\sin \alpha = \frac{R'}{R}. \tag{7}$$

Therefore, the expression for α is

$$\alpha = \arcsin \frac{R'}{R}. \tag{8}$$

Therefore, the area of sector ABD is calculated as

$$A_{S_{ABD}} = \frac{2\pi R^2}{360} \cdot \arcsin \frac{R'}{R}. \tag{9}$$

In summary, the area of zone 1 is obtained as

$$A_1 = A_{S_{ABD}} - A_{T_{ABD}} = \left(\frac{2\pi R^2}{360} \cdot \arcsin \frac{R'}{R} \right) - \left(R' \times \sqrt{R^2 - R'^2} \right). \tag{10}$$

Based on the same analysis method, the area of zone 2 can be obtained as

$$A_2 = A_{S_{CBD}} - A_{T_{CBD}}. \tag{11}$$

IC can be expressed as

$$IC = AC - AI = R - \sqrt{R^2 - R'^2}. \tag{12}$$

The relationship between angle β and angle α is given by

$$\beta = 90° - \alpha. \tag{13}$$

Therefore, the area of the sector CBD is obtained as

$$A_{S_{CBD}} = \frac{2\pi R'^2}{360} (90° - \alpha). \tag{14}$$

The area of triangle *CBD* is given by

$$A_{T_{CBD}} = R\prime \times \left(R - \sqrt{R^2 - R'^2}\right). \tag{15}$$

According to Equations (14) and (15), the area of zone 2 is

$$A_2 = A_{S_{CBD}} - A_{T_{CBD}} = \frac{2\pi R'^2}{360} \cdot (90 - \alpha) - R' \times \left(R - \sqrt{R^2 - R'^2}\right). \tag{16}$$

Therefore, the corrosion pit area can be expressed as

$$\Delta A_s = A_1 + A_2 = 2\pi \left[\frac{R'^2}{360} \cdot (90 - \alpha) + \frac{R^2}{360} \cdot \alpha\right] - R' \cdot \left[\sqrt{R^2 - R'^2} + \left(R - \sqrt{R^2 - R'^2}\right)\right]. \tag{17}$$

Equation (17) can be simplified as

$$\Delta A_s = 2\pi \left[\frac{R'^2}{360} \cdot (90° - \alpha) + \frac{R^2}{360} \cdot \alpha\right] - RR'. \tag{18}$$

where ΔA_s is the area of the corrosion pit and denotes the loss of cross-sectional area of the reinforcement, R' is the corrosion pit radius and R is the initial radius of the reinforcement. In general, the degree of corrosion of reinforced concrete piles is expressed as the percentage mass loss of reinforcement ρ, and ρ can be calculated as

$$\rho = \frac{M_1 - M_2}{M_1} \tag{19}$$

where M_1 and M_2 are the masses of the reinforcement bars before corrosion and the mass after corrosion, respectively. Equation (19) can be rewritten as [23]

$$\rho = \frac{M_1 - M_2}{M_1} = \frac{\rho_{rebar} \cdot \Delta A_s}{\rho_{rebar} \cdot A_{rebar}} = \frac{\Delta A_s}{A_{rebar}}. \tag{20}$$

According to Equations (18) and (20), the following equation can be obtained as

$$\rho A_{rebar} = 2\pi \left[\frac{R'^2}{360}(90° - \alpha) + \frac{R^2}{360}\alpha\right] - RR' \tag{21}$$

where ρ is the degree of corrosion of the reinforcement, and A_{rebar} is the initial area of the reinforcement. Equation (21) involves two unknown parameters ρ and R', when the degree of corrosion ρ is known, the radius of pit radius R' can be obtained for different corrosion degrees. However, since the unknown parameter R' is present in the trigonometric function, the radius of the corrosion pit cannot be solved by simple reduction. Therefore, the discretization method can be used to solve the problem. The corrosion pit radius R' is ranging from 0 to 5. Combined with relevant software programming, when the corrosion degree ρ is determined, the corrosion pit radius is discretized into countless points within its feasible range, and the point that makes both sides of the equation infinitely close is found. This point is the optimal solution we are looking for. It is inevitable that using the discrete method to solve the corrosion pit model can result in certain calculation errors E, so this article also analyzes the discrete error.

2.3. Analysis of Crack Width

After the onset of cracking in RC structures, the corrosion products on the surface can first fill the local pores and then be deposited at the interface between the reinforcement and the concrete. The depth of the corrosion pit required for reinforcement cracking is calculated by [24]

$$x_0 = 7.53 + 9.32 \frac{c}{D_0} \qquad (22)$$

where x_0 is the corrosion pit depth required for cracking, c is the concrete cover thickness (mm) and D_0 is the initial diameter of the reinforcement. The above equation shows that the depth of the corrosion pit required for cracking is related to the ratio of the concrete cover thickness and the initial diameter of the reinforcement. The relationship between the depth of the corrosion crater and reinforcement cross-section loss can be obtained by [15]

$$x = \frac{D_0}{\alpha} \left[1 - \sqrt{1 - \frac{\Delta A_s}{A_{rebar}}} \right] \cdot 10^3 \qquad (23)$$

where α is the pitting factor, is the ratio of localized corrosion depth to uniform corrosion depth, $\alpha = 2$, for uniform corrosion, $4 < \alpha < 8$, for localized corrosion, x is the depth of the corrosion pit (μm), D_0 is the initial diameter of the reinforcement (mm), A_{rebar} is the initial area of the reinforcement and ΔA_s is the loss of reinforcement cross-section (mm^2). According to Equations (22) and (23), the local steel section loss at the beginning of cracking can be obtained as [17]

$$\Delta A_{s0} = A_{rebar} \left[1 - \left[1 - \frac{\alpha}{D_0} \left(7.53 + 9.32 \frac{c}{D_0} \right) \cdot 10^{-3} \right]^2 \right]. \qquad (24)$$

It should be noted that the local reinforcement section loss at the onset of cracking is determined by Equation (24) without taking into account the properties of the concrete. Based on the experimental data, a linear regression formula obtained by Vidal et al. [15] is used for predicting crack width as

$$w = K(\Delta A_s - \Delta A_{s0}) \qquad (25)$$

where ΔA_{s0} is the loss of the local steel cross-section at the beginning of cracking, and K is the slope of the curve ($K = 0.0575$). According to the limit crack criterion, when the crack width exceeds the limit crack value, the reinforced concrete pile loses stability.

3. Analysis Results

In this paper, the selected limit crack width is $w_{lim} = 0.4$ mm. The values of model parameters used in this analysis are shown in Table 1. According to Equation (22), the pit depth at the beginning of reinforcement cracking is 0.054 mm, and according to Equation (24), the local steel section loss at the beginning of cracking is 4.73 mm^2. The relationship between the radius of the corrosion pit and the loss of reinforcement cross-section with the degree of corrosion is shown in Figure 2. When the corrosion degree increases, the corrosion pit radius gradually increases, and the reinforcement section loss gradually increases. This is consistent with the findings of the experiment by Li et al. [12]. The error analysis caused by solving the corrosion pit radius with the discrete method is shown in Figure 3. The error varies randomly and irregularly as the corrosion level increases from 1% to 16%, while each error is very small; therefore, the influence of the error formed by the discrete method on the results can be ignored. Figure 4 shows the relationship between corrosion crater radius and corrosion crater depth. The depth of the corrosion pit increases with increasing pit radius.

Table 1. Values of main parameters used in the analysis.

Parameters	Unit	Values	Description	Reference
a	mm	190	Inner radius of pile	-
b	mm	300	Outer radius of pile	-
D_0	mm	10	Initial diameter of steel bar	-
c	mm	50	Concrete cover thickness	-
w_{lim}	mm	0.4	Limit crack width	[25]
α	-	5.65	Pitting factor	[15]
K	-	0.0575	Slope of curve	[26]

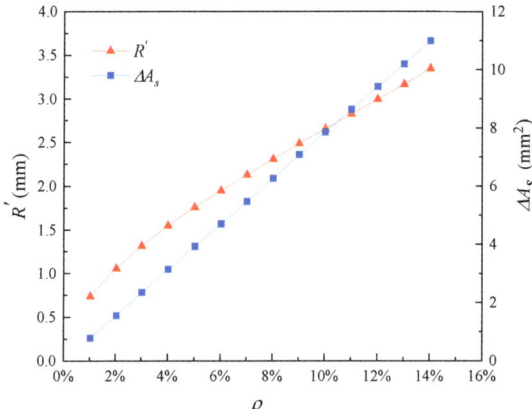

Figure 2. Influence of corrosion degree on radius of corrosion pit and cross-sectional area loss of reinforcement.

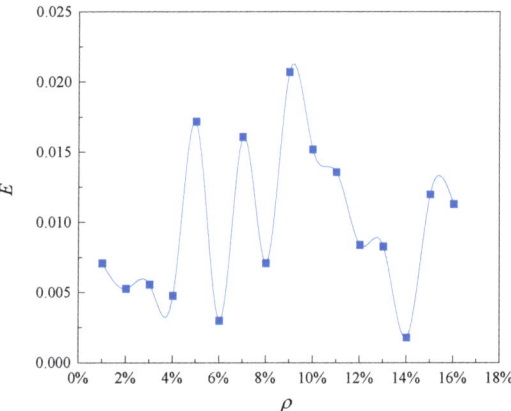

Figure 3. Calculation error of corrosion pit radius.

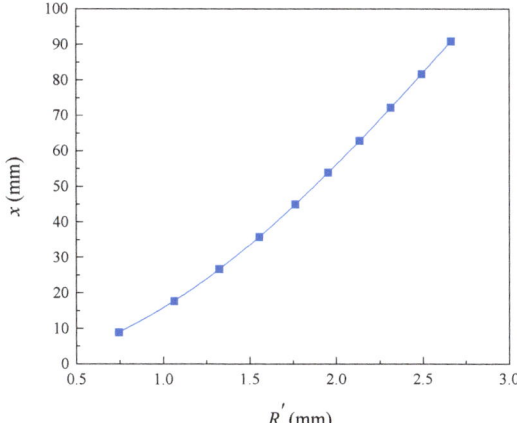

Figure 4. Influence of pit radius variation on corrosion pit depth.

The depth of corrosion pits under the influence of different corrosion degrees and pitting factors is shown in Table 2. When the corrosion degree is 8%, the pit depth decreases from 115.50 mm to 57.75 mm as the pitting factor increases from 4 to 8. The crater depth decreases with the increase in pitting factor. This is because when the corrosion degree is certain, the cross-sectional loss of reinforcement is also a fixed value, and the depth of the corrosion pit is inversely proportional to the pitting factor. When the pitting factor is 6, the pit depth increases from 50.81 mm to 76.99 mm as the corrosion degree increases from 6% to 9%. The corrosion degree increases from 6% to 7%, from 7% to 8%, and from 8% to 9%, and the pit depth increases by 16.53%, 14.69%, and 13.27%, respectively. The depth of the corrosion crater increases slowly. This is because, with the increase in corrosion level, the local corrosion of reinforcement becomes more and more serious. The corrosion products generated by corrosion attach to the surface of reinforcement and block the flow of media required for corrosion such as chlorine ion and oxygen, thus slowing down the development of corrosion to a certain extent.

Table 2. Depth of corrosion pits under different degrees of corrosion and pitting factors (mm).

ρ \ α	4	5	6	7	8
6%	76.20	60.97	50.81	43.55	38.10
7%	88.82	71.06	59.21	50.75	44.41
8%	101.96	81.57	67.97	58.26	50.98
9%	115.50	92.40	76.99	66.00	57.75

The crack width under the influence of different corrosion degrees and pitting factors is shown in Table 3. When the corrosion degree is 9%, the crack width reduces from 0.21 mm to 0.03 mm as the pitting factor increases from 4 to 8. The crack width decreases with the increase in pitting factor. When the pitting factor is 5, the crack width increases from 0.03 mm to 0.17 mm as the corrosion degree increases from 6% to 9%. The crack width increases 4.67 times. The crack width increases with the increase in corrosion degree. This is because the progress of corrosion, air, oxygen, and other corrosive media is sufficient, and corrosion can develop rapidly at the beginning. The crack width can expand rapidly before reaching the limit crack width, but after the crack width reaches the threshold, the corrosion rate can slow down because the accumulation of corrosion products restricts the flow of air and so on.

Table 3. Crack width under different degrees of corrosion and pitting factors (mm).

ρ \ α	4	5	6	7	8
6%	0.08	0.03	-	-	-
7%	0.12	0.07	0.03	-	-
8%	0.16	0.12	0.07	0.03	-
9%	0.21	0.17	0.12	0.07	0.03

The influence of the concrete cover thickness on the required pit depth at the initiation of reinforcement cracking is shown in Figure 5. When concrete cover thickness increases from 40 mm to 60 mm, the corrosion pit depth at the beginning of reinforcement cracking increases from 44.81 mm to 63.45 mm. The depth of the pit required for the initiation of reinforcement cracking increases with the concrete cover thickness. This is because increasing the concrete cover thickness not only increases the diffusion path of the chloride but also decreases the tensile stresses on the outer surface of the concrete cover due to the radial expansion of the corroded reinforcement, thus delaying the initiation of cracking the concrete cover.

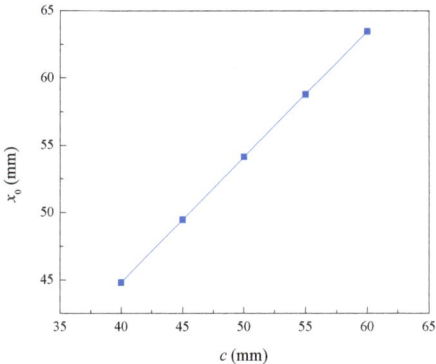

Figure 5. Influence of concrete cover thickness on pit depth during cracking.

The influence of the thickness of concrete cover on crack width is shown in Table 4. When the corrosion degree is 8%, the crack width decreases from 0.14 mm to 0.04 mm as the concrete cover thickness increases from 40 mm to 60 mm. The crack width decreases with the increase in concrete cover thickness. This is because the increase in the thickness of the concrete cover improves the impermeability, corrosion resistance, and other durability of the reinforced concrete, so the crack expansion is improved. Generally speaking, the greater the thickness of the concrete cover, the longer it takes for the external corrosive medium to reach the surface of the steel, the less likely it is for the reinforcement to rust, and the better the durability of the concrete. However, the concrete cover should not be too thick, as the shrinkage and temperature stresses are not well controlled during the hardening process.

Table 4. Crack width under different concrete cover thicknesses and corrosion levels (mm).

ρ \ c	40	45	50	55	60
6%	0.05	0.02	-	-	-
7%	0.09	0.07	0.04	0.02	-
8%	0.14	0.11	0.09	0.07	0.04
9%	0.18	0.16	0.14	0.11	0.09

The effect of the initial diameter of reinforcement on the radius of the corrosion pit and the loss of the reinforcement section is shown in Figure 6. When the corrosion degree is 6%, the radius of the corrosion pit increases from 1.56 mm to 3.12 mm with the reinforcement diameter increasing from 8 mm to 16 mm. At the same corrosion degree, the radius of the corrosion pit tends to increase with the increase in the initial diameter of reinforcement. When the initial diameter is 10 mm, the corrosion pit radius increases from 1.95 mm to 2.49 mm as the corrosion level increases from 6% to 9%. At the same initial diameter, the radius of the corrosion pit tends to increase with the increase in corrosion degree. This shows that the corrosion radius and the loss of cross-sectional area of reinforcement increase with the increase in the initial diameter of reinforcement.

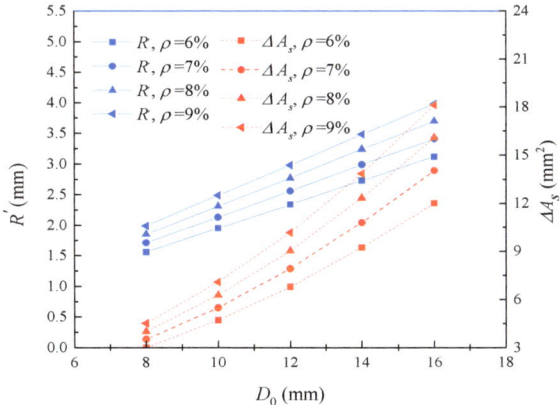

Figure 6. Effect of initial diameter on corrosion pit radius and loss of reinforcement cross-section.

Figure 7 shows the influence of the initial diameter of reinforcement on the depth of the corrosion pit at the beginning of reinforcement cracking and in the process of corrosion. When the corrosion degree is 6%, 7%, 8%, and 9%, the corrosion pit depth increases from 43.16 mm to 86.32 mm, from 50.61 mm to 100.71 mm, from 57.85 mm to 115.70 mm, and from 65.31 mm to 130.61 mm, as the initial diameter of reinforcement increases from 8 mm to 16 mm. At the same corrosion degree, the pit depth increases with the increase in the initial diameter of the reinforcement. When the initial diameter of the reinforcement increases from 8 mm to 16 mm, the required corrosion pit depth for the cracking of the reinforcement decreases from 65.78 mm to 36.66 mm. The depth of the corrosion pit required for reinforcement cracking reduces as the initial diameter of the reinforcement increases.

The localized steel section loss at the initiation of rebar cracking is related to the ratio of concrete cover thickness to rebar diameter and the pitting factor. As shown in Figure 8, when the pitting factor is 6, the local steel section loss at the beginning of cracking increases from 4.12 mm^2 to 5.87 mm^2 as the ratio of concrete cover to steel diameter increases from 4 to 6. In addition, the local steel section loss at the beginning of cracking increases with the increase in pitting factor. This is because the increase in pitting factor means that the ratio of local corrosion degree to uniform corrosion degree increases, the local corrosion degree increases, and then the loss of cross-sectional area of the reinforcement at the beginning of reinforcement cracking increases.

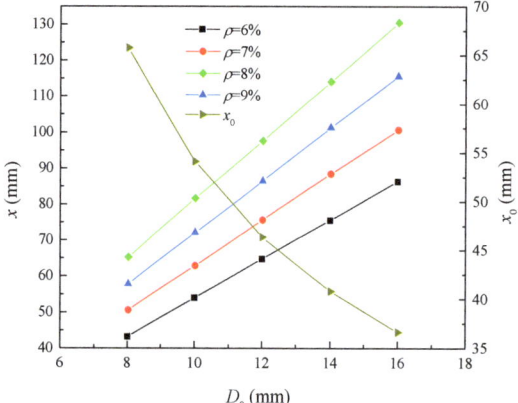

Figure 7. Influence of initial diameter on the pit depth at the initiation of cracking and the pit depth during corrosion.

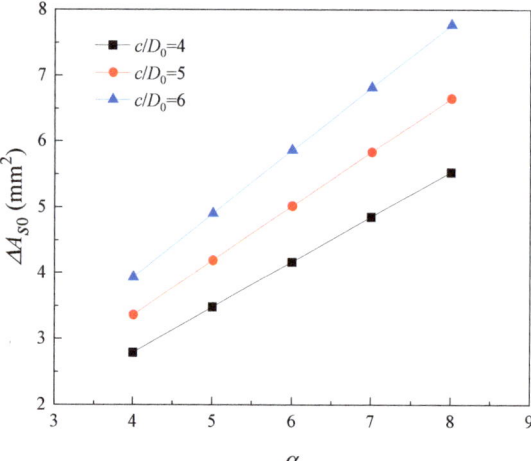

Figure 8. Factors affecting localized steel section loss at crack initiation.

The influence of corrosion degree on crack width is shown in Figure 9a. When the corrosion level is approximately 15%, the crack width reaches the ultimate crack width of 0.4 mm, and the structure becomes unstable. The influence of the loss of cross-sectional area of reinforcement on the crack width is shown in Figure 9b. When the crack width reaches the limit crack width, the loss of cross-sectional area of reinforcement is about 11.79 mm^2. The crack width increases with the increase in corrosion degree. This is because the reinforcement corrosion becomes more serious with the increase in corrosion degree. The crack gradually expands with the reinforcement corrosion until the structure is unstable.

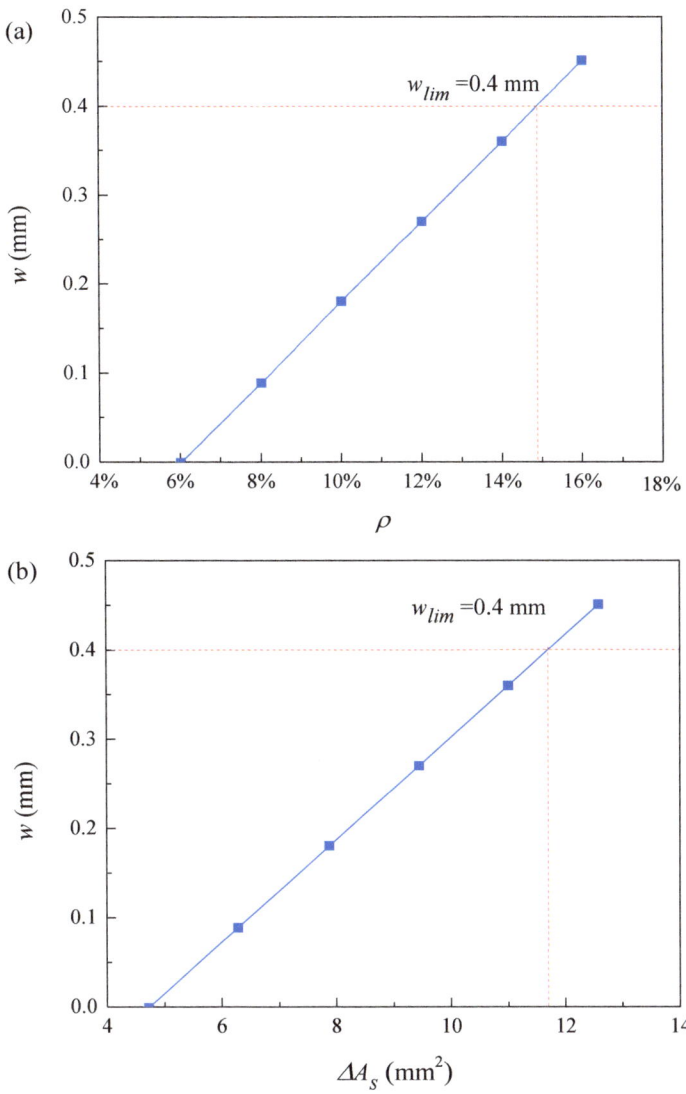

Figure 9. Effect of different factors on crack width: (**a**) corrosion degree and (**b**) loss of reinforcement cross-sectional area.

Figure 10 shows the influence of the initial diameter of reinforcement on the model error of the corrosion pit radius. As the corrosion level increases from 1% to 16%, the error curves follow very similar trends for initial steel diameters of 8 mm, 10 mm, and 12 mm. The error varies within the range from 0 to 2%. Therefore, it can be concluded that the initial diameter of reinforcement has little effect on the variation trend of the error of the pit radius model, and the error caused by the solution of the pit radius model can be ignored.

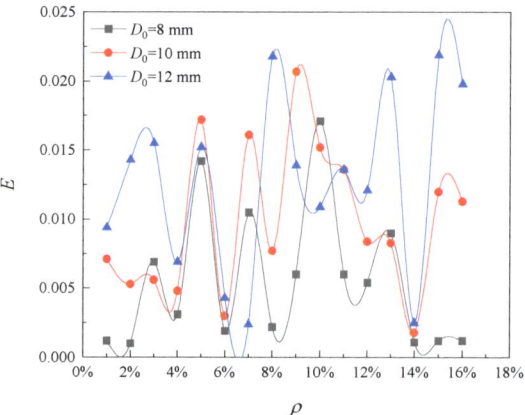

Figure 10. Effect of diameter change on error.

4. Discussion

The corrosion of steel bars can generally be divided into two types: uniform corrosion and local corrosion. The former assumes uniform corrosion on the surface of steel bars, uniform expansion stress generated by corrosion, and uniform distribution of corrosion products and depth along the surface of steel bars. The impact of uniform corrosion on piles is mainly proactive damage. The latter focuses on the study of local corrosion on the surface of steel bars, which is defined as local corrosion on the metal surface. It refers to the local dissolution of the metal caused by the rupture of the surface passivation protective film, which is limited to a single point or small area, and ultimately forms a form of voids (i.e., corrosion pits or cracks), which is a passive injury. Under the assumption of uniform corrosion, the research generally takes the time of corrosion initiation or the time of the first crack appearance as the standard for judging the durability of piles. Under the assumption of local corrosion, the time when cracks develop into severe cracks is used as the criterion for judging the instability of concrete piles. Kim and Kim [11] carried out a numerical analysis of localized corrosion (two-dimensional corrosion) of reinforced concrete structures. The results show that the depth of localized corrosion is 3.5 times greater than the depth of uniform corrosion after 3.5 years from the onset of corrosion. The concrete surface cracking time in localized corrosion mode is earlier and the vertical surface displacement is much greater than in uniform corrosion [21]. It has been shown that in the localized corrosion mode, cracks appear earlier on the concrete surface and the pressure to crack the concrete surface is much less than in the uniform corrosion mode. In addition, localized corrosion requires less loss of reinforcement section and shorter periods to reach the limit state of the concrete structure [27].

Table 5 shows the predicted durability life of piles under two different corrosion mode assumptions. It should be noted that the durability life of piles obtained under both assumptions is based on the theoretical model. In order to reduce errors, we ensure that the same parameter values are consistent for both corrosion modes and use the same judgment criteria to predict the durability life of piles. From Table 5, it can be seen that the predicted durability life of piles under uniform corrosion mode is much smaller than that under local corrosion mode. This is because the assumption of uniform corrosion greatly simplifies the modeling process, while in fact, the predicted durability life based on the assumption of uniform corrosion only indicates the beginning of corrosion, and the pile has not yet been adversely affected by potential inhibition of its function and performance. In general, corrosion of steel bars occurs simultaneously with uniform corrosion and local corrosion, but the harm of local corrosion to reinforced concrete structures is much greater than that caused by uniform corrosion. According to the short plate effect, the first cause of instability of RC structures mostly depends on the degree of local corrosion. Local corrosion causes passive damage to RC structures. When RC piles are subjected to local corrosion, the surface of the reinforced concrete expands, and the passivation film on the surface ruptures locally, causing concrete cracking. The effective area of the steel bar surface continues to decrease, the bonding between the steel bar and the concrete protective layer deteriorates, and the strength of the pile structure continues to deteriorate, which may lead to local instability of the pile and a decrease in its integrity and safety.

Table 5. Prediction results of durability life of piles under different corrosion modes.

Corrosion Mode	Method	Model	Durability Life (Years)
Uniform corrosion [28]	Determination method	$F(t) = C_{th} - C(r,t)$	34.8
	Reliability method	$P_f(t) = P[F(t) \leq 0] = P[C_{th} \leq C(r,t)]$, $T_i = [P_f(t) \geq P_{fmax}]$	27.4
Localized corrosion [20]	Determination method	$T_{sp,lim} = T_i + T_{cr} + T_{cp}$	77.60

In conclusion, it is too conservative to use the uniform corrosion model to study the corrosion of reinforced concrete structures, and there is a gap in the actual engineering conditions. The cracking time of the concrete surface, the crack initiation time, and the time when the cracks reach the ultimate crack are much longer in the uniform corrosion model than in the localized corrosion model. Therefore, in order to be more relevant to the actual marine engineering, the crack analysis of reinforced concrete structures in localized corrosion mode should be developed as much as possible. This paper proposes a new method for predicting cracks in concrete piles under localized corrosion of chloride in the marine environment, and the predicted results in this paper are in conformity with many experimental results. However, the shape, size, and spatial distribution of the specific corrosion pits and cracks need to be investigated in more depth. For example, by enhancing the identification technology of corrosion pits, the size and shape of corrosion pits at different positions on the surface of steel bars can be identified, and the impact of corrosion pit growth of different shapes on the durability of piles can be further studied.

5. Conclusions

This article is focused on the corrosion pits and their effect on the development of cracks in RC piles. In this paper, a novel prediction model for crack development of RC piles with local chloride corrosion in the marine environment is proposed. The generation and development of cracks in RC piles subjected to localized corrosion by chloride ions are studied and a sensitivity analysis of the important parameters of the cracking phase is carried out. Therefore, the following conclusions can be obtained:

(1) The radius of the corrosion pit, loss of cross-sectional area of reinforcement, and corrosion pit depth increase as the corrosion level increases. The corrosion depth decreases with the increase in pitting factor and increases with the increase in corrosion pit radius and initial diameter of steel bars. The crack width decreases with the increase in pitting factor and protective layer thickness. During the initial period of the crack, the development rate of corrosion pits is relatively fast, but before reaching the limit crack, the development of the corrosion pits becomes slow.

(2) The required pit depth for reinforcement cracking increases with the increase in the ratio of concrete cover thickness to diameter. The loss of the cross-sectional area of reinforcement and the radius of the corrosion pit increase with the increase in the initial diameter of reinforcement. The loss of the local reinforcement section at the beginning of reinforcement cracking increases with the increase in the ratio of concrete cover thickness to initial diameter and increases with the increase in the pitting factor.

(3) According to the limit crack criterion, the maximum corrosion degree of the reinforced concrete pile is about 15%. The error caused by the discrete method can be ignored.

The above analysis results are obtained without considering the influence of concrete characteristics on steel section loss.

(4) The prediction of the durability life of chloride ion corroded piles under the assumption of uniform corrosion is too conservative. In fact, the predicted durability life based on the assumption of uniform corrosion only indicates the beginning of corrosion, and the pile has not yet been adversely affected by the potential inhibition of its function and performance. In order to more accurately predict the time when the crack reaches the limit crack, the durability prediction of piles should be carried out under the assumption of local corrosion, rather than uniform corrosion. When evaluating the durability life of RC piles in actual service, engineers should pay more attention to the local corrosion on the surface of RC piles. This paper proposes a new method to predict the cracking of concrete piles by localized corrosion of chloride in the marine environment. However, the impact of different shapes and spatial distributions of corrosion pits on crack width is currently unclear. If advanced corrosion pit identification technology can be combined to study the impact of different shapes and spatial distributions of corrosion pits on the development of crack width, it will provide effective measures for the durability evaluation of future RC structures.

Author Contributions: Conceptualization, W.S. and D.S.; methodology, W.S. and X.H.; writing—original draft, X.H.; formal analysis, X.H. and W.Z.; investigation, D.S.; visualization, D.S.; writing—review and editing, W.S. and W.Z.; supervision, W.S. and W.Z.; funding acquisition, W.S. All authors have read and agreed to the published version of the manuscript.

Funding: This research was financially supported by the National Natural Science Foundation of China (Grant No. 52078289), the Shanghai International Science and Technology Cooperation Project (Grant No. 19520744100), State Key Laboratory of Hydraulic Engineering Simulation and Safety (Tianjin University) (Grant No. HESS-2321), and State Key Laboratory of Coastal and Offshore Engineering (Dalian University of Technology) (Grant No. LP2111).

Institutional Review Board Statement: Not applicable.

Informed Consent Statement: Not applicable.

Data Availability Statement: The data presented in this study are available on request from the corresponding author.

Conflicts of Interest: The authors declare no conflict of interest.

References

1. Zhao, Y.X.; Wu, Y.Y.; Jin, W.L. Distribution of millscale on corroded steel bars and penetration of steel corrosion products in concrete. *Corros. Sci.* **2013**, *66*, 160–168. [CrossRef]
2. Chao, Z.M.; Dang, Y.B.; Pan, Y.; Wang, F.Y.; Wang, M.; Zhang, J.; Yang, C.X. Prediction of the shale gas permeability: A data mining approach. *Geomech. Energy Environ.* **2023**, *33*, 100435. [CrossRef]
3. Jang, B.S.; Oh, B.H. Effects of non-uniform corrosion on the cracking and service life of reinforced concrete structures. *Cem. Concr. Res.* **2010**, *40*, 1441–1450. [CrossRef]
4. Cao, C. 3D simulation of localized steel corrosion in chloride contaminated reinforced concrete. *Constr. Build. Mater.* **2014**, *72*, 434–443. [CrossRef]
5. Chen, E.; Berrocal, C.G.; Fernandez, I.; Löfgren, I.; Lundgren, K. Assessment of the mechanical behaviour of reinforcement bars with localized pitting corrosion by digital image correlation. *Eng. Struct.* **2020**, *219*, 110936. [CrossRef]
6. Moreno, E.; Cobo, A.; Palomo, G.; González, N. Mathematical models to predict the mechanical behavior of reinforcements depending on their degree of corrosion and the diameter of the rebars. *Constr. Build. Mater.* **2014**, *61*, 156–163. [CrossRef]
7. Tang, F.J.; Lin, Z.B.; Chen, G.D.; Yi, W.J. Three-dimensional corrosion pit measurement and statistical mechanical degradation analysis of deformed steel bars subjected to accelerated corrosion. *Constr. Build. Mater.* **2014**, *70*, 104–117. [CrossRef]
8. Liu, J.; Wang, Z.G.; Zhang, R.B.; Du, X.L. Mesoscopic simulation on flexural behavior of single-way reinforced concrete slab with rebars subjected to localized corrosion. *Structures* **2021**, *31*, 815–827.
9. Xia, N.; Ren, Q.W.; Liang, R.Y.; Payer, J.; Patnaik, A. Nonuniform corrosion-induced stresses in steel-reinforced concrete. *J. Eng. Mech.* **2012**, *138*, 338–346. [CrossRef]
10. Zhao, Y.X.; Hu, B.Y.; Yu, J.; Jin, W.L. Non-uniform distribution of rust layer around steel bar in concrete. *Corros. Sci.* **2011**, *53*, 4300–4308. [CrossRef]

11. Kim, C.-Y.; Kim, J.-K. Numerical analysis of localized steel corrosion in concrete. *Constr. Build. Mater.* **2008**, *22*, 1129–1136. [CrossRef]
12. Li, D.W.; Ren, W.; Li, L.; Guan, X.T.; Mi, X.M. Pitting corrosion of reinforcing steel bars in chloride contaminated concrete. *Constr. Build. Mater.* **2019**, *199*, 359–368. [CrossRef]
13. Darowicki, K.; Mirakowski, A.; Krakowiak, S. Investigation of pitting corrosion of stainless steel by means of acoustic emission and potentiodynamic methods. *Corros. Sci.* **2003**, *45*, 1747–1756. [CrossRef]
14. Kioumarsi, M.M.; Hendriks, M.A.N.; Kohler, J.; Geiker, M.R. The effect of interference of corrosion pits on the failure probability of a reinforced concrete beam. *Eng. Struct.* **2016**, *114*, 113–121. [CrossRef]
15. Vidal, T.; Castel, A.; François, R. Analyzing crack width to predict corrosion in reinforced concrete. *Cem. Concr. Res.* **2004**, *34*, 165–174. [CrossRef]
16. Zhang, W.B.; Shi, D.D.; Shen, Z.Z.; Zhang, J.; Zhao, S.; Gan, L.; Li, Q.M.; Chen, Y.S.; Tang, P. Influence of chopped basalt fibers on the fracture performance of concrete subjected to calcium leaching. *Theor. Appl. Fract. Mech.* **2023**, *125*, 103934. [CrossRef]
17. Zhang, R.J.; Castel, A.; François, R. Concrete cover cracking with reinforcement corrosion of RC beam during chloride-induced corrosion process. *Cem. Concr. Res.* **2010**, *40*, 415–425. [CrossRef]
18. Li, C.Q.; Lawanwisut, W.; Zheng, J.J.; Kijawatworawet, W. Crack width due to corroded bar in reinforced concrete structures. *Int. J. Mater. Struct. Reliab.* **2005**, *3*, 87–94.
19. Zhao, Y.X.; Yu, J.; Jin, W.L. Damage analysis and cracking model of reinforced concrete structures with rebar corrosion. *Corros. Sci.* **2011**, *53*, 3388–3397. [CrossRef]
20. Shao, W.; He, X.Q.; Shi, D.D. Durability life prediction of RC piles subjected to localized corrosion in chloride environments. *Eng. Fail. Anal.* **2022**, *136*, 106184. [CrossRef]
21. Guzmán, S.; Gálvez, J.C. Modelling of concrete cover cracking due to non-uniform corrosion of reinforcing steel. *Constr. Build. Mater.* **2017**, *155*, 1063–1071. [CrossRef]
22. Eliass, E.A.; Fekak, F.E.; Garibaldi, L.; Ahmed, E. A numerical study of pitting corrosion in reinforced concrete structures. *J. Build. Eng.* **2021**, *43*, 102789.
23. Sun, J.; Ding, Z.H.; Huang, Q. Corrosion fatigue life prediction for steel bar in concrete based on fatigue crack propagation and equivalent initial flaw size. *Constr. Build. Mater.* **2019**, *195*, 208–217. [CrossRef]
24. Alonso, C.; Andrade, C.; Rodriguez, J.; Diez, J.M. Factors controlling cracking of concrete affected by reinforcement corrosion. *Mater. Struct.* **1998**, *31*, 435–441. [CrossRef]
25. Stewart, M.G.; Mullard, J.A. Spatial time-dependent reliability analysis of corrosion damage and the timing of first repair for RC structures. *Eng. Struct.* **2007**, *29*, 1457–1464. [CrossRef]
26. Stewart, M.G. Spatial variability of pitting corrosion and its influence on structural fragility and reliability of RC beams in flexure. *Struct. Saf.* **2004**, *26*, 453–470. [CrossRef]
27. Zhao, Y.X.; Karimi, A.R.; Wong, H.S.; Hu, B.Y.; Buenfeld, N.R.; Jin, W.L. Comparison of uniform and non-uniform corrosion induced damage in reinforced concrete based on a Gaussian description of the corrosion layer. *Corros. Sci.* **2011**, *53*, 2803–2814. [CrossRef]
28. Shao, W.; Nie, Y.H.; Liang, F.Y.; Shi, D.D. A novel comprehensive evaluation method for the corrosion initiation life of RC hollow piles in chloride environments. *Constr. Build. Mater.* **2020**, *249*, 118801. [CrossRef]

Article

On the Change in Hydrogen Diffusion and Trapping Behaviour of Pearlitic Rail Steel at Different Stages of Production

Matthias Eichinger [1,*], Bernd Loder [1], Michael Tkadletz [2], Holger Schnideritsch [3], Gerald Klösch [3] and Gregor Mori [1]

[1] General and Analytical Chemistry, Montanuniversitaet Leoben, Franz Josef-Strasse 18, 8700 Leoben, Austria
[2] Functional Materials and Materials Systems, Montanuniversitaet Leoben, Franz Josef-Strasse 18, 8700 Leoben, Austria
[3] Voestalpine Stahl Donawitz GmbH, Kerpelystraße 199, 8700 Leoben, Austria
* Correspondence: matthias.eichinger@unileoben.ac.at

Abstract: To avoid hydrogen flaking in rail production, it is of crucial importance to understand the differences in hydrogen diffusion and trapping between different production steps. Therefore, as-cast unfinished material was compared with two finished rails, hot-rolled and head-hardened, using electron backscattered diffraction (EBSD), electrochemical permeation, and thermal desorption spectroscopy (TDS). A significant increase in dislocation density was in the head-hardened rail compared with the other material states. This leads to an effective hydrogen diffusion coefficient of 5.8×10^{-7} cm^2/s which is lower by a factor of four than the diffusion coefficients examined in the other states. Thermal desorption spectroscopy analyses show a clear difference between unfinished and finished rail materials. While a peak in activation energy between 32 and 38 kJ/mol is present at all states, only as-cast unfinished material shows a second peak with an activation energy of 47 kJ/mol, which is related to microvoids. The results show that in the investigated material, the effect of increasing dislocation density has a stronger influence on the effective diffusion coefficient than the presence of a second active trapping site.

Keywords: pearlitic rail steel; hydrogen diffusion; hydrogen trapping; electrochemical permeation; thermal desorption spectroscopy

Citation: Eichinger, M.; Loder, B.; Tkadletz, M.; Schnideritsch, H.; Klösch, G.; Mori, G. On the Change in Hydrogen Diffusion and Trapping Behaviour of Pearlitic Rail Steel at Different Stages of Production. *Materials* **2023**, *16*, 5780. https://doi.org/10.3390/ma16175780

Academic Editor: Ming Liu

Received: 25 July 2023
Revised: 14 August 2023
Accepted: 22 August 2023
Published: 23 August 2023

1. Introduction

The international movement of goods is constantly increasing as globalization progresses. Trains are a cost-effective way of transporting large quantities of goods over land. To construct new and maintain existing railway tracks, more than 10 million tons of railway steel are produced per year.

Rail steel classically has a pearlitic microstructure, although rail steel with a bainitic structure is also produced on the basis of modern alloy concepts and heat treatments [1–3]. It is known that the applied steels are prone to hydrogen embrittlement (HE), especially to hydrogen flaking [4,5]. This phenomenon describes hydrogen-based cracks which appear in the material hours or even days after production. These cracks lead to a drastic decrease in the material's mechanical properties and fatigue strength [6–8]. Hydrogen uptake during the production process, in combination with internal stresses (resulting from phase transformations, temperature differences, or deformation) and a prone microstructure, has been identified as the main reason for hydrogen flaking [7]. To reduce the hydrogen content, a vacuum treatment is mandatory for most rail steel before casting. The only production unit between vacuum degassing and continuous casting and, therefore, the place where most hydrogen is taken up by the steel bath is the tundish, where hydrogen can be provided due to the humidity of refractory lining or casting powder [5,9]. During solidification near eutectoid, melts traverse several phase transformations with different hydrogen solubilities. The liquid metal has the highest solubility for hydrogen, followed by the face-centered cubic austenitic lattice structure, which can hold significantly higher amounts of hydrogen compared with a body-centered cubic ferritic one [10]. As the solidification of the bloom progresses from the outside to the inside during continuous casting, hydrogen, which can no longer be dissolved in the solidified outer layers, is absorbed by the liquid phase in the center of the bloom. This results in an enrichment of hydrogen from the outside to the inside of the bloom. Over the course of the rail rolling process, the bloom is reshaped in such a way that the former center of the bloom (high hydrogen content) is located at the transition from the rail neck to the rail head and, thus, at the area with the narrowest cross-section. Moreover, this section is subjected to the fastest cooling rates during heat treatment. Hydrogen diffuses as a function of temperature, time, and microstructure in the matrix of the material and accumulates at trapping sides such as dislocations, grain boundaries, phase boundaries, vacancies, microvoids, or non-metallic inclusions. If a sufficiently high enrichment of hydrogen atoms takes place at such trapping sides, the lattice cannot dissolve the hydrogen anymore, and it precipitates as molecular hydrogen. This process is linked to a high volume expansion resulting in high internal stresses that can lead to crack formation [11].

Although the problem of hydrogen flaking has been present in the manufacturing of rail steel for a long time, detailed studies about hydrogen diffusion and trapping behavior at different steps in the production of rail steel are rare in the literature. Deep knowledge of these processes can significantly improve the production of rail steel, especially regarding the heat treatments applied to reduce the material's hydrogen content. In this context, this study aims to provide a detailed analysis of hydrogen interactions in eutectoid rail steels with a special focus on the changes in hydrogen diffusion and trapping concerning different production steps and heat treatments. To achieve this objective, three material conditions corresponding to different production steps (pre-cast material, hot-rolled rail, and head-hardened rail) are investigated using a combination of high-resolution electron microscopy, electrochemical permeation, and thermal desorption spectroscopy (TDS).

2. Materials and Methods

2.1. Materials

The investigated material is a eutectoid low alloyed steel with a chemical composition of 0.79 wt.% C, 0.30 wt.% Si, 1.06 wt.% Mn, 0.016 wt.% P, 0.014 wt.% S, 0.051 wt.% Cr and 0.0005 wt.% Ti. In secondary metallurgy, the steel was degassed to reduce the dissolved hydrogen content below production limits prior to tundish treatment and continuous casting. To avoid hydrogen flaking of the finished rails, blooms were stored in effusion boxes to reduce their hydrogen content. Thereafter, the material was heated up to 1000 °C and rolled into rails. To analyze the diffusion and trapping behavior of the as-cast (As-cast) bloom, specimens were taken at 75% of the bloom's height. In the course of this study, the finished rail materials were examined in hot-rolled (HR) as well as in heat-treated (HT) conditions, i.e., head-hardened conditions. Therefore, samples were manufactured out of the rail head. Figure 1 gives the exact sampling positions (highlighted in red) for all states. The abbreviations in brackets indicate the further designation of the different material conditions in the text.

2.2. Materials Characterization

The microstructure of all samples was characterized across scales using a light optical microscope (LOM) (Olympus, Tokyo, Japan) and a scanning electron microscope equipped with an electron backscattered diffraction (EBSD) detector. The investigations using the LOM were conducted on an Olympus AX70. The investigated specimens were hot embedded in epoxy resin and ground stepwise with SiC abrasive paper of grits 120, 240, 500, and 1000. Thereafter, samples were polished with 1 μm diamond paste and etched using 3% nital solution for 10 (As-cast, HR) to 30 (HT) seconds. Additionally, a cross-section of the As-cast conditions bloom was prepared by etching it with a mixture of 50% HCl and 50%

H_2O at 60 °C. Furthermore, the material's hardness (HV1) was determined using a fm-300 (Future Tech, Kanagawa, Japan) hardness testing device.

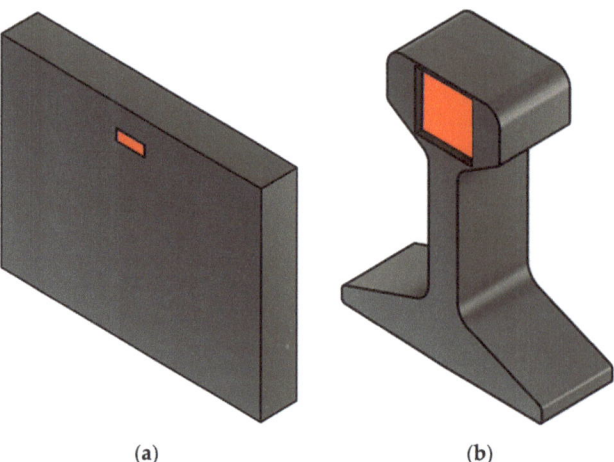

(a)   (b)

Figure 1. Positions of sample machining (red) of (**a**) As-cast and (**b**) HR and HT material condition.

In order to avoid mechanical deformations caused by the sample preparation, the cross sections for the EBSD investigations were produced using an IonSlicer (Hitachi IM4000+, Tokyo, Japan). For sample preparation, an excitation voltage of 6 kV and an oscillation of 15°/6 per minute were applied for 60 min. EBSD analyses were carried out using a GeminiSEM 450 (Zeiss, Jena, Germany) field emission SEM coupled with an Symmetry EBSD detector (Oxford Instruments, Abingdon, UK). A step size of 200 nm was chosen for the investigations. Imaging was conducted using the software AZtec 5.1 as well as AZtecCrystal 2.1.

The dislocation densities of all material states were determined using X-ray diffraction (XRD) with a Bruker D8 Advance Davinci (Bruker, Billerica, MA, USA) diffractometer, which was equipped with a Cu-K_α-X-ray tube. The measuring procedure and the selected parameter settings are described in detail in [12]. To evaluate data regarding the dislocation density, whole powder pattern modeling was conducted using a Topas (Bruker, Billerica, MA, USA) macro based on the theory of Krivoglaz and Wilkins [13–15].

2.3. Electrochemical Permeation Testing

For the permeation measurements, square-shaped samples with a cross-section of 40×40 mm and a thickness of 1.2 mm were manufactured. The thickness of these specimens was further reduced upon grinding with SiC abrasive paper with grit 1000 to a final thickness of 1.000 ± 0.005 mm. A 100 nm thick palladium layer was applied on one side of the sample using physical vapor deposition. The specimen was inserted into a double cell, according to Devanathan and Stachurski [16], in such a way that the palladinated side was facing the cell's oxidation side. The electrolytes used for the permeation experiment were 0.1 M NaOH on the oxidation side and 3.5% NaCl solution with the addition of 1 g/L $SC(NH_2)_2$ on the charging side, respectively. To ensure a minimum amount of dissolved oxygen, both electrolytes were continuously purged with Ar (99.999%). The experiments were conducted at 25 °C, and a charging current of 1 mA/cm^2 was applied. The potential on the oxidation side was kept constant at +540 mV against a standard hydrogen electrode, equal to +300 mV against a standard calomel electrode. Two charging cycles were performed on each material condition.

These measured data were evaluated using the standardized time-lag method according to Equation (1) [17]:

$$D_{eff} = \frac{L^2}{6 \cdot t_{lag}}$$ (1)

where D_{eff} is the effective diffusion coefficient of hydrogen, L is the sample thickness, and t_{lag} is the time between applying the charging current and reaching 63% of the oxidation currents plateau value.

2.4. Thermal Desorption Spectroscopy

TDS spectra of all materials were recorded using a Bruker Galileo G8 (Bruker, Billerica, MA, USA) coupled with an IR07 infrared furnace and an IPI quadrupole mass spectrometer to determine the active traps in the sample materials. For this purpose, specimens with a geometry of $30 \times 6 \times 1.2$ mm were manufactured and electrolytically charged with hydrogen. These specimens acted as the cathode and were placed in the center of a platinum mesh electrode which served as the anode to secure homogeneous hydrogen charging. Charging was conducted at 10 mA/cm^2 for one hour at 25 °C, where 0.5 M H_2SO_4 with the addition of 1 g/L SC(NH$_2$)$_2$ served as electrolyte. After charging, the specimens were flushed with acetone and immediately cooled in liquid nitrogen. Prior to recording the TDS spectra, the specimens were held for ten minutes at room temperature (25 °C) in air to remove hydrogen adsorbed onto the surface. This was performed to prevent interference with the measured signal originating from the hydrogen effusing from the traps. The system was calibrated with a calibration gas consisting of nitrogen and 5.05% H_2. For the measurements, N_2 (99.999%) was used as the carrier gas. The recorded TDS spectra were conducted under the application of four different heating rates (200, 400, 800, and 1200 K/h), during which the specimens were heated from 25 to 450 °C. To ensure no deep traps resulting in high-temperature peaks were missed, selected tests were heated up to 900 °C. Peak deconvolution was performed using the PeakFit v4.12 software, applying a Gaussian distribution for all single peaks. The activation energies for the different peaks of TDS spectra were calculated according to Kissinger's approach after Equation (2) [18,19]:

$$\frac{d\left(\ln \frac{\phi}{T_p^2}\right)}{d\left(\frac{1}{T_p}\right)} = -\frac{E_A}{R}$$ (2)

where ϕ is the heating rate, T_p is the temperature of the peak center from the TDS spectrum, E_A is the activation energy for the desorption of hydrogen from the specific trap, and R is the universal gas constant.

3. Results

3.1. Materials Characterization

Figure 2 gives the microstructure of As-cast (a), HR (b), and HT (c) conditions analyzed using LOM at a 100-fold magnification. Although a purely pearlitic microstructure is visible for all conditions, the appearance of the pearlite colonies changes significantly depending on the material's condition. The As-cast state has the coarsest microstructure of all the materials examined; its pearlite colonies are oriented toward the dendritic solidification and have a diameter of several hundred micrometers. A decrease in the average pearlite colony diameter compared to the As-cast condition is observed for the HR (around 100 µm) and the HT (<100 µm) conditions. Furthermore, the pearlite colonies in the HR as well as in the HT state are oriented randomly and, therefore, independent from the material's prior solidification direction. In Figure 3, the bloom's cross-section is provided. As indicated by the full and dashed lines, the cross-section can be divided into three regimes marked as I to III. Regime I represents the bloom center where big blowholes and pores are present. The size and density of these defects decrease with increasing distance from the bloom center so that in Regime II, minor pores, blowholes, and microvoids were observed. Regime III, closest to the surface, is free of pores and blowholes.

(a)   (b)   (c)

Figure 2. Microstructure of the (**a**) As-cast, (**b**) HR, and (**c**) HT states observed using LOM.

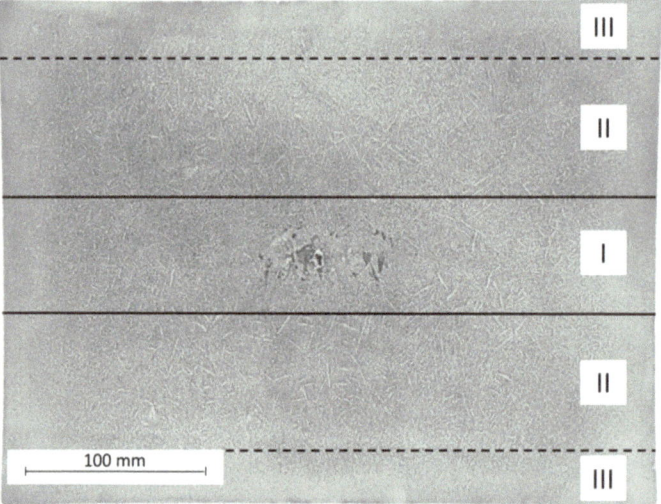

Figure 3. Cross-section of the As-cast bloom with indications of different regimes (I to III) regarding the presence of blowholes, pores, and microvoids. Regime I represents the blooms center with large sized blowholes and pores. In regime II minor pores, blowholes and microvoids are present. Regime III is free of blowholes, pores and microvoids.

The hardness measurements show a significant increase for the HT condition (364 $\pm$ 8 HV) compared with the As-cast (278 $\pm$ 3 HV) and the HR (285 $\pm$ 7 HV) states.

The results of the EBSD investigations are given in Figure 4 in the form of inverse pole figure (IPF) maps in the x-direction ((a), (c), and (e)) and Kernel average misorientation (KAM) maps (b), (d) and (f)) for all samples. Additionally, the average KAM factors are given in Figure 4a–c. The crystallographic orientation distribution of the As-cast material (Figure 4a) shows a relevant orientation texture over large parts of its cross-section. For HR (c), the IPF maps show a significantly decreased fraction of textured microstructure compared to the As-cast state. The finest grain size with a completely random distribution of grain orientation is shown for the HT (e) material. The amount of zones with low KAM factor (blue) decreases from the As-cast (b) to HR (d) condition and further nearly disappear in the HT (f) state. In the HT material, there is a significantly higher proportion of areas with a high KAM factor (red) compared with other conditions. This results in average KAM factors of 0.66 for the As-cast, 0.74 for HR, and 1.10 for the HT conditions. Figure 4b shows that the zones with the lowest KAM factors are located in sample areas of highly oriented texture. The black areas in Figure 4 are due to insufficient resolution in these sample areas. Apart from IPF and KAM, the fractions of body-centered cubic (bcc) and face-centered cubic (fcc) lattice structures were determined using EBSD, resulting in 100% bcc and 0% fcc microstructures for all investigated materials. The fraction of low-angle grain boundaries (LAGB) and high-angle grain boundaries (HAGB) are comparable for the HR and HT conditions and resulted in about 80% of LAGB (with an angle $\leq 15°$) and 20% HAGB (with an angle $\geq 15°$) respectively. In comparison, the As-cast condition has a lower proportion of HAGB at 11.8%.

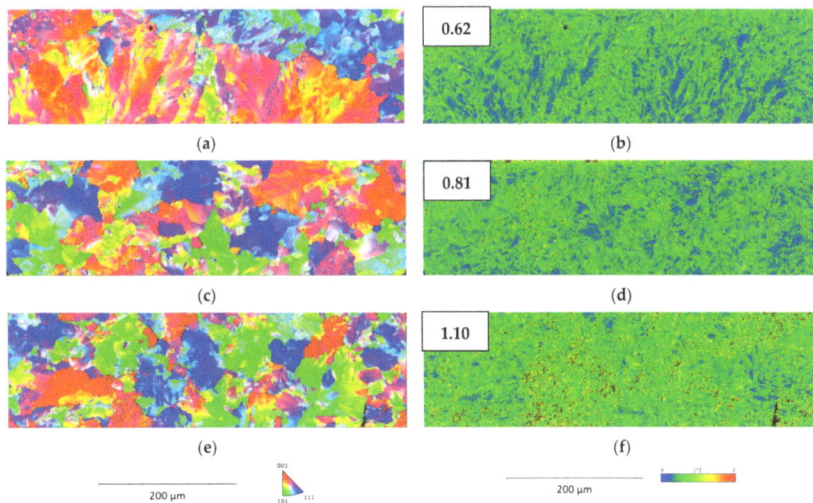

Figure 4. IPF maps in the x-direction of (**a**) As-cast, (**c**) HR, (**e**) HT state, and KAM factor map with an indication of the average KAM factors of (**b**) As-cast, (**d**) HR and (**f**) HT condition.

XRD measurements prove a significant increase in the dislocation density from 2.1×10^{14} m^{-2} for the As-cast material to 3.9×10^{14} m^{-2} for the HR and 9.2×10^{14} m^{-2} for the HT conditions.

Summing up, one can conclude that significant differences between the microstructures of the As-cast, HR, and HT conditions are present. The processes in rail production lead to an increasing refinement of the microstructure from the As-cast to the HR and HT state. This simultaneously leads to a reduction in texture and an increase in hardness as well as the KAM factor and the dislocation density. In addition, the annealing and forming treatments close blowholes, pores, and microvoids. A qualitative summary and comparison of the microstructural features of all investigated material states is provided in Table 1.

Table 1. Summary of microstructural features present in the investigated material conditions.

Microstructural Feature	Material Condition		
	As-Cast	HR	HT
Pearlite colony size	coarse	fine	very fine
Texture	dendritic oriented	none	none
Hardness [HV]	278 ± 3	285 ± 7	364 ± 8
Microvoids	yes	no	no
KAM	0.62	0.81	1.10
Dislocation density [m^{-2}]	2.1×10^{14}	3.9×10^{14}	9.3×10^{14}
Lattice structure	bcc	bcc	bcc
LAGB [%]	88.2	78.5	80.3
HAGBs [%]	11.8	21.5	19.7

3.2. Electrochemical Permeation Measurement

To determine the influence of hot rolling and different heat treatments on hydrogen diffusion, electrochemical permeation tests were conducted on all material conditions. Figure 5a shows the permeation transients of all states during the first charging cycle. It is shown that the permeation transients of the As-cast and HR material do not differ significantly as the current rises as a function of time. There are minor differences in the steepness of corresponding transients for these two material states. In comparison, the HT state shows different behavior in terms of both criteria. The time lag was two times higher compared with the As-cast and HR conditions. Apart from this, the slope of the HT permeation transient is less steep compared with the ones from the As-cast and the HR state. The effective diffusion coefficients of the first (D_{eff1}) and second (D_{eff2}) charging cycles obtained from electrochemical permeation experiments are shown in Table 2 for all states. It can be seen that D_{eff1} is equal for As-cast and HR. Compared with that, the HT material provides a four times lower effective diffusion coefficient in the first charging cycle. The effective diffusion coefficient of the As-cast condition increases by nearly a factor of two between the first and second charging cycles. To illustrate this behavior, Figure 5b gives the permeation transients for the first and second charging cycles of the As-cast state. It is shown that the time lag until an increase in oxidation current occurs significantly decreases from the first to the second charging cycle. In terms of HR, just a slight increase in the effective diffusion coefficient was measured for its second charging cycle, and the effective diffusion coefficient of the HT state was unaffected by the number of applied charging cycles. The permeation transients of HR and HT do not show a change between the first and second charging cycles comparable to the one of the As-cast state.

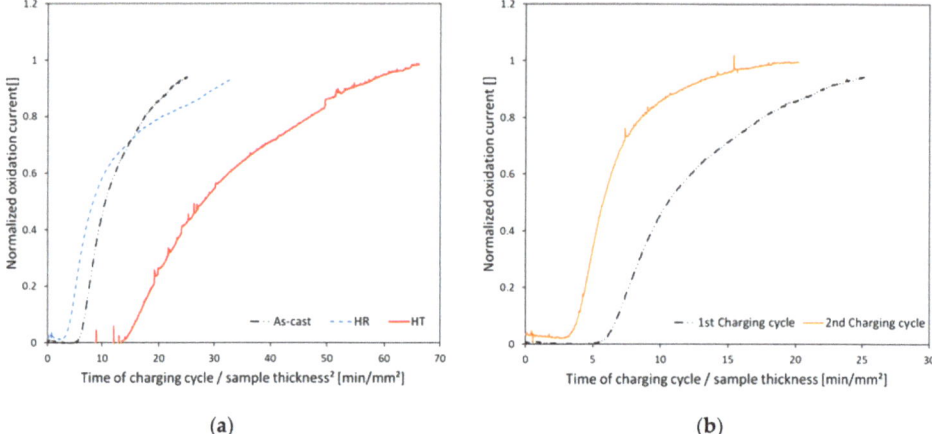

(a)　　　　　　　　　　　　　　　　　(b)

Figure 5. (**a**) First charging cycles permeation transients of all material conditions, (**b**) Comparison of permeation transients of the As-cast materials first and second charging cycle.

Table 2. Diffusion coefficients of the investigated materials.

Material Condition	D_{eff1} [cm^2/s]	D_{eff2} [cm^2/s]
As-cast	2.3×10^{-6}	4.3×10^{-6}
HR	2.3×10^{-6}	3.0×10^{-6}
HT	5.8×10^{-7}	5.8×10^{-7}

3.3. Thermal Desorption Spectroscopy

Figure 6 shows the desorption spectra of the As-cast (a), the HR (c), and the HT (e) material obtained by applying all four heating rates and the corresponding peak deconvolutions at a heating rate of 400 K/h for As-cast (b), HR (d) and HT (f) condition. The obtained desorption rates increase, with the heating rate reaching its maximum values at a heating rate of 1200 K/h. For the As-cast and HR states, they are close to each other and in the range of 0.0085 to 0.0095 wt.-ppm/s. The HT condition, compared with the other material conditions, has a slightly increased maximum desorption rate of 0.012 wt.-ppm/s. Considering the peak deconvolution, it can be seen that the measured sum curve of the As-cast condition (b) can be fitted by two peaks and reaches baseline at 250 °C. Therefore, 250 °C can be considered the temperature where all the hydrogen absorbed during charging has effused from the specimen. In contradiction to the As-cast state, the spectra of HR (d) and HT (f) consist of one single peak and reach their baselines at about 150 °C. Based on the spectra of peak deconvolution, it can be seen that the maxima of the As-cast material's first peak and the peaks from HR as well as HT all occur in a close temperature range between 60 and 70 °C. The peak maximum of the As-cast material's second peak is shifted towards higher temperatures and located at 130 °C. The plots of $\ln(\phi/T_p^2)$ as a function of $1/T_p$ are given in Figure 7, where ϕ represents the heating rate, and T_p is the temperature at which the peak maximum occurs. The plots depicted in Figure 7 were used for the calculation of the activation energies corresponding to the peaks shown in Figure 6. The slope's steepness corresponds to the activation energy of the specific trapping site. The peak activation energies gained by the application of Kissinger's approach are summarized in Table 3. These plots result in equal correlation coefficients between 0.9800 and 0.9915 for the first peak of the As-cast state and the peaks of the HR and the HT conditions. In comparison, data obtained for the second peak of the As-cast condition show a lower correlation coefficient of 0.9327 which further leads to a larger scatter of the peak's activation energy. It can be seen that the activation energies obtained for the first peak of the As-cast and the peaks of HR and HT conditions are between 34 ± 5 (HR and HT) and 38 ± 4 kJ/mol (As-cast). Due to the metrological uncertainty of ± 3 to 4 kJ/mol, the activation energies of these peaks do not differ significantly. For the second peak of the As-cast condition, an activation energy of 47 ± 8 kJ/mol was determined. Although the scattering ranges of the activation energies of both peaks deconvoluted from the As-cast state's TDS spectrum are slightly overlapping, Figure 6a,b prove the presence of two clearly distinguishable peaks.

Table 3. Summary of the trap activation energies obtained by TDS measurements.

Material Condition	Activation Energy Peak 1 (E_{A1}) [kJ/mol]	Activation Energy Peak 2 (E_{A2}) [kJ/mol]
As-cast	38 ± 4	47 ± 8
HR	34 ± 4	
HT	32 ± 3	

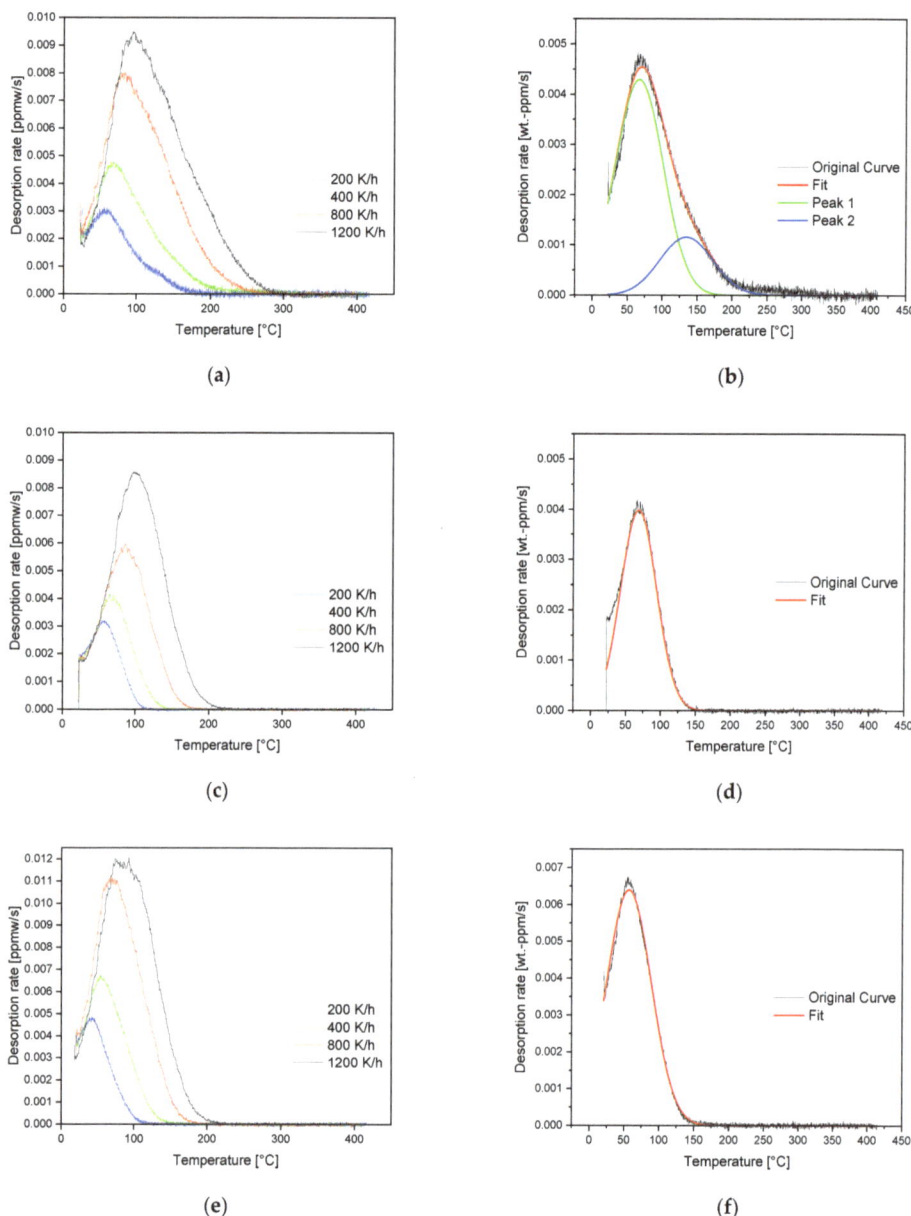

Figure 6. Desorption rates according to the applied heating rates for (**a**) As-cast, (**c**) HR, (**e**) HT and peak deconvolution of desorption spectra obtained at 400 K/h for (**b**) As-cast, (**d**) HR, and (**f**) HT condition.

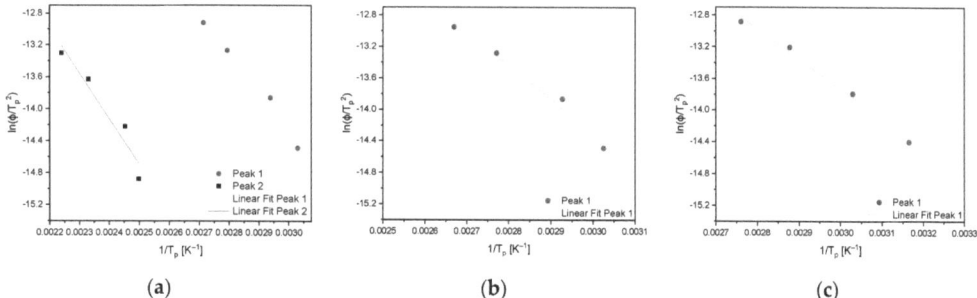

Figure 7. Plots of $\ln(\phi/T_p^2)$ as a function of $1/T_p$ for (**a**) As-cast, (**b**) HR, and (**c**) HT material.

4. Discussion

4.1. Materials Characterization

The As-cast material solidifies from the melt. Its microstructure is dominated by coarse pearlite colonies oriented in the direction of solidification. Reheating the steel above its austenitization temperature, followed by hot rolling, leads to a significant change in its microstructure due to recrystallization processes. According to Table 1, the pearlite structure becomes finer when transitioning from the As-cast state to the HR and HT states. Microstructure and mechanical properties of pearlitic steels depend to a large amount on the cooling rate, which is indirectly proportional to the interlamellar spacing. The dependency of the mechanical properties on the interlamellar spacing is usually described using the Hall–Petch relationship so that a decrease in the interlamellar spacing results in an increase in yield strength, ultimate tensile strength, and hardness for non-oriented pearlite colonies [20–23]. This explains why the HT condition, which is head-hardened (quenched) and, therefore, provides the finest microstructure, shows the highest hardness values among the investigated conditions.

The oriented solidification structure of the As-cast state is modified during the heat treatment. The HR and HT conditions show the isotropic orientation of the pearlite colonies, as indicated by the IPF maps in Figure 3. Apart from the differences in orientation between the material conditions, EBSD data do not indicate any retained austenite, which could act as a hydrogen trap in any of the investigated materials.

According to Calcagnotto et al. [24], KAM is directly related to the density of geometrical necessary dislocations (GND) following Equation (3).

$$\rho_{GND} = \frac{2 \cdot \vartheta}{u \cdot b} \tag{3}$$

where ρ_{GND} gives the GND density, ϑ the misorientation angle, u the unit length, and b the magnitude of the Burgers vector. It has to be mentioned that KAM is based on GNDs and does not provide any information about statistically stored dislocations. Although this fact limits the derivation of the dislocation density, KAM can be considered a qualitative measure of the dislocation density [24–27]. As can be seen in the KAM map of the As-cast material (Figure 4b), the misorientation and, therefore, the dislocation density is lowest in big pearlite colonies with uniform orientation. Thus refinement of the pearlitic structure leads to an increase in dislocation density, as can be observed when comparing the KAM maps of the As-cast and the HT material states. The HT condition shows a more uniform distribution of zones with increased KAM values and an increased area where the misorientation angle is about 2°. When analyzing the HR condition, it has to be considered that after hot rolling, the cooling of the material is carried out on a cooling bed. While on this cooling bed, the material is cooled by the surrounding air. Therefore, a significantly lower cooling rate is applied compared with the HT condition. The lower cooling rate results in a significantly longer time for dislocations to heal and annihilate and, therefore, leads to a decrease in dislocation density compared to the HT state. The continuous increase in the dislocation density from the As-cast over the HR to the HT material state observed quantitatively by the comparison of the KAM factors was further confirmed using XRD measurements.

4.2. Electrochemical Permeation Measurement

Van den Eckhout et al. describe in [28] the influence of microstructural characteristics on the hydrogen permeation of steels. They conclude that trap density is detrimental to diffusion behavior because a high trap density leads to low diffusion coefficients and, therefore, to low diffusion rates. According to this, a high trap density results in longer lag times and shifts the increase in diffusion transient slopes towards longer charging times.

The effective hydrogen diffusion coefficients related to the As-cast and the HR material condition are both equal to 2.3×10^{-6} cm^2/s and about four times higher than the effective hydrogen diffusion coefficient related to the HT (5.8×10^{-7} cm^2/s) state. Due to the fact that dislocations act as hydrogen trapping sites in steel [29,30], the increasing dislocation density from the As-cast and HR to the HT condition was identified as a possible source for the difference in diffusion coefficients. The direct dependency of the effective hydrogen diffusion coefficient and the dislocation density has been well described in the literature [12,31–33]. Thereby, an increase in dislocation density leads to a decrease in the effective diffusion coefficient. Considering the breakthrough time of the As-cast and HR diffusion transients, a similar dislocation density can be assumed, which is also confirmed by the EBSD data and hardness measurements provided in Section 3.1.

As described in [34], irreversible or so-called deep traps are filled up with hydrogen prior to reversible traps. Additionally, to the preferential uptake of hydrogen by deep traps, the high binding energies of these trapping sites are responsible for stronger hydrogen bonding. Therefore, hydrogen can hardly diffuse through the lattice at ambient temperature. Thus, apart from the trap density, the binding energy of the traps also influences the permeation transients in a way that a large concentration of irreversible traps, which are considered traps with an activation energy > 60 kJ/mol [28] and lead to an increase in the steepness of the permeation transients slope.

When comparing the slopes of the diffusion transients (Figure 5a), it can be seen that the steepness of the HTs diffusion transient is significantly lower compared to the diffusion transients of the other material conditions. This leads, in combination with the increased breakthrough time, to the assumption that the HT material contains the highest trap density and the lowest ratio of deep to shallow traps. The slope of the As-cast material's diffusion transient (Figure 5a) is slightly steeper than the one of HR, which indicates a higher ratio of deep to shallow traps in this material state. This assumption is supported by the results of the comparison between the first and second charging cycles. For the second charging cycle of the As-cast material, a shift in breakthrough time towards lower times is shown in Figure 4b. This indicates, as described in [35], that the material contains a significant amount of deep traps from which hydrogen cannot be removed during discharging. Therefore, these traps are already occupied with hydrogen prior to the second charging cycle, which leads to faster diffusion in the material and, thus, higher effective diffusion coefficients as well as lower breakthrough times. In the case of the HR and HT material conditions, this behavior was not observed, and diffusion transients, as well as effective diffusion coefficients, are not differing significantly for the first and second charging cycles. Because almost complete hydrogen effusion occurs during specimen discharging of HR and HT conditions, it can be assumed that trapping sites with lower activation energy predominate in both the HR and HT states compared with the As-cast state.

4.3. Thermal Desorption Spectroscopy

Analyzing the TDS spectra, a clear difference regarding the hydrogen trapping behavior of the tested material states can be distinguished. In principle, the investigated materials can be divided into two groups. The first group consists of the HR as well as the HT material condition, and the resulting sum curves of TDS spectra can be deconvoluted into a single peak. The As-cast state represents the second group, whereby its TDS spectrum can be deconvoluted into two peaks. Therefore, it can be assumed that the HR and the HT condition contain less active trapping sites than the As-cast material state. Table 4 provides a summary of literature data on the activation energies of trapping sites that can be present in the investigated material.

Table 4. Summary of literature data on activation energies of hydrogen traps.

Trapping Side	Material	E_A [kJ/mol]	Reference
Dislocations	Pure iron, Low alloyed steel	20–34	[36–41]
Dislocation cores	Pure iron	~42	[42]
Grain boundaries	Pure iron, Low alloyed steel	17–33	[19,37,43,44]
Fe/Fe$_3$C interface unstrained	Pearlite, Low alloyed steel	~24–31	[40,45,46]
Fe/Fe$_3$C interface strained	Pearlite	90–105	[47,48]
Microvoids	Pure iron, Low alloyed steel	40–50	[49,50]
MnS	Low alloyed steel	~72	[49]

The calculation of the activation energy related to the first peak for all three states results in energies between 32 and 38 ± 4 kJ/mol. Considering the spectra given in Figure 6, it can be seen that the first peak of all material conditions has its maximum in a close temperature range between 60 and 70 °C (for an applied heating rate of 400 K/h). Due to this observation, it can be concluded that the hydrogen responsible for the first peak originates from the same trapping sites. It is unlikely that just one single active trap is present in HR and HT. It is reasonable to assume that different reversible traps with very similar activation energies are present next to each other, providing only one joint cumulative peak. Since the shape of the peak corresponds to the Gaussian normal distribution curve on which peak-fitting is based, it is not possible to further deconvolute this sum peak. Considering the calculated activation energy and the microstructural artifacts, dislocations, dislocation cores, grain boundaries, and the interface between ferrite and Fe$_3$C are considered possible trapping sites contributing to the first peak. Activation energies for dislocations given in the literature are mostly between 20 and 34 kJ/mol [36–41]. Taketomi et al. show in their work [42] that the activation energy in an elastic stress field caused by a dislocation increases significantly with decreasing distance from the dislocation core. They report activation energies up to 42 kJ/mol directly at the dislocation core. To release hydrogen trapped at grain boundaries, an activation energy very similar to dislocations, which is specified to lie between 17 and 33 kJ/mol [19,37,43,44], is needed. For the activation energy of hydrogen trapped at the ferrite/Fe$_3$C interface, it has to be clearly distinguished between the level of strain which was applied to the material because, in unstrained pearlitic steels, the ferrite/Fe$_3$C interface shows activation energies between 23.5 and 31 kJ/mol [40,45,46]. However, when pearlitic material is deformed plastically, e.g., by cold drawing, an additional peak with activation energies between 90 and 105 kJ/mol occurs. This is related to hydrogen trapped at the strained ferrite/Fe$_3$C interface [47,48]. When comparing the spectra of HR and HT, it can be seen that with an increasing amount of dislocations also, the peak height increases. This finding confirms the above-described contribution of dislocations to the cumulative peak obtained by TDS measurements.

Identification of the second peak's activation energy of the As-cast material results in 47 kJ/mol and is in good correlation with the value for microvoids, which was found to be 48.3 kJ/mol for low alloyed steel by Lee et al. [49]. In addition, the continuous casting process with its typical microvoids from the solidification process indicates a relation to microvoids. Near the center of the cast bloom section, there is a large number of microvoids. As indicated in Figure 3, the amount of microvoids decreases with increasing distance from the bloom's center. Figure 8 compares the deconvoluted TDS spectra of specimens taken from as-cast bloom material at 50 (a) and 75% (b) of its height. It can be seen that the signal of the second peak is significantly more pronounced in the spectrum of the specimen taken at 50% of the bloom's height compared with one of the specimens taken at 75% of its height. This is an indication that the amount of hydrogen in the trapping site regarding the second peak decreases over the bloom's cross-section as the number of microvoids decreases. Microvoids are closed during hot-rolling, which can explain why a second peak can solely be observed in the As-cast condition. This is supported by the work of Laureys et al. [51], who showed by means of TDS analysis that microvoids can be annihilated by the application of annealing treatment. Therefore, the second peak of the As-cast material can be related to microvoids, responsible for the differences in diffusion coefficients of the first and second charging cycles described in the section above.

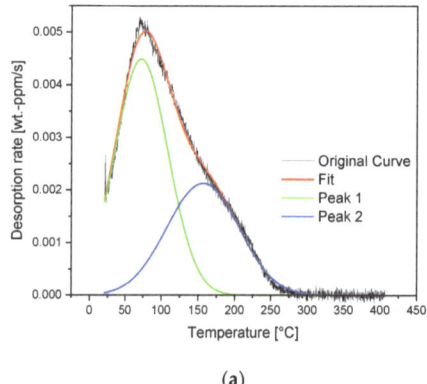 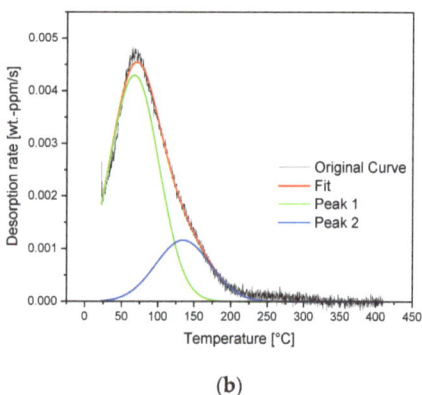

(a)   (b)

Figure 8. Deconvoluted TDS spectra were obtained from a heating rate of 400 K/h of As-cast bloom material at (**a**) 50% and (**b**) 75% of the peak height.

Non-metallic inclusions, especially MnS, are known to be initiation sites for hydrogen flaking [4,5,52]. In the present study, TDS spectra do not show evidence of hydrogen trapping directly at MnS, which leads to a hydrogen desorption peak with an activation energy of about 72 kJ/mol [49]. To act as an effective trapping site, inclusions must be small and finely dispersed; if MnS particles are relatively large in size, unevenly distributed, and in low density, the contribution to a materials desorption spectrum can be below the detection limit of TDS analysis.

5. Conclusions

The present study shows that hydrogen diffusion and trapping in pearlitic rail steel are not influenced to a significant amount by the material's crystallographic orientation. Quenching of the material leads to a finer pearlite structure resulting in higher hardness. Furthermore, higher cooling rates result in a substantial increase in dislocation density confirmed using EBSD, XRD and TDS. The increase in dislocation density due to the applied heat treatment caused a decrease in the effective hydrogen diffusion coefficient by a factor of 4 and was determined to be around 5.8×10^{-7} cm^2/s. According to data from electrochemical permeation and considering the TDS spectra, it can be assumed that hydrogen diffusion and trapping are controlled mainly by dislocation density in the finished rail materials (HR and HT). TDS analysis confirms the presence of a second peak related to an activation energy of 47.5 kJ/mol for the As-cast material that microvoids act as strong hydrogen traps. Although non-metallic inclusions, especially MnS, are discussed as possible initiation sides for hydrogen flaking in literature, TDS analysis did not show any evidence for MnS acting as a hydrogen trapping site in the investigated materials.

Author Contributions: M.E.: Conceptualization, Methodology, Investigation, Writing—Original Draft; B.L.: Investigation, Visualization, Writing—Review and Editing; M.T.: Investigation; H.S.: Conceptualization, Resources; G.K.: Conceptualization, Resources; G.M.: Conceptualization, Writing—Review and Editing, Supervision, Resources, Funding Acquisition. All authors have read and agreed to the published version of the manuscript.

Materials **2023**, 16, 5780

Funding: This research received no external funding.

Data Availability Statement: The data presented in this study are available on request from the corresponding author.

Acknowledgments: The authors would like to express their deep gratitude to voestalpine Stahl Donawitz GmbH for the project funding and the provision of sample material.

Conflicts of Interest: The authors declare no conflict of interest.

References

1.   Dean, S.W.; Sahay, S.S.; Mohapatra, G.; Totten, G.E. Overview of Pearlitic Rail Steel: Accelerated Cooling, Quenching, Microstructure, and Mechanical Properties. *J. ASTM Int.* **2009**, 6, 102021. [CrossRef]
2.   Kuziak, R.; Zygmunt, T. A New Method of Rail Head Hardening of Standard-Gauge Rails for Improved Wear and Damage Resistance. *Steel Res. Int.* **2012**, 84, 13–19. [CrossRef]
3.   Panda, B.; Balasubramaniam, R.; Moon, A. Microstructure and mechanical properties of novel rail steels. *Mater. Sci. Technol.* **2009**, 25, 1375–1382. [CrossRef]
4.   Laureys, A.; Van Stappen, J.; Depover, T.; Cnudde, V.; Verbeken, K. Electrochemical hydrogen charging to simulate hydrogen flaking in pressure vessel steels. *Eng. Fract. Mech.* **2019**, 217, 106546. [CrossRef]
5.   Ravichandar, D.; Balusamy, T.; Balachandran, G. Behaviour of Hydrogen During the Manufacture of Rail Steels. *Trans. Indian Inst. Met.* **2019**, 72, 3285–3294. [CrossRef]
6.   Uyama, H.; Yamada, H.; Hidaka, H.; Mitamura, N. The Effects of Hydrogen on Microstructural Change and Surface Originated Flaking in Rolling Contact Fatigue. *Tribol. Online* **2011**, 6, 123–132. [CrossRef]
7.   Gao, N.; Wei-Xun, Y.; Yin-Zhi, C. Flakes in low carbon high strength low alloy steel. *Mater. Charact.* **1992**, 28, 15–21. [CrossRef]
8.   Voronenko, B.I. Hydrogen and flakes in steel. *Met. Sci. Heat Treat.* **1997**, 39, 462–470. [CrossRef]
9.   Ravichandar, D.; Balusamy, T.; Gobinath, R.; Balachandran, G. Behaviour of Hydrogen in Industrial Scale Steel Melts. *Trans. Indian Inst. Met.* **2018**, 71, 2505–2515. [CrossRef]
10. Pillot, S.; Coudreuse, L. Hydrogen-Induced Disbonding and Embrittlement of Steels Used in Petrochemical Refining. In *Gaseous Hydrogen Embrittlement of Materials in Energy Technologies*; Gangloff, R.P., Somerday, B.P., Eds.; Woodhead Publishing: Sawston, UK, 2012; pp. 51–93. ISBN 9781845696771.
11. Fan, J.; Chen, H.; Zhao, W.; Yan, L. Study on Flake Formation Behavior and Its Influence Factors in Cr5 Steel. *Materials* **2018**, 11, 690. [CrossRef]
12. Drexler, A.; Siegl, W.; Ecker, W.; Tkadletz, M.; Klösch, G.; Schnideritsch, H.; Mori, G.; Svoboda, J.; Fischer, F. Cycled hydrogen permeation through Armco iron—A joint experimental and modeling approach. *Corros. Sci.* **2020**, 176, 109017. [CrossRef]
13. Wilkens, M. The determination of density and distribution of dislocations in deformed single crystals from broadened X-ray diffraction profiles. *Phys. Status Solidi* **1970**, 2, 359–370. [CrossRef]
14. Krivoglaz, M.A. Influence of correlation in position of dislocations on x-ray diffraction by deformed crystals. *Phys. Met. Met.* **1984**, 55, 1–12.
15. Krivoglaz, M.A.; Ryaboshapka, K.P. Theory of X-ray scattering by crystals containing dislocations, screw and edge disloca-tions randomly distributed throughout the crystal. *Fiz. Met. Met.* **1963**, 15, 18–31.
16. Devanathan, M.A.V.; Stachurski, Z. The adsorption and diffusion of electrolytic hydrogen in palladium. *Proc. R. Soc. Lond. Ser. A Math. Phys. Sci.* **1962**, 270, 90–102. [CrossRef]
17. Austrian Standards Institute. *Elektrochemisches Verfahren zur Messung der Wasserstoffpermeation und zur Bestimmung von Wasserstoffaufnahme und -Transport in Metallen*; Austrian Standards Institute: Vienna, Austria, 2014.
18. Kissinger, H.E. Reaction Kinetics in Differential Thermal Analysis. *Anal. Chem.* **1957**, 29, 1702–1706. [CrossRef]
19. Choo, W.Y.; Lee, J.Y. Thermal analysis of trapped hydrogen in pure iron. *Met. Trans. A* **1982**, 13, 135–140. [CrossRef]
20. Toribio, J.; González, B.; Matos, J.-C.; Ayaso, F.-J. Influence of Microstructure on Strength and Ductility in Fully Pearlitic Steels. *Metals* **2016**, 6, 318. [CrossRef]
21. Modi, O.; Desmukh, N.; Mondal, D.; Jha, A.; Yegneswaran, A.; Khaira, H. Effect of interlamellar spacing on the mechanical properties of 0.65% C steel. *Mater. Charact.* **2001**, 46, 347–352. [CrossRef]
22. Marder, A.R.; Bramfitt, B.L. The effect of morphology on the strength of pearlite. *Met. Trans. A* **1976**, 7, 365–372. [CrossRef]
23. Elwazri, A.; Wanjara, P.; Yue, S. The effect of microstructural characteristics of pearlite on the mechanical properties of hypereutectoid steel. *Mater. Sci. Eng. A* **2005**, 404, 91–98. [CrossRef]
24. Calcagnotto, M.; Ponge, D.; Demir, E.; Raabe, D. Orientation gradients and geometrically necessary dislocations in ultrafine grained dual-phase steels studied by 2D and 3D EBSD. *Mater. Sci. Eng. A* **2010**, 527, 2738–2746. [CrossRef]
25. Guglielmi, P.O.; Ziehmer, M.; Lilleodden, E.T. On a novel strain indicator based on uncorrelated misorientation angles for correlating dislocation density to local strength. *Acta Mater.* **2018**, 150, 195–205. [CrossRef]
26. Jiang, J.; Britton, T.; Wilkinson, A. Evolution of dislocation density distributions in copper during tensile deformation. *Acta Mater.* **2013**, 61, 7227–7239. [CrossRef]

27. Li, H.; Hsu, E.; Szpunar, J.; Utsunomiya, H.; Sakai, T. Deformation mechanism and texture and microstructure evolution during high-speed rolling of AZ31B Mg sheets. *J. Mater. Sci.* **2008**, *43*, 7148–7156. [CrossRef]
28. Eeckhout, E.V.D.; Depover, T.; Verbeken, K. The Effect of Microstructural Characteristics on the Hydrogen Permeation Transient in Quenched and Tempered Martensitic Alloys. *Metals* **2018**, *8*, 779. [CrossRef]
29. Choo, W.Y.; Lee, J.Y. Hydrogen trapping phenomena in carbon steel. *J. Mater. Sci.* **1982**, *17*, 1930–1938. [CrossRef]
30. Shi, R.; Chen, L.; Wang, Z.; Yang, X.-S.; Qiao, L.; Pang, X. Quantitative investigation on deep hydrogen trapping in tempered martensitic steel. *J. Alloys Compd.* **2021**, *854*, 157218. [CrossRef]
31. Luppo, M.; Ovejero-Garcia, J. The influence of microstructure on the trapping and diffusion of hydrogen in a low carbon steel. *Corros. Sci.* **1991**, *32*, 1125–1136. [CrossRef]
32. Eeckhout, E.V.D.; Laureys, A.; Van Ingelgem, Y.; Verbeken, K. Hydrogen permeation through deformed and heat-treated Armco pure iron. *Mater. Sci. Technol.* **2017**, *33*, 1515–1523. [CrossRef]
33. Siegl, W.; Ecker, W.; Klarner, J.; Kloesch, G.; Mori, G.; Drexler, A.; Winter, G.; Schnideritsch, H. Hydrogen trapping in heat treated and deformed Armco iron. In Proceedings of the CORROSION 2019, Nashville, TX, USA, 24–28 March 2019.
34. Dadfarnia, M.; Sofronis, P.; Neeraj, T. Hydrogen interaction with multiple traps: Can it be used to mitigate embrittlement? *Int. J. Hydrogen Energy* **2011**, *36*, 10141–10148. [CrossRef]
35. Chen, L.; Antonov, S.; Su, Y.; Qiao, L. Dislocation cell walls with high dislocation density as effective hydrogen traps in Armco iron. *Mater. Corros.* **2022**, *73*, 346–357. [CrossRef]
36. Hirth, J.P. Effects of hydrogen on the properties of iron and steel. *Met. Trans. A* **1980**, *11*, 861–890. [CrossRef]
37. Pressouyre, G.M. A classification of hydrogen traps in steel. *Met. Trans. A* **1979**, *10*, 1571–1573. [CrossRef]
38. Takai, K.; Chiba, Y.; Noguchi, K.; Nozue, A. Visualization of the hydrogen desorption process from ferrite, pearlite, and graphite by secondary ion mass spectrometry. *Met. Mater. Trans. A* **2002**, *33*, 2659–2665. [CrossRef]
39. Wei, F.-G.; Tsuzaki, K. Response of hydrogen trapping capability to microstructural change in tempered Fe–0.2C martensite. *Scr. Mater.* **2005**, *52*, 467–472. [CrossRef]
40. Pinson, M.; Claeys, L.; Springer, H.; Bliznuk, V.; Depover, T.; Verbeken, K. Investigation of the effect of carbon on the reversible hydrogen trapping behavior in lab-cast martensitic Fe C steels. *Mater. Charact.* **2022**, *184*, 111671. [CrossRef]
41. Moshtaghi, M.; Loder, B.; Safyari, M.; Willidal, T.; Hojo, T.; Mori, G. Hydrogen trapping and desorption affected by ferrite grain boundary types in shielded metal and flux-cored arc weldments with Ni addition. *Int. J. Hydrogen Energy* **2022**, *47*, 20676–20683. [CrossRef]
42. Taketomi, S.; Matsumoto, R.; Miyazaki, N. Atomistic study of hydrogen distribution and diffusion around a {112}<111> edge dislocation in alpha iron. *Acta Mater.* **2008**, *56*, 3761–3769. [CrossRef]
43. Wallaert, E.; Depover, T.; Arafin, M.; Verbeken, K. Thermal Desorption Spectroscopy Evaluation of the Hydrogen-Trapping Capacity of NbC and NbN Precipitates. *Met. Mater. Trans. A* **2014**, *45*, 2412–2420. [CrossRef]
44. Vandewalle, L.; Konstantinović, M.J.; Verbeken, K.; Depover, T. A combined thermal desorption spectroscopy and internal friction study on the interaction of hydrogen with microstructural defects and the influence of carbon distribution. *Acta Mater.* **2022**, *241*, 118374. [CrossRef]
45. Hagi, H. Effect of Interface between Cementite and Ferrite on Diffusion of Hydrogen in Carbon Steels. *Mater. Trans. JIM* **1994**, *35*, 168–173. [CrossRef]
46. Truschner, M.; Pengg, J.; Loder, B.; Köberl, H.; Gruber, P.; Moshtaghi, M.; Mori, G. Hydrogen resistance and trapping behaviour of a cold-drawn ferritic–pearlitic steel wire. *Int. J. Mater. Res.* **2023**, *114*, 439–452. [CrossRef]
47. Yu, S.-H.; Lee, S.-M.; Lee, S.; Nam, J.-H.; Lee, J.-S.; Bae, C.-M.; Lee, Y.-K. Effects of lamellar structure on tensile properties and resistance to hydrogen embrittlement of pearlitic steel. *Acta Mater.* **2019**, *172*, 92–101. [CrossRef]
48. Doshida, T.; Takai, K. Dependence of hydrogen-induced lattice defects and hydrogen embrittlement of cold-drawn pearlitic steels on hydrogen trap state, temperature, strain rate and hydrogen content. *Acta Mater.* **2014**, *79*, 93–107. [CrossRef]
49. Lee, J.L.; Lee, J.Y. Hydrogen trapping in AISI 4340 steel. *Met. Sci.* **1983**, *17*, 426–432. [CrossRef]
50. Lee, J.-Y.; Lee, J.-L. A trapping theory of hydrogen in pure iron. *Philos. Mag. A* **1987**, *56*, 293–309. [CrossRef]
51. Laureys, A.; Claeys, L.; Pinson, M.; Depover, T.; Verbeken, K. Thermal desorption spectroscopy evaluation of hydrogen-induced damage and deformation-induced defects. *Mater. Sci. Technol.* **2020**, *36*, 1389–1397. [CrossRef]
52. De Bruycker, E.; De Vroey, S.; Huysmans, S.; Stubbe, J. Phenomenology of Hydrogen Flaking in Nuclear Reactor Pressure Vessels. *Mater. Test.* **2014**, *56*, 439–444. [CrossRef]

MDPI AG
Grosspeteranlage 5
4052 Basel
Switzerland
Tel.: +41 61 683 77 34

Materials Editorial Office
E-mail: materials@mdpi.com
www.mdpi.com/journal/materials

www.ingramcontent.com/pod-product-compliance
Lightning Source LLC
LaVergne TN
LVHW072338090526
838202LV00019B/2438